NEW COLLOID AND SURFACE SCIENCE RESEARCH

New Colloid and Surface Science Research

Lawrence V. Schwartz
Editor

Nova Science Publishers, Inc.
New York

Library of Congress Cataloging-in-Publication Data

Schwartz, Lawrence V.
New colloid and surface science research / editor, Lawrence V. Schwartz.
p. cm.
Includes index.
ISBN-13: 978-1-60021-594-0 (hardcover)
ISBN-10: 1-60021-594-7 (hardcover)
1. Colloids. 2. Surface che8stry. I. Title.
QD549.S375 2007
05-07 541,.345—dc22

2007001426

Published by Nova Science Publishers, Inc. ✣ New York

Contents

PREFACE

Colloid and surface science research spans a wide range of topics including biological interactions at surfaces, molecular assembly of selective surfaces, role of surface chemistry in microelectronics and catalysis, tribology, and colloidal physics in the context of crystallization and suspensions; fluid interfaces; adsorption; surface aspects of catalysis; dispersion preparation, characterization and stability; aerosols, foams and emulsions; surfaces forces; micelles and microemulsions; light scattering and spectroscopy; nanoparticles; new material science; detergency and wetting; thin films, liquid membranes and bilayers; surfactant science; polymer colloids; rheology of colloidal and disperse systems; electrical phenomena in interfacial and disperse systems. This book presents new research in this field from around the globe.

Chapter 1 - The peculiar behavior of stable and unstable convection that is produced when a single liquid drop or a periodic array of evenly spaced droplets dissolve in a surrounding liquid is investigated by means of non-invasive optical techniques and computer simulations. Numerical computations are used to capture additional insights into the physics of the problem as well as to discern the role played by buoyancy forces, surface tension gradients and the interaction of both in determining dissolution kinetics and instability mechanisms. The considered system is intended to model the typical phenomena occurring during the thermal processing of liquid-liquid systems exhibiting a miscibility gap (the so-called "immiscible alloys"). These alloys undergo sedimentation of the separated heavier phase to the bottom of the container under normal gravity conditions. Droplets in non-equilibrium conditions on the bottom are responsible for the occurrence of still poorly known phenomena. The present analysis shows that even a single dissolving droplet can behave as an intriguing pattern-forming dynamical system of exceptional complexity leading to a wealth of different spatio-temporal modes of convection when the imposed temperature gradient is increased. The case of a multi-droplet configuration is even more complex; the distribution of solute depends on the multicellular structure of the convective field and on associated 'pluming phenomena'. Significant adjustments in such patterns take place as time passes. The structure of the velocity field and the number of rising solutal plumes exhibit sensitivity to the number of droplets and to the possible presence of surface Marangoni effects. New classes of possible instability mechanisms (pulsating, traveling, erratic) are identified, described and compared with other canonical phenomena traditionally considered in the literature as reference problems for the study of the evolution a system may exhibit from the stationary state to the fully developed turbulent regime (i.e. the instabilities of pool fires and of gas mixtures, the classical Rayleigh-Bénard problem with the related "wind of turbulence", etc.).

Chapter 2 – In general, driving forces for the fluid transport through micropores include dynamic pressure differences between the two ends of a capillary pore (convection), concentration differences of an impermeable solute between the two bulk solutions outside the pores (osmosis), and tangential electric fields that interact with the electric double layer adjacent to a charged pore wall (electroosmosis). Another driving force for the flow of liquid solutions in a capillary pore, which has commanded less attention, involves concentration gradients of a permeable solute along the capillary that interacts with the pore wall. The fluid motion associated with this mechanism is termed diffusioosmosis.

The steady diffusioosmotic flows of an electrolyte solution along a charged plane wall, in a capillary slit, and in a capillary tube generated by an imposed tangential concentration gradient are theoretically examined in this chapter. The charged walls may have either a constant surface potential or a constant surface charge density of an arbitrary quantity. The electric double layers adjacent to the charged walls may have an arbitrary thickness and their electrostatic potential distributions are determined by the Poisson-Boltzmann equation. Solving a modified Navier-Stokes equation with the constraint of no net electric current arising from the cocurrent diffusion, electric migration, and diffusioosmotic convection of the electrolyte ions, the macroscopic electric field and the fluid velocity in the tangential direction induced by the imposed electrolyte concentration gradient are obtained as functions of the lateral position for various cases. The results indicate that the effect of the deviation of the local induced tangential electric field inside the double layer from its bulk-phase quantity and the relaxation effect on the diffusioosmotic velocity of the fluid are significant in most practical situations, even for the case of very thin double layer.

Chapter 3 - Biological products are often produced as or included in particles, such as inclusion bodies, crystals, cell organelles, viruses and virus-like particles. Selective harvesting of these specific particles is crucial for efficient product recovery. It requires separation techniques that have the ability to separate the target particles from the liquid phase and in general also from other particles, such as cells, cell debris, by-products or immobilised catalysts. In many cases the target particles contain the product already at high purity, making selective particle recovery advantageous. Conventional separation methods such as centrifugation and filtration are inefficient in many particle-particle separations due to the small density differences and overlapping particle size distributions of the particles that require separation. New physical separation methods are therefore desirable.

Particles can adsorb to fluid interfaces. Under specific conditions this process is selective for one type of particle in a particle mixture, which makes the phenomenon useful for particle-particle separation. Selectivity in the adsorption process can be created in several ways.

This chapter deals with the use of fluid interfaces for particle separation both from an experimental and theoretical point of view. It shows that interfacial separation is very valuable for separation of biotechnological particle mixtures. Two principles for creating the fluid interfaces are discussed: 1) injection of the dispersed phase, and 2) emulsification. Selectivity in these processes can be created by selective particle-interface collision, selective particle attachment, selective particle detachment, or a combination of these sub-processes. Experimental examples from literature are given for all of these sub-processes. For example, ampicillin and phenylglycine crystals were separated by making use of solvent-water interfaces (Jauregi et al., 2001; 2002; Hoeben et al., 2005). In this case the crystals showed competition for the interface and under certain conditions partitioning of the phenylglycine crystals occurred from one liquid phase to the other. A second example is the separation of polyhydroxyalkanoate inclusion bodies from ruptured fermentation broth making use of the air-water interface (Van Hee et al., 2006b). The selectivity was created by selective aggregation of the inclusion bodies, which enhanced their transport to the interface (more favourable hydrodynamics) relative to that of the other particles in the mixture. Finally, the flotation behaviour of synthetic micrometer-sized bioparticles was investigated by determining the relation between the properties of the surface groups of the particles and their adsorption behaviour to the air-water interface (Van Hee et al., 2006c). Particle aggregation again showed to be essential for these small particles. In addition, the floatability of the particles was dominated by the adsorption kinetics of their surface groups. Besides these examples, many other examples from literature are used to illustrate the mechanisms that can be applied for interfacial bioparticle separation.

Chapter 4 - Magnetic colloids are complex fluids composed by magnetic and non-magnetic particles dispersed in liquid carriers. These fluids can develop intense rheological changes upon application of an external magnetic field. This remarkable property makes possible the use of these fluids in a wide range of technological applications and, therefore, they center the attention of both scientists and engineers interested in field-responsive materials. Usually, magnetic colloids are classified into: (a) ferrofluids (FF) and (b) magnetorheological (MR) fluids. FF are dispersions of magnetic nanoparticles in a carrier liquid while MR fluids contain micron-sized particles of ferro- or ferrimagnetic materials. In the case of FF, although thermal energy prevents particle sedimentation and aggregation by magnetic attraction, it does not avoid coagulation due to van der Waals forces. Therefore, the use of adequate surfactants and/or thixotropic agents is needed to overcome this problem. In MR fluids, high density magnetic particles are dispersed in low density oil carriers and, consequently, the suspensions suffer from excessive gravitational settling, which can hinder the MR response of the fluid.

In this chapter, the preparation of fluids belonging to both categories (FF and MR fluids) is described. Furthermore, the effect of different stabilizing additives (thixotropic agents, surfactants, magnetic nanoparticles) and carrier liquids in both the sedimentation and the redispersibility of these fluids is analyzed. These studies are conducted using different experimental techniques: (i) optical and electromagnetic induction techniques to monitor the stability of dilute and concentrated suspensions; (ii) contact angle and pendant drop measurements to explain the observed stability in different carriers; and (iii) steady-state rheological tests to monitor the redispersibility of the suspensions. The results of these experiments are analyzed in detail and magnetic fluids with improved stability properties are prepared accordingly. Finally, an extensive study on the rheological properties of these fluids, in absence and presence of magnetic field, is described. Particularly, the effect of additive concentration and magnetic field intensity on the magnetorheological behavior is studied in detail. For this purpose, both controlled-stress and controlled-rate rheometers are employed.

Chapter 5 - Although the DLVO theory is a fundamental theory for explaining the stability of charged emulsion and colloidal particles in water, the treatments of the theory developed so far have a fundamental imperfection. The theory shows that the force preventing the coagulation and coalescence of charged particles is the double layer force that is generated by the overlap of diffuse electric double layers around the particles. The double layer force depends on the surface charge density and the separation distance between the surfaces of adjacent particles, whereas the surface charge density depends on the separation distance. The self consistency should therefore be attained between the surface charge density and the separation distance. However, except for our recent study, no attempt has been made for attaining the self consistency.

Here we have obtained the double layer potential energy for the front surfaces of approaching two charged particles by taking the migration of surface charge into account. The potential energy including the van der Waals energy is given by

$$U = \frac{\varepsilon_l \kappa \varsigma^2 e^{-\kappa w}}{8(h+1)} \left[\frac{[4(h+1)+3h(a_1-a_2)E_0 \cos\eta/\varsigma]^2}{h+1+(h-1)e^{-\kappa w}} - \frac{[3h(a_1+a_2)E_0 \cos\eta/\varsigma]^2}{h+1-(h-1)e^{-\kappa w}} \right] - \frac{A_H}{12\pi w^2}$$

where $h = z_s e\zeta (s_{+,0} + s_{-,0}) / [kT (s_{+,0} - s_{-,0})]$; z_s, the valence of surface ions; e, the unit charge; $s_{+,0}$ and $s_{-,0}$, number densities of positive and negative surface ions; kT, the thermal energy at temperature T; ε_l, the dielectric constant of water; κ, the Debye reciprocal length; ζ, the zeta potential; w, the separation distance between the front surfaces of approaching two spheres with radii a_1 and a_2; E_0, the intensity of an external electric field; η, the angle between the axis of approach and the external field; and A_H, the Hamaker constant of the particles, respectively. The above equation indicates that emulsion and colloidal systems can be destabilized by applying external electric fields. Experimental evidences of the efficient destruction of an oil-in-water emulsion by low external electric fields are also shown.

Chapter 6 - A first-order reversible kinetic model and a general isotherm model describing Cd^{2+} and Cu^{2+} biosorption by aerobic granules were developed and verified by biosorption data obtained at different initial metal and aerobic granules concentrations. Compared to the second-order kinetic model, the proposed first-order kinetic model could provide a more satisfactory description of experimental data. The general isotherm model derived from a thermodynamic approach could be reduced to the well known Langmuir and Freundlich isotherm equations in some special cases.

The overall biosorption rate constants and biosorption capacities of Cd^{2+} and Cu^{2+} by aerobic granules were determined by the first-order kinetic model, and were found to be closely related to initial metal and granules concentrations, i.e., a high initial metal concentration would favor biosorption, while an increased initial granule (biosorbent) concentration would hinder bisorption. To interpret these experimental observations, two theoretical models were further developed from the proposed kinetic and isotherm models. The model simulation is in good agreement with experimental data obtained at different initial granule, Cd^{2+} and Cu^{2+} concentrations. It is expected that the modeling system developed can offer in-depth sights into biosorption of heavy metals by aerobic granules.

In: New Colloid and Surface Science Research
Editor: Lawrence V. Schwartz, pp. 1-59
ISBN: 978-1-60021-594-0

Chapter 1

SINGLE- AND MULTI-DROPLET CONFIGURATIONS OUT OF THERMODYNAMIC EQUILIBRIUM: PULSATING, TRAVELING AND ERRATIC FLUID-DYNAMIC INSTABILITIES

*Marcello Lappa**
MARS (Microgravity Advanced Research and Support Center)
Via Gianturco 31 - 80146, Napoli, Italy,
Also, Via Salvator Rosa 53, San Giorgio a Cremano, 80046, Napoli, Italy

ABSTRACT

The peculiar behavior of stable and unstable convection that is produced when a single liquid drop or a periodic array of evenly spaced droplets dissolve in a surrounding liquid is investigated by means of non-invasive optical techniques and computer simulations. Numerical computations are used to capture additional insights into the physics of the problem as well as to discern the role played by buoyancy forces, surface tension gradients and the interaction of both in determining dissolution kinetics and instability mechanisms. The considered system is intended to model the typical phenomena occurring during the thermal processing of liquid-liquid systems exhibiting a miscibility gap (the so-called "immiscible alloys"). These alloys undergo sedimentation of the separated heavier phase to the bottom of the container under normal gravity conditions. Droplets in non-equilibrium conditions on the bottom are responsible for the occurrence of still poorly known phenomena. The present analysis shows that even a single dissolving droplet can behave as an intriguing pattern-forming dynamical system of exceptional complexity leading to a wealth of different spatio-temporal modes of convection when the imposed temperature gradient is increased. The case of a multi-droplet configuration is even more complex; the distribution of solute depends on the multicellular structure of the convective field and on associated 'pluming phenomena'. Significant adjustments in such patterns take place as time passes. The structure of the

* E-mail: lappa@marscenter.it, marlappa@unina.it

velocity field and the number of rising solutal plumes exhibit sensitivity to the number of droplets and to the possible presence of surface Marangoni effects. New classes of possible instability mechanisms (pulsating, traveling, erratic) are identified, described and compared with other canonical phenomena traditionally considered in the literature as reference problems for the study of the evolution a system may exhibit from the stationary state to the fully developed turbulent regime (i.e. the instabilities of pool fires and of gas mixtures, the classical Rayleigh-Bénard problem with the related "wind of turbulence", etc.).

1. Introduction

When a liquid system with a miscibility gap is cooled below the so-called "critical temperature" (T_c), it demixes into two liquids of different compositions and densities. Above T_c both components are completely miscible in the liquid state, but below this temperature two immiscible liquids coexist (they are usually referred to as the "majority" and "minority" phases).

Many alloys of technological interest exhibit such a feature; there are, in fact, several interesting and relevant examples in both the inorganic and organic realms (sulfides and silicates systems, glasses, many metallic composites used as superconductors, magnetic materials and high-performance electrical contacts). The properties of these materials are linked to the internal structure, which develops as the alloy solidifies from its molten state. Therefore, research in this area is primarily concerned with advancing the understanding of the related processing so that the structure and, ultimately, the properties of the alloy, can be controlled as the materials are originally formed.

In the Earth's gravitational field the usual differences in density cause rapid spatial separation of the components of these alloys through sedimentation or flotation. This prevents homogeneous distribution of particles in the matrix using simple and inexpensive casting processes (Ratke, 1992). Superimposed on this is the fact that during the solidification process, the melt is subject to temperature gradients. In a gravity field, they lead in most cases to buoyancy driven convective flows and regardless of the presence of gravity (i.e. also under microgravity conditions) they can produce Marangoni stresses at the interface between the two liquids.

These stresses have been found to be directly or indirectly responsible for several additional complex and intriguing phenomena. In microgravity conditions they act as the driving force for the so-called "Marangoni migration": due to the surface Marangoni stress distribution induced by thermal effects free liquid drops surrounded by an immiscible liquid migrate towards the hot region (see, e.g., the experimental results of Langbein and Heide, 1984, 1987; Braun et al., 1993 and the numerical simulations of Bassano, 2003; Lappa, 2005a and Esmaeeli 2005). On the ground, the presence of thermal Marangoni surface stresses can be the source of fascinating fluid-dynamic instabilities driven by the direct opposition of thermal to solutal (dissolution) effects (Lappa et al., 2004a,b).

Even if, as explained before, the phases behave as immiscible liquids when they are formed, droplets, formed in some regions by cooling of the considered alloy, can move, due to the aforementioned gravitational sedimentation, to regions with different temperature, e.g., towards the bottom of the container that is typically colder than the top (the system is usually cooled from below to avoid the onset of massive thermal buoyancy convection). Since the phases, on the basis of the phase diagram, have different equilibrium concentrations at different temperatures, a droplet that is in equilibrium (i.e. immiscible) with the surrounding liquid in a certain region at a certain temperature will no longer be in such a stable condition if it rapidly moves to another zone with a different temperature. Therein, in fact, it will tend to return to equilibrium exchanging mass with the surrounding phase and leading to the onset of solutal convection.

Owing to this behavior, the interaction between liquid drops and the surrounding fluid becomes very complex. It is manifested by a fluid motion brought about by different coexisting mechanisms: Marangoni convection, generated by the interfacial stresses due to temperature gradients along the drop surface (thermocapillary convection) and gravitational convection driven by the volumetric buoyancy forces caused by thermally and/or solutally generated density variations in the bulk of the fluid. As outlined above, the delicate interplay among these effects can lead to the onset of oscillatory phenomena.

These instabilities represent an additional aspect of the problem that should be taken into account in addition to the aforementioned migration and sedimentation mechanisms pertaining to the droplets and other well-known relevant phenomena of different spatial extensions and temporal durations (droplet coagulation and coalescence prevention mechanisms, see, e.g., Lappa 2005b,c for a review). Many of these effects are still poorly known, due to the experimental difficulties encountered when trying to discern the separate contributions provided by many concurrent and interacting mechanisms (the aforementioned droplet nucleation and ensuing growth, sedimentation, Marangoni migration, coalescence, dissolution and related interplay of all these aspects). During material processing the different effects cannot be separated; they all act together and the related mutual interference can be very complex.

Some of these limitations, however, can be removed if a single stationary droplet is considered. This fundamental situation is considered in the first part of the present analysis (Section 5). Conditions corresponding to the aforementioned lack of equilibrium (in terms of surface concentration distribution) are reproduced by directly injecting the pure liquid representing the minority phase into the majority phase at a temperature below the critical one. This is accomplished by means of a needle and, accordingly, a single dissolving droplet anchored to the top of the needle is obtained. Drop coagulation events cannot occur since a single drop is present and drop motion is prevented by the fact that the droplet is locked to its initial position.

Two different fundamental thermal situations are considered: isothermal (Sect. 5.1) and non-isothermal liquid (Sect. 5.2, vertical temperature gradient imposed across the container). The latter one better reproduces conditions corresponding to real industrial processes (where, as explained before, the materials are formed by means of a thermal quench of the initially homogeneous mixture exhibiting a miscibility gap from temperatures above the gap to lower temperatures inside the gap). In such a context the onset of oscillatory convection is investigated for different possible initial volumes of the droplet. These results are refined in Sect. 5.3 that reports a variety of possible scenarios as a function of the applied temperature gradient.

It is worth anticipating that in the light of these studies, beyond technological aspects, the considered single-drop configuration also exhibits the features of an intriguing patter-forming dynamical system. As a consequence, in addition to its technological and industrial significance, it becomes clear that the problem under consideration in this chapter also exhibits a theoretical kinship with the complex of problems that come under the heading "route to chaos". Hence, it can be also regarded as an interesting and relevant subject of investigation within the framework of fundamental studies focused on the possible bifurcations of complex systems.

The second part of the chapter (Section 6) extends the investigation to the more realistic situation in which a relatively large number of droplets are uniformly distributed on the bottom of the container (i.e. a discrete layer of N drops for different values of N). It is shown how the convective structures originated from the different droplets can interact in unexpected ways and, therefore, "ensemble behaviors" with many interacting droplets, should be regarded as an important aspect of the problem (Lappa, 2006). Due to their geometric complexity, such configurations have not been studied in great detail. They are still a challenging task for numerical simulation. Nevertheless, numerical simulations are of fundamental importance within the context of these studies. From an experimental point of view, in fact, for N>1 the concentration pattern (interferometric fringes, see Sect. 2 for additional details) around the droplets becomes so complex that a coherent re-construction of its structure is very difficult (often impossible). Accordingly the first part of the present chapter (dealing with the single-droplet configuration) proceeds with the support of both experimental and numerical investigations whereas the second part (multi-droplet configuration) is mostly based on computer simulations.

2. Liquids, Geometrical Parameters and Configuration Under Investigation

2.1. Liquid-Liquid System

According to the major difficulty in investigating liquid metals for their opaqueness, organic liquids are often used as model substances. For instance, methanol and cyclohexane have been selected and investigated both in Earth and in space laboratories since they are transparent liquids exhibiting a miscibility gap in the phase diagram. They are used as model liquids for the present analysis as well (see Table I for the related physical properties).

Table I. Physical properties of the cyclohexane-methanol system ((1) pure methanol, (2) pure cyclohexane)

Diffusion coefficient D [$m^2\ s^{-1}$]	2×10^{-9}
Density $\rho_{1(o)}$ [$kg\ m^{-3}$]	0.782×10^{3}
Density $\rho_{2(o)}$ [$kg\ m^{-3}$]	0.769×10^{3}
Kinematic viscosity ν_1 [$m^2\ s^{-1}$]	1.03×10^{-6}
Kinematic viscosity ν_2 [$m^2\ s^{-1}$]	1.275×10^{-6}
Solutal expansion coefficient β_{c1} [-]	-4.62×10^{-2}
Solutal expansion coefficient β_{c2} [-]	4.95×10^{-2}
Thermal expansion coefficient β_{T1} [K^{-1}]	1.14×10^{-3}

Table I. (Continued)

Thermal expansion coefficient β_{T2} [K^{-1}]	$1.14x10^{-3}$
Specific heat C_{p1} [$J\ kg^{-1}K^{-1}$]	$2.51x10^{3}$
Specific heat C_{p2} [$J\ kg^{-1}K^{-1}$]	$1.86x10^{3}$
Thermal conductivity λ_1 [$J\ m^{-1}s^{-1}K^{-1}$]	$2.02x10^{-1}$
Thermal conductivity λ_2 [$J\ m^{-1}s^{-1}K^{-1}$]	$1.25x10^{-1}$
Surface tension derivative σ_T [$dyne\ m^{-1}K^{-1}$]	17.42

As anticipated, conditions corresponding to lack of equilibrium (induced by the gravitational shift of droplets formed in the bulk by the spontaneous thermodynamic decomposition process, to zones with different temperature) are reproduced here by simply considering initial conditions with droplets of pure methanol surrounded by pure cyclohexane at a temperature below the critical one (the fluids are miscible in such a condition). The droplets are attached at the tip of the needles used to inject them into the matrix from the bottom of the container.

2.2. Experimental Technique

The experiments have been performed using a cell made of quartz, 1 [cm] x 1 [cm] x 4 [cm], filled with liquid. The generic droplet is formed at the tip of a capillary. This tube is made of glass (SIGMA P-6679) and has external diameter d =1 [mm].

Within the test cell filled with cyclohexane, an isothermal or non-isothermal temperature field can be established. Two Peltier elements at the top and bottom walls can be controlled, in fact, to obtain a constant temperature inside the enclosure and/or to introduce a temperature gradient parallel to the axis of the aforementioned capillary tube used to inject and sustain the droplet. In the latter case (non isothermal conditions), the test cell is heated from above in order to prevent the onset of large-scale thermal buoyancy convection in the liquid bulk.

The experimental procedure includes the direct visual observation of droplets and an interferometric analysis performed to analyze the concentration field and therefore the mass transfer at the liquid-liquid interface. In particular a Wollaston lateral shearing interferometer [Malacara] has been used to analyze refraction index gradients: fringes correspond to the locations having equal values of the component of gradient along the shearing direction. The variation of gradient along the shearing direction related to consecutive fringes is given by $\lambda/L\delta$, λ being the wavelength, L the cell thickness and δ the shearing distance. By rotating the Wollaston prism by 90° horizontal and vertical shearing can be obtained; for the experiments with single drops, in practice the direction is horizontal, being the problem axisymmetric along the capillary axis. The light source is a diode laser, λ=638.6 [nm] (Micro Laser Systems) and the wave-front deformation of the cell's quartz walls amounts to λ in order to reduce optical aberrations.

A digital camera (Hamamatsu) has been also used, having a sensor area of 8.58 x 8.58 [mm^2], and a resolution of 1024 x 1280 pixels and providing a direct visualization in the direction perpendicular to the axis of the capillary tubes.

3. Numerical Method

A number of mathematical models and methods have appeared for numerical/theoretical analysis of the fundamental mechanisms related to phase separation in alloys with a miscibility gap when they are disjoint or partially combined. For the sake of simplicity, non-miscible liquids (like oil and water) have often been selected as model fluids for fundamental studies concerning the sedimentation and migration phenomena; these particular aspects have been treated within the framework of Volume of Fluid ("VOF" also referred to as "Volume Tracking") techniques.

VOF-based simulations, however, are still very rare in literature (rare and excellent efforts have been provided by Mehdi-Nejad et al., 2003 for the case of droplet sedimentation; and Haj-Hariri et al., 1997 for the case of Marangoni migration).

As discussed by Rider and Kothe, 1998, in fact, these techniques yet possess solution algorithms that are too often perceived as being heuristic and without mathematical formalism. Part of this misperception lies in the difficulty of applying standard hyperbolic PDE (Partial Differential Equations) numerical analysis tools, which assume algebraic formulations, to methods that are largely geometric in nature (hence, the more appropriate term volume tracking). To some extent the lack of formalism in volume tracking methods, manifested as an obscure underlying methodology, has impeded progress in evolutionary algorithmic improvements and application in particular to the case of systems with a miscibility gap (classical Volume of Fluid methods need to be modified (re-formulated) to account for interface motion (expansion or contraction) induced by solute absorption or release).

An alternative numerical approach for the analysis of the dissolution kinetics and fluid-dynamic instabilities mentioned in the introduction, could be based, in principle, on the so-called linear stability analysis. This approach allows one to avoid the straightforward integration in time that would be required by Volume of Fluid methods as well as the solution of the Navier-Stokes equations in their complete and non-linear form. The related mathematical treatment of the problem generally proceeds along the following lines: an initial flow that represents a stationary state of the system is considered. By supposing that the various physical variables describing the flow suffer small (infinitesimal) increments, the equations governing these increments are obtained. In obtaining these equations from the relevant equation of motion, all products and powers (higher than the first) of the increments are neglected while retaining only terms that are linear in them. Since stability means stability with respect to all possible (infinitesimal) disturbances, for an investigation on stability to be complete it is necessary that the reaction of the system to all possible disturbances be examined. This is accomplished by expressing an arbitrary disturbance as a superposition of certain basic possible modes and examining the stability of the system with respect to each of these modes. In practice, this approach considers all possible infinitesimal perturbations and usually provides better understanding of the instability mechanisms as well as more precise critical values of the governing parameters.

Despite the important capabilities offered to the investigators by linear stability analyses, it is opinion of the author of this chapter, however, that the direct straightforward integration in time of the governing balance equations offers even more interesting advantages for the case under investigation here. This philosophy can provide, in principle, a wealth of additional aspects that could not be obtained on the basis of linear stability computations: The effective amplitude of the disturbances when they are saturated; the behavior of the system far from the threshold point; a quite exhaustive picture of the subsequent stages of temporal evolution, etc.

The effective capability of the VOF methods to capture fluid-dynamic instabilities driven by surface processes (with these methods the interface separating the two liquids must be smeared over a region with a non-zero thickness) has been recently proven (the DDVOD method, see Lappa, 2005d). The basic features of the DDVOF are discussed in Sect. 3.1.

3.1. DDVOF - Volume of Fluid Method for Dissolving Drops

The DDVOF is a single-region formulation and allows a fixed-grid solution to be undertaken. It is therefore able to utilize standard solution procedures for the fluid flow and species equations directly, without resorting to mathematical manipulations and transformations (e.g., body-fitted curvilinear coordinates).

The model is based on the mass-balance equations. The diffusion of the species is governed by the equations (drop=phase 1→ϕ=1, matrix=phase 2→ϕ=0):

$$\frac{\partial C_1}{\partial t} = \phi\left[-\underline{\nabla}\cdot\left(\underline{V}C_1\right) + D\,\nabla^2 C_1\right] \tag{1}$$

$$\frac{\partial C_2}{\partial t} = (1-\phi)\left[-\underline{\nabla}\cdot\left(\underline{V}C_2\right) + D\,\nabla^2 C_2\right] \tag{2}$$

where D is the interdiffusion coefficient, C_1 and C_2 are the concentrations (mass fraction) of one of the two components (for instance, methanol) in the minority (drop) and majority (matrix) phases, respectively, and ϕ is the well-known phase variable traditionally used in the context of VOF methods.

The species equations (1) and (2) are solved throughout the computational domain including the drop and the matrix. The presence of the terms ϕ and (1-ϕ) ensures, in fact, that each equation characterizes a different phase. The initial constant concentrations in the generic drop and in the matrix are denoted by $C_{1(o)}$ and $C_{2(o)}$, respectively (for the case under investigation: $C_{1(o)}$ = 1, i.e. drop of pure methanol and $C_{2(o)}$ = 0, i.e. matrix of pure cyclohexane). As time passes, owing to molecular diffusion through the interface, the drop tends to be enriched with cyclohexane (i.e. C_1 becomes <1), correspondingly, the majority external phase tends to be enriched with methanol (i.e. C_2 becomes >0).

The behavior of the two phases is coupled through the equilibrium concentration values imposed on the two sides of the liquid-liquid interface (this means that when a drop of pure methanol is injected into the cyclohexane liquid matrix, it is assumed that equilibrium at the liquid/liquid interface is attained instantaneously, see Perez De Ortiz and Sawistowski, 1973a). The aforementioned values come from the miscibility law (see Figure 1) according to the local value of the temperature of the interface (for isothermal conditions, of course, the equilibrium values are constant along the interface).

At the interface ($0<\phi<1, |\underline{\nabla}\phi| \neq 0$), the concentrations must satisfy the equilibrium conditions:

$$C_1\big|_i = C_{1(e)}(T) \quad (3a)$$

$$C_2\big|_i = C_{2(e)}(T) \quad (3b)$$

i.e. the equilibrium methanol concentrations at the consolute points of the miscibility diagram (Figure 1a).

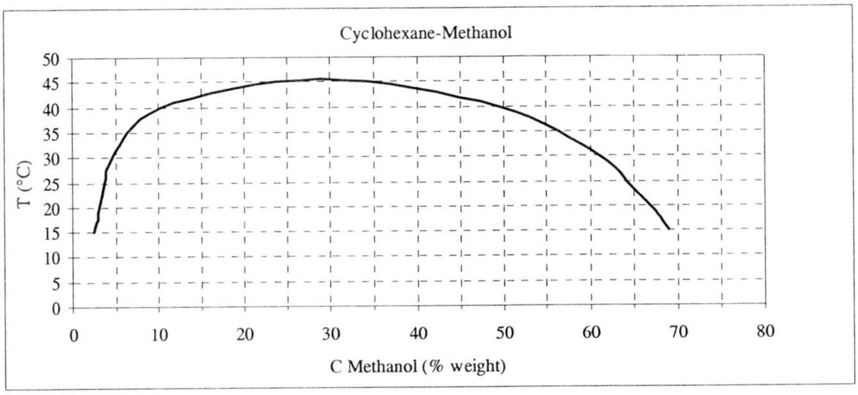

a)

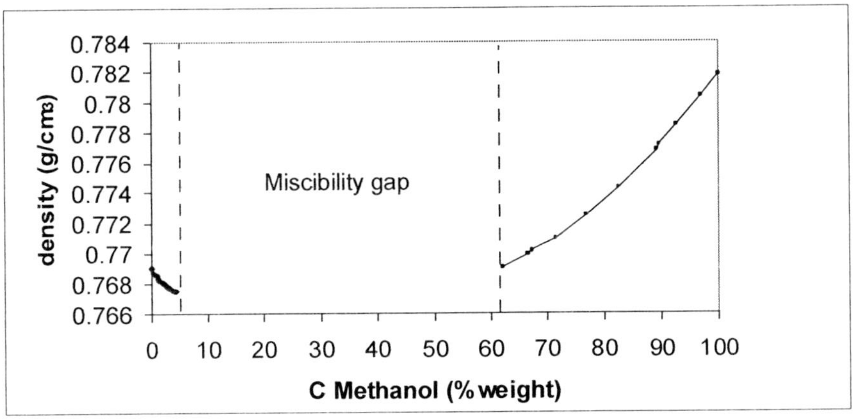

b)

Figure 1. Phase diagram (a) and density versus methanol concentration (b) for the cyclohexane-methanol system.

Equations (3) behave as 'moving boundary conditions'. Note that in place of the 'sharp' transition from one phase to the other that would characterize a multiple-region formulations (in that case the interface separating the bulk phases is a mathematical boundary of zero thickness), here (following the VOF philosophy) the phase field varies smoothly but rapidly through an interfacial region whose thickness is not zero. This region, defined by the mathematical conditions $|\underline{\nabla}\phi| \neq 0, 0<\phi<1$, moves through the computational domain according to the behavior of the different phases (i.e. according to the behavior of ϕ).

The flow is governed by the continuity, and Navier-Stokes equations:

$$\underline{\nabla} \cdot \underline{V} = 0 \quad (4)$$

$$\frac{\partial(\rho \underline{V})}{\partial t} = -\underline{\nabla} p - \underline{\nabla} \cdot [\rho \underline{V}\underline{V}] + \underline{\nabla} \cdot \left[\mu\left(\underline{\nabla}\underline{V} + \underline{\nabla}\underline{V}^T\right)\right] + \underline{F}_g + \underline{F}_{Ma} \quad (5)$$

where, in turn, the density ρ and the dynamic viscosity μ read

$$\rho = \rho_{1(o)}\phi + \rho_{2(o)}(1-\phi) \tag{6a}$$

$$\mu = \mu_1\phi + \mu_2(1-\phi) \tag{6b}$$

moreover:

$$F_g = +\{\rho g\lfloor\beta_{C1}(C_1 - C_{1(o)})\rfloor\phi + \rho g\lfloor\beta_{C2}(C_2 - C_{2(o)})\rfloor(1-\phi) + \rho g\lfloor\beta_{T1}(T - T_{ref})\rfloor\phi + \rho g\lfloor\beta_{T2}(T - T_{ref})\rfloor(1-\phi)\ \} \tag{6c}$$

the Boussinesq approximation being used to model the buoyancy forces(β_{C1} and β_{C2} are the solutal expansion coefficients related to the mutual interpenetration of the phases (1) and (2), β_{T1} and β_{T2}, the thermal expansion coefficients, T_{ref} a reference temperature).

The temperature distribution is governed by the energy equation:

$$\frac{\partial \rho C_P T}{\partial t} = \left[-\underline{\nabla}\cdot(\rho C_P \underline{V}T) + \underline{\nabla}\cdot(\lambda\ \underline{\nabla}T)\right] \tag{7}$$

where the thermal conductivity λ and the specific heat C_P are computed as

$$\lambda = \lambda_1\phi + \lambda_2(1-\phi) \tag{8a}$$

$$C_P = C_{P1}\phi + C_{P2}(1-\phi) \tag{8b}$$

The core of the DDVOF method is its technique for adjourning ϕ. The tracking of the interface between the phases is accomplished by the solution of a special continuity integral equation for the volume of each liquid drop (taking into account the release or absorption of solute through its interface). Within this context the generic liquid drop is assumed to be a sphere of radius R_i increasing due to growth or decreasing due to dissolution (the assumption of spherical shape is reliable when the Bond number $Bo = \frac{\Delta\rho g R^2}{\sigma}$ is sufficiently small, i.e. for small droplets and/or relatively large values of the surface tension σ).

If ζ denotes the velocity with which the interface moves through the reference frame, the overall jump mass balance reads $\delta\left[\rho(V_n - \zeta)\right] = 0$ where V_n is the fluid velocity component orthogonal to the free surface. Assuming $\rho_1 \cong \rho_2$ (see Table I) the above equation reduces to $(V_{n1} - \zeta) = (V_{n2} - \zeta) \rightarrow V_{n1} = V_{n2}$. Moreover, $V_n = 0$ on both sides of the interface since there is no local macroscopic convective transport of fluid across the drop surface (the inward or outward motion of the interface being only based on molecular diffusion). Thus the specie balance jump equation $\delta\left[\rho(V_n - \zeta)C - \rho D\frac{\partial C}{\partial n}\right] = 0$ reads (see also Bassano, 2003):

$$-D\left.\frac{\partial C_2}{\partial n}\right|_i + C_2\big|_i \dot{m} = -D\left.\frac{\partial C_1}{\partial n}\right|_i + C_1\big|_i \dot{m} \tag{9a}$$

Where $\dot{m}$ is the mass flow rate through the interface (n points towards the matrix; if the generic drop dissolves, $\dot{m}$ is positive and ζ is negative, i.e. $\dot{m} = -\zeta$ with $\zeta \cong \frac{dR_i}{dt}$); therefore

$$D\left(\left.\frac{\partial C_2}{\partial n}\right|_i - \left.\frac{\partial C_1}{\partial n}\right|_i\right) = \zeta \ \Delta C \text{ with } \Delta C = C_1|_i - C_2|_i \tag{9b}$$

i.e. the difference between the diffusive solutal fluxes computed at the two different sides of the dissolving liquid-liquid surface is balanced by the convective contribution due to the interface motion. In conclusion integrating over the surface of the generic drop, one obtains for the time evolution of its radius:

$$\frac{dR_i}{dt} = D\frac{1}{S}\oint \frac{1}{C_1|_i - C_2|_i}\left(\left.\frac{\partial C_2}{\partial n}\right|_i - \left.\frac{\partial C_1}{\partial n}\right|_i\right) dl \tag{10}$$

where S is the surface of the drop, and

$$\frac{\partial C}{\partial n} = \underline{\nabla} C \cdot \underline{\hat{n}} = \alpha \frac{\partial C}{\partial x} + \beta \frac{\partial C}{\partial y} \tag{11}$$

$$\underline{\hat{n}} = -\frac{\underline{\nabla}\phi}{|\underline{\nabla}\phi|} = (\alpha, \beta) \tag{12}$$

$$\alpha = -\frac{\partial \bar{\phi}}{\partial x} / \sqrt{\left(\frac{\partial \bar{\phi}}{\partial x}\right)^2 + \left(\frac{\partial \bar{\phi}}{\partial y}\right)^2}, \quad \beta = -\frac{\partial \bar{\phi}}{\partial y} / \sqrt{\left(\frac{\partial \bar{\phi}}{\partial x}\right)^2 + \left(\frac{\partial \bar{\phi}}{\partial y}\right)^2} \tag{13}$$

As shown in eq. (13), in practice, the unit vector $\underline{\hat{n}}$ perpendicular to the interface results from the gradient of a smoothed phase field $\bar{\phi}$, where the transition from one phase to the other takes place continuously over several cells (typically 8 or 10 for a grid having 400x400 points) . The smoothed phase field $\bar{\phi}$ is obtained by convolution of the unsmoothed field ϕ with an interpolation function. The interface orientation depends on the direction of the volume fraction gradient of the phase $\bar{\phi}$ within the cell, and that of the neighbor cell (or cells) sharing the face in question. Depending on the interface's orientation and on the side (phase (1) or phase (2)) on which computations are performed, concentration gradients can be discretized by forward or backward schemes. After the computation of the radius at the new instant (n+1), then the distribution of the phase variable ϕ is reinitialized to take into account the volume change, i.e. $\phi=1$ for $r<R_i$ and $\phi=0$ for $r>R_i$.

Equations (1), (2), (4), (5), (7) and (10) for each drop represent a system of five partial differential equations and N ordinary differential equations (N=number of droplets) whose solution governs the nonlinear behavior of the physical system under investigation.

It is worth recalling that the present mathematical model and related numerical technique can be regarded as a very hybrid volume-tracking method.

In the "classical" VOF methods, the phase field variable ϕ is 'advected' solving an appropriate partial differential transport equation; as mentioned before, this formulation is traditionally used for the solution of typical problems dealing with the migration of bubbles or drops in liquids. It relies on the fact that the fluids are not interpenetrating. For the present method the equation governing the evolution of ϕ comes from mass-balance conditions rather than from transport. Moreover, interpenetration of the different fluids is allowed according to the coupled behavior of eqs. (1), (2) and (10).

Moreover, as previously pointed out, the present model, assumes the drop shape to be spherical. In the classical VOF methods the shape of the interface is unknown: ϕ is computed solving a separate equation, then the interface is reconstructed within the framework of a special geometrical algorithm (e.g., SLIC or PLIC, see Lappa, 2005c). In the present case, it is the opposite. The interface shape is assumed to be known. Its radius is computed solving the separate integral balance equation (10) and finally the phase variable is defined on the basis of the current value of the radius. The introduction of the phase variable is still necessary since, from a computational point of view, it is used for the definition of the materials physical properties as shown by eqs. (6) and (8). As illustrated before, it is also used as a practical geometrical variable for the definition of the unit vector perpendicular to the fluid/fluid interface.

Equation (10) provides a necessary coupling among the species and momentum equations. The density and the dynamic viscosity of the liquid in eq. (5), in fact, are computed according to the instantaneous distribution of ϕ. Additional coupling between the species and momentum equations is due to the volume force term (Boussinesq approximation) in eq. (5).

The additional source term $\underline{F}_{Ma}$ in eq. (5) takes into account the other driving force involved in the phenomena under investigation, i.e. the Marangoni effect. The strategy used herein to account for this force, in fact, is based on the Continuum Surface Force (CSF) and Stress (CSS) models (the philosophy underlying this approach is quite complex and is not treated here for the sake of brevity; the reader will find all the necessary details in the excellent study of Haj-Hariri et al., 1997); with this model surface tension gradients along the interface (Marangoni stress) are incorporated as body forces per unit volume in the momentum equations rather than as boundary conditions: Thus the second source term in eq. (5) reads

$$\underline{F}_{Ma} = \left[\frac{2\sigma}{R_i} \underline{\hat{n}} - \frac{\partial \sigma}{\partial T} \left(\underline{\underline{I}} - \underline{\hat{n}}\underline{\hat{n}} \right) \cdot \underline{\nabla} T \right] |\underline{\nabla} \phi| \tag{14}$$

where σ is the surface tension, σ_T its temperature derivative and $\underline{\underline{I}}$ the identity matrix. Note that following Warren and Webb (1969) the sole (experimentally determined) parameter σ_T is needed to account for the dependence of the surface tension on both T and C (the surface equilibrium concentrations, in fact, are not free parameters but, in their turn, functions of T as shown by Figure 1 and Eqs. (3)).

The Navier-Stokes equations (4) and (5) together with the species and energy equations can be solved within the framework of a standard projection method (see, e.g., Brown et al., 2001; Armfield and Street, 2002; Guermond et al., 2006). These methods are not described here since they are well known.

4. Plume Dynamics

Before embarking on the detailed description of the convective structures originated from single- or multi-droplet configurations and related instabilities, it is worth opening a brief discussion about the dynamics of plumes originated from localized sources of buoyancy (regions of finite extent with a constant solute concentration). At this stage of this chapter the brief treatment of this classical subject must be regarded as a necessary pre-requisite introduced to support the reader in the reading and understanding of the next sections (where concepts such as "plume", "plume cap", "jet", etc. are at the basis of the discussions).

The attention is focused on the development of a single plume from a line source of solute. A single laminar plume, also known as a free or unbounded plume, develops as the fluid near a point or line solute source begins to rise. As the lighter fluid rises, it pushes aside the material above it and is, itself, in turn deflected. The rising material produces a stalk, while the deflected fluid produces a cap on top. As the pushing and deflection continue, the edge of the cap may further fold over. The result is something that looks like a mushroom.

As outlined above, there are two types of commonly studied free plumes: a 2D line source plume that is produced by placing a horizontal source of finite extent in a box, and an axisymmetric plume created by a point source. Theory for steady laminar plumes is well-established. Scaling arguments indicate that the vertical velocity is constant (Batchelor, 1954). Numerical results are available for different cases (e.g. Fujii 1963; Brand and Lahey 1967; Worster 1986). These analyses are valid for the plume stem far from the leading edge (the cap) and do not specify the cap behavior.

Studies about the starting (cap) behavior are due to Moses et al., (1993), Kaminski and Jaupart, (2003). It is well-known since 1954, however, that as a plume rises, it widens due to the diffusion of both solute and momentum in the lateral plane (Batchelor, 1954). A line source plume forms a wedge, with the angle controlled by the rate of entrainment (Batchelor, 1954).

It is also known that different regimes of plume growth (categorized in terms of relative thickness of solutal and velocity boundary layers along the plume body) are possible in the Sc-Ra space (where $Sc=\nu/D$ is the Schmidt number, ν being the liquid kinematic viscosity and $Ra=g\beta_c(\Delta C)L^3/\nu D$ is the solutal Rayleigh number, L being a reference length and ΔC the solute gradient between the source and the surrounding liquid).

These distinct regimes of plume growth can be delineated by varying systematically the Rayleigh and Schmidt numbers. They, as mentioned above, are basically defined by the growth of solutal and velocity boundary layers with the plume height. The variation in the thickness of these boundary layers leads to partition the Ra–Sc number space into three regimes of plume growth.

The *diffusive-viscous regime (DV)* is defined by both wide solutal and velocity boundary layers. Plumes in this regime have thick plume stems with no distinct plume heads and no strong vortices.

The strength of the diffusive component is clearly evident in the concentration map. It is responsible for the homogenization of the concentration field around the plume, which leads, as mentioned above, to a thick plume stem without a separate plume head. The velocity field for a plume with Sc=1 and $Ra=10^5$ is shown in Figure 2a. There are cells of opposite vorticity on the right and left halves of the plume. The cells are relatively diffuse and no strong spiraling vortices are seen.

This regime is characterized by low Rayleigh numbers ($O(10^5)$) and Schmidt numbers of order unity. Due to the lack of fine-scale turbulent structures, numerical simulations of this regime can be conducted with relatively coarse grids.

The *viscous-nondiffusive regime (VND)* is defined by thin concentration boundary layers and thick velocity boundary layers. It has thin plumes with sharp features and no strong vortices. Figure 2b shows that the plume shape is clear and distinct with a thin stem. The plume has a well-defined head with side lobes. With time the plume lobes remain relatively distinct and are not significantly deformed by vortex structures. The vorticity field is featured by two broad regions of opposing vorticity on each side of the plume without sharp vortices. The thick velocity boundary layers seen in this flow result in the velocity being dissipated by viscous friction rather than by vortices.

This regime occurs typically at $Ra=O(10^7)$ and higher Schmidt number ($Sc \geq O(10)$). Since it is characterized by small, thin plume structures due to the low diffusion coefficient, fine grids are required to properly capture the related dynamics.

The *inviscid-nondiffusive (IVND)* regime is defined by both thin concentration and velocity boundary layers. The nondiffusive nature of the regime can be clearly observed in the concentration field (See Figure 2c). The plume has a thin, sharp stem with a well-defined cap and lobes that are significantly deformed by vortex structures.

This regime takes place for high Rayleigh numbers ($Ra>O(10^6)$) and Schmidt numbers $=O(1)$. This is the most turbulent of the possible regimes. It is characterized by fine, turbulent structures that cover a wide range of small scales. This makes the regime the most difficult one from a computational point of view.

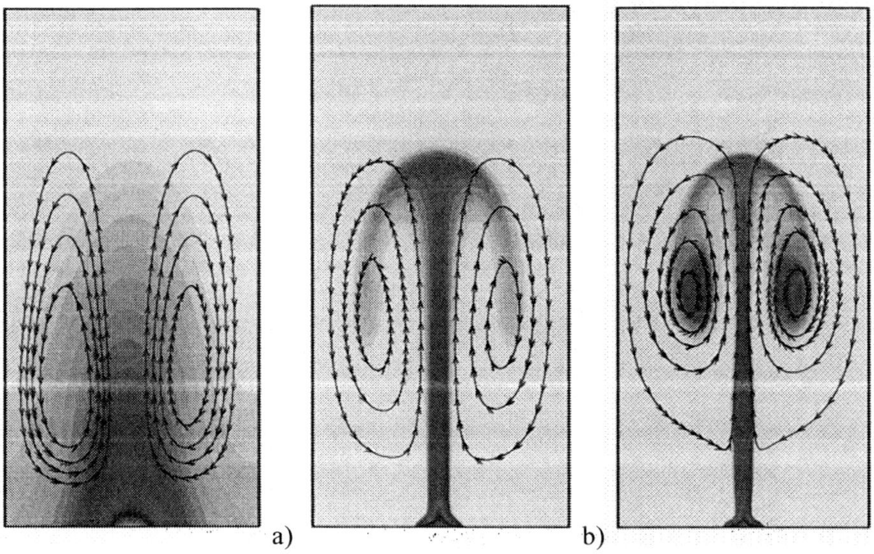

a) b) c)

Figure 2. Example of plumes (concentration contour map and streamlines) for various possible regimes (the plume is obtained by placing a line solute source at the center of the bottom surface of a box): a) Viscous-diffusive regime (DV) $Sc=O(1)$ and $Ra=O(10^5)$, b) Viscous-nondiffusive regime (VND) $Sc=O(10)$ and $Ra=O(10^7)$, c) Inviscid-nondiffusive regime (IVND) $S=O(1)$ and $Ra \geq O(10^7)$.

In the DV regime, the diffusion of solute away from the plume into the surrounding fluid tends to result in broader, more large-scale structures, whereas the VND and the IVND regimes, are characterized by a wide range of small-scale structures.

The regimes related to the experimental configuration described in Sect. 2 and depicted in detail in Sect. 5 and 6 correspond to the VND regime (Sc=O(10^3), Ra=O(10^8)).

5. Single-Droplet Configuration

Hereafter, some experimental results (supported by numerical simulations in the framework of the DDVOF method) are discussed as relevant examples of the curious dissolution phenomena that may be encountered when dealing with a single dissolving droplet.

First, the attention is focused on the case of an isothermal configuration, then the presence of a temperature gradient is considered and compared with the isothermal condition. It is worthwhile to stress how, in the case of constant temperature throughout the system, the onset of Marangoni stresses at the drop surface is prevented. In this case, in fact, neither temperature nor solutal gradients can exist at the interface separating the fluids (the equilibrium concentration values outside and inside the droplet, respectively, are constant along the surface). The comparison between isothermal and nonisothermal conditions, allows the role played by the Marangoni effect in the dissolution process and on the associated convective mechanism to be discerned.

5.1. Dissolution in Isothermal Conditions

According to the experimental results, if a droplet of methanol is formed on the tip of a capillary immersed into the cyclohexane (Figure 3), in spite of the small differences in density between the drop and the surrounding matrix, a convective plume is clearly observed; a jet originates at the tip of the drop and is directed upward (Figure 4).

In earlier experiments with different liquid-liquid systems (aniline/water, isobutanol/water, ethilacetate/water, see Agble and Mendes Tatsis, 2000) similar convective plumes were also observed. In these studies, the systems basically consisted of sitting drops or of pendant drops. It was shown that the convective plume originated from the drop is directed downward or upward according to whether the liquid is heavier or lighter than that used for the matrix.

For the system under investigation, the droplet (methanol) is heavier than the surrounding matrix. Nevertheless, the experiments performed in this case (isothermal conditions) show an upward methanol flow, starting from upper point of the drop. This flow corresponds to the classical solutal plume typically observed in the buoyant phenomena illustrated in Sect. 4. This means that for the case treated here a rising plume is created in the host liquid (cyclohexane) despite this liquid exhibiting a smaller density with respect to the drop (methanol). Such an apparent counterintuitive behavior, in reality, is a consequence of the fact that the density of the cyclohexane-methanol system does not necessarily behave as an increasing function of the methanol concentration and can be a decreasing function for small values of this concentration (Figure 1b). In practice, at the early stages of the dissolution process, the density of the majority phase (rich in cyclohexane) tends to become smaller even though methanol is heavier than cyclohexane.

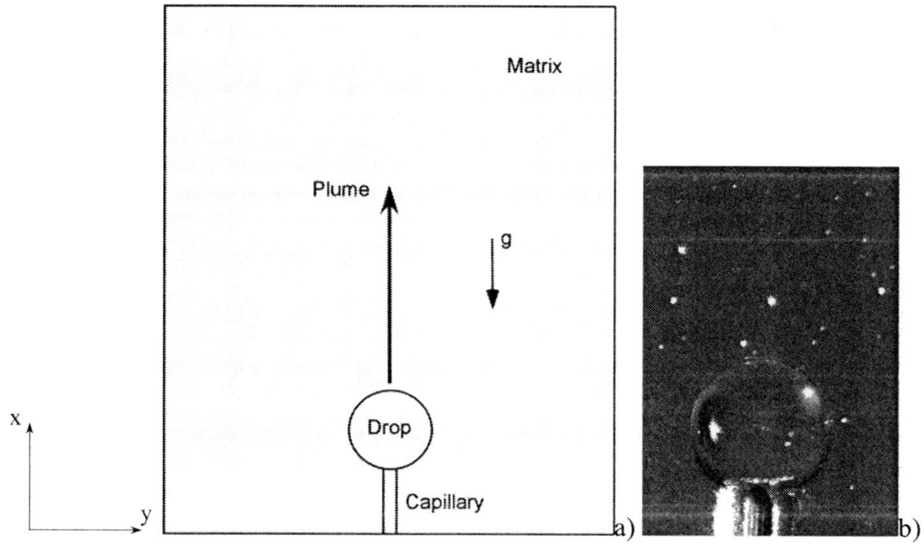

Figure 3. a) Sketch of the geometrical configuration; b) Liquid drop at the tip of the capillary tube.

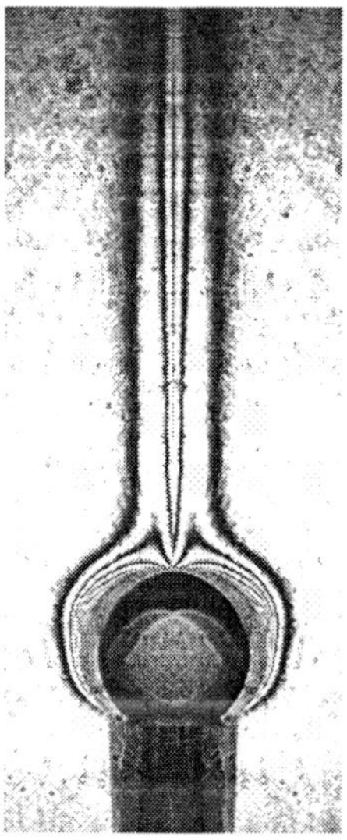

Figure 4.

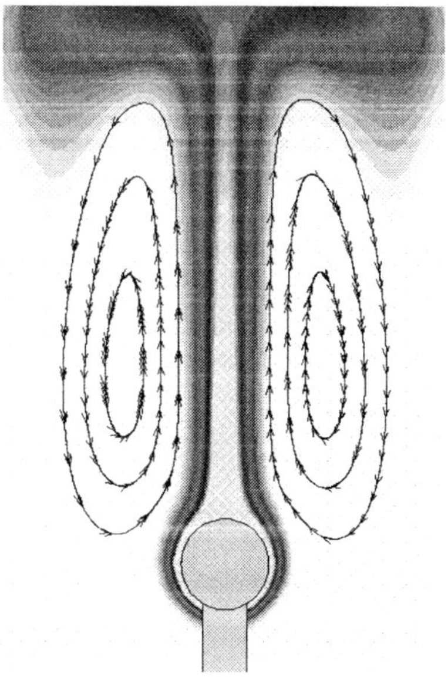

Figure 5.

Rising plume from a droplet of methanol dissolving in cyclohexane (uniform temperature distribution T= 25 [°C], initial drop volume 3 [μl]): Figure 4, experimental results, Figure 5, numerical results.

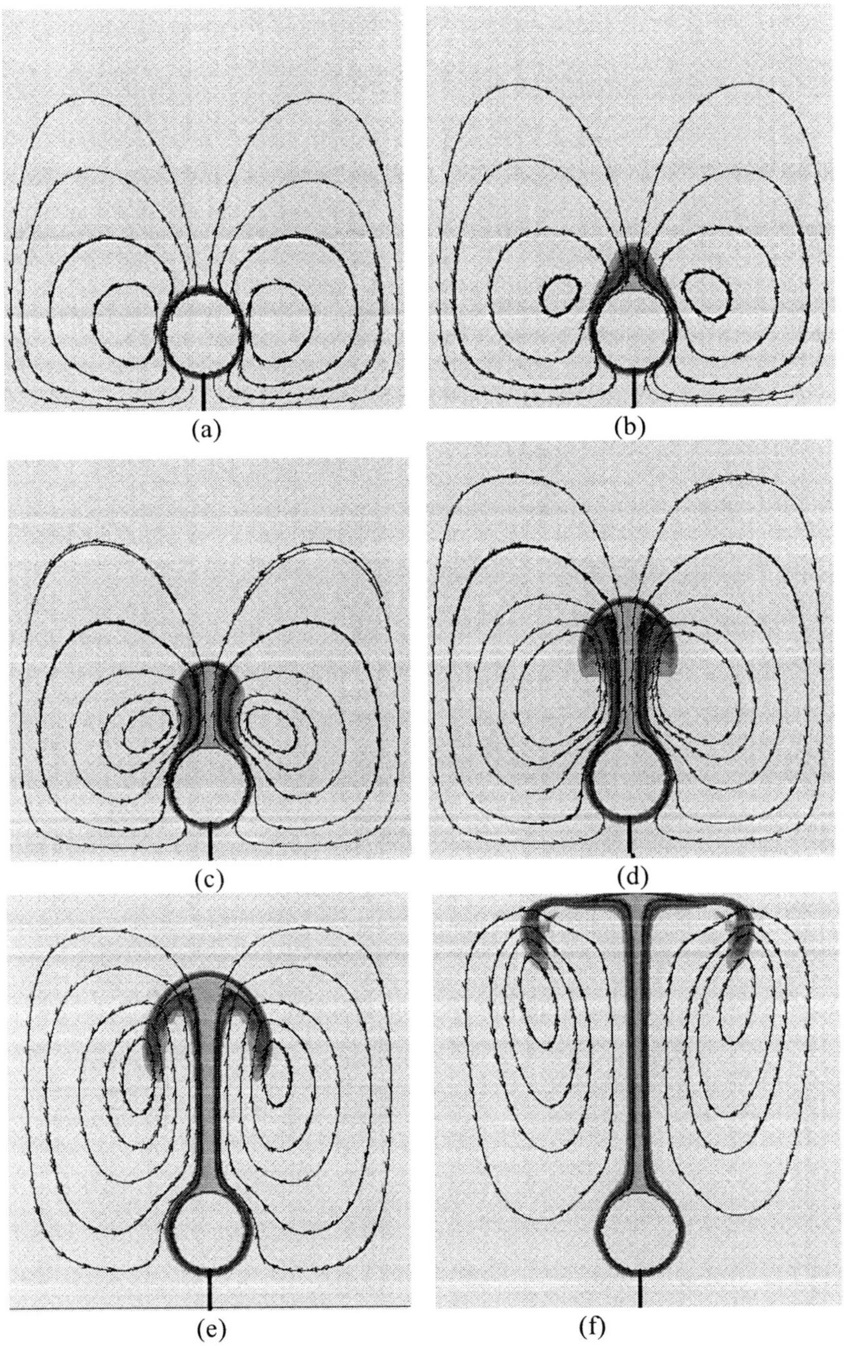

Figure 6.Computed evolution of the concentration and flow-field patterns at the beginning of the dissolution process (droplet of methanol dissolving into a matrix of cyclohexane, initial volume of the droplet 3 [μl]). The time intervals are 8 [s].

As shown in Figure 6, after an initial transient time, a quite stable buoyant rising jet continuously carrying methanol towards the top of the test cell is created (Figure 5).

This case is particularly significant and relevant within the context of this chapter since it demonstrates (in principle) that, owing to the particular density behavior of systems with a miscibility gap, drops sedimented to the bottom of the enclosure containing the immiscible alloy can give rise to rising solutal jets even though they are heavier than the surrounding liquid.

5.2. Dissolution in Non-Isothermal Conditions

If a temperature gradient ($\partial T/\partial x$) is applied to the liquid matrix, by heating the top of the test cell and cooling the bottom, the regular release of methanol in the rising jet discussed in Sect. 5.1, is taken over by a new spatiotemporal behavior (Lappa et al., 2004a).

For sufficiently small values of the temperature gradient ($\partial T/\partial x$), the transport of methanol in the rising jet is similar to that obtained in the case of isothermal conditions, but if $\partial T/\partial x$ exceeds a certain critical value (for instance, for a droplet having an initial volume of 3 [μl], the critical $\partial T/\partial x$ is about 1 [°C cm^{-1}]), the aforementioned mechanism undergoes a transition to an oscillatory axisymmetric complex flow pattern.

Instead of a steady regular plume, release of solute occurs in the form of discrete events separated by certain time (and thus space) intervals.

The instability, in fact, leads somehow to a periodic shooting of a packet of lighter fluid in the surrounding liquid (Figure 7).

This behavior appears in the generic meridian plane in the form of a couple of curls that are initially formed close to the drop surface and then are transported upward by buoyancy effects. In practice, the phenomenon behaves as a disturbance traveling towards the top (i.e. a perturbation of the solutal and flow field periodically rising along the symmetry axis).

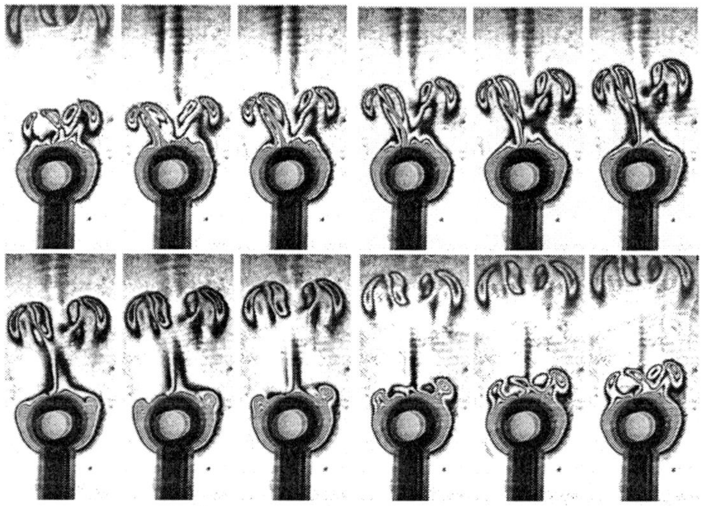

Figure 7. Experimental results (patterns of interference fringes): oscillatory instability of the axisymmetric plume rising from a drop of methanol dissolving in a cyclohexane matrix with $\partial T/\partial x$ = 2.3 [°C cm^{-1}] ($T_{average}$=27 [°C]). The initial volume of the drop is 3 [μl]. The images are taken at time intervals of about 1 [s]. One period of the oscillating behavior is shown.

This scenario is confirmed by the numerical simulations (Figure 8). The mathematical model and the associated numerical algorithm introduced in Sect. 3 are able to capture the complex time-dependent phenomena and to elucidate some experimental behaviors. For instance, the numerical simulations in Figure 8 show that the curls in the interference-fringe pattern correspond to the presence of a toroidal convection roll periodically released at the peak of the dissolving drop.

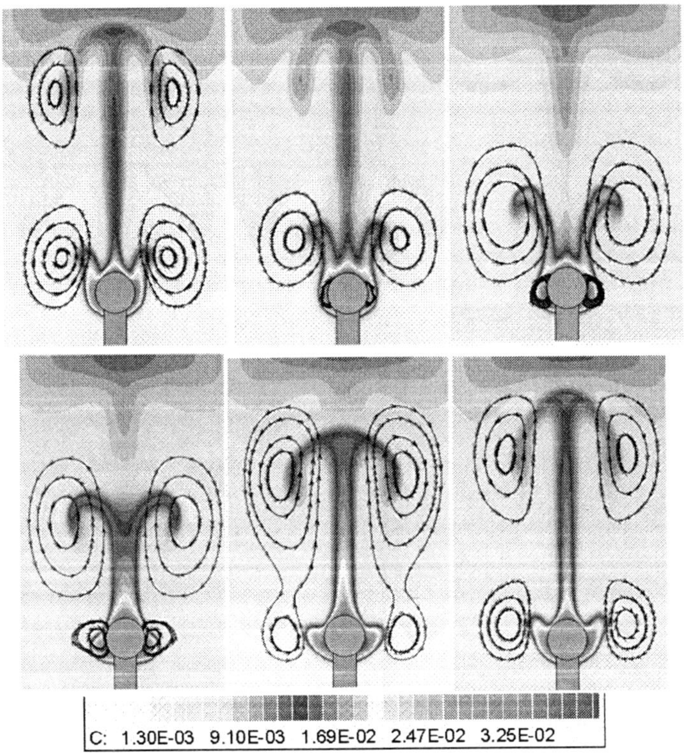

Figure 8. Computed evolution of the unstable solutal plume: concentration and flow field patterns for a 3 [μl] droplet of methanol dissolving into a matrix of cyclohexane with $\partial T/\partial x$=2.3 [°C cm^{-1}] ($T_{average}$=27 [°C]). One period of the oscillating behavior is shown (the period τ has been divided into 6 parts and the concentration and velocity fields are shown in figs a,b,c,d,e,f corresponding to t=0, τ/6, τ/3, τ/2, 2τ/3, 5τ/6, respectively, with τ≅19 [s]).

5.3. Higher Modes of the Unstable Convection

If the analysis is extended to higher values of the imposed temperature gradient, as expected, the phenomena exhibit an increasing degree of complexity (Lappa and Piccolo, 2004).

The periodic release of the aforementioned convection roll tends, in fact, to become very intricate. Figures 11 and 12 show intriguing computed stages of evolution of the solutal plume in the case of a droplet having initial volume 5 [μl] and $\partial T/\partial x$= 5 [°C cm^{-1}] and 8 [°C cm^{-1}], respectively.

The behaviors in these figures provide a clear evidence for the high sensitivity that these systems exhibit to the applied temperature difference. Very complex scenarios arise. The rising jet is not only broken into discrete temporal events, but loses its single-finger structure (shown in Figures 7 and 8). In the generic meridian plane, in fact, two somewhat wave-shaped branches appear (see, in particular, Figure 9).

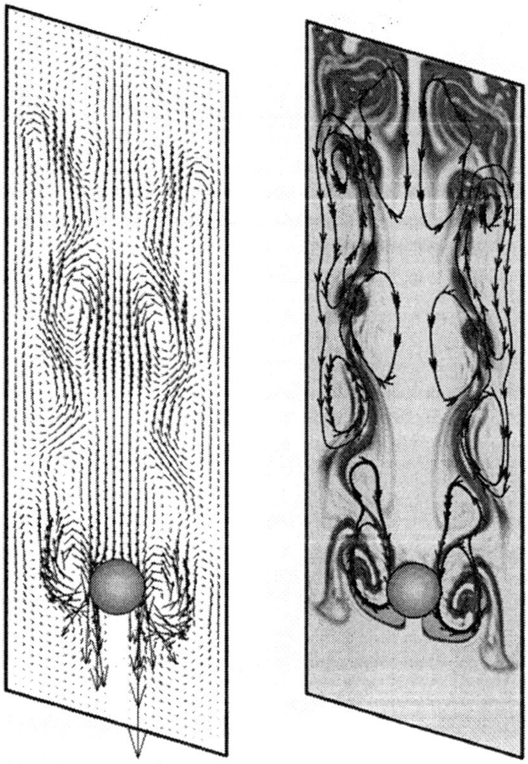

Figure 9. Computed vector velocity field and solutal plume: concentration and flow-field patterns for a 5 [μl] droplet of methanol dissolving into a matrix of cyclohexane with $\partial T/\partial x=5$ [°C cm^{-1}] ($T_{average}=27$ [°C], t= 88.0 [s]).

The frequency with which the toroidal vortices are released increases as a function of $\partial T/\partial x$ and, accordingly, many interacting rolls can simultaneously affect flow and solute distribution in the test cell. Figure 9 show, in particular, that the multicellular structure along the symmetry axis is featured by a periodic alternance of toroidal rolls with small and large diameter. This effect is responsible for the ondulations in the aforementioned two opposed branches (visible in the concentration distribution for the generic meridian plane).

Table II. Modes of convection as they appear in the generic meridian plane (5 [μl] droplet of methanol dissolving into the matrix of cyclohexane)

$\partial T/\partial x$ [°Ccm^{-1}]	Regime
<1.2	Stable plume
1.5	Pulsating rising jet - single finger structure
5	Periodic shooting of solutal packets (sinusoidally shaped branches)
8	Quasiperiodic shooting of solutal packets (sinusoidally shaped branches)

Progression from spatial "order" to "disorder" versus $\partial T/\partial x$ is shown in Figures 10, 11 and 12. Superimposed on these aspects is the evolutionary transition from the initial steady state to a final quasi-periodic fluctuating flow regime that appears if the phenomenon is considered from a temporal point of view.

When $\partial T/\partial x$ is sufficiently small a well-defined frequency (f_0) can be identified but other incommensurable frequencies superimpose on the principal one when $\partial T/\partial x$ is increased.

Table II summarizes the prominent spatial and temporal structure of the rising solutal jet as a function of the increasing $\partial T/\partial x$: "stable plume", "pulsating rising jet - single finger structure", "periodic shooting of solutal packets - sinusoidally-shaped branches", "quasiperiodic shooting of solutal packets - sinusoidally-shaped branches" are the different possible regimes of evolution of the system shown, respectively, in Figures 5, 10, 11 and 12.

This time-dependent oscillatory convection is not only of inherent phenomenological significance, but it is also of prime practical interest to ground-based immiscible alloys processing. The oscillatory thermosolutal convection generated by the drop, in fact, could influence solute segregation and interfere with solidification mechanisms in a significant manner.

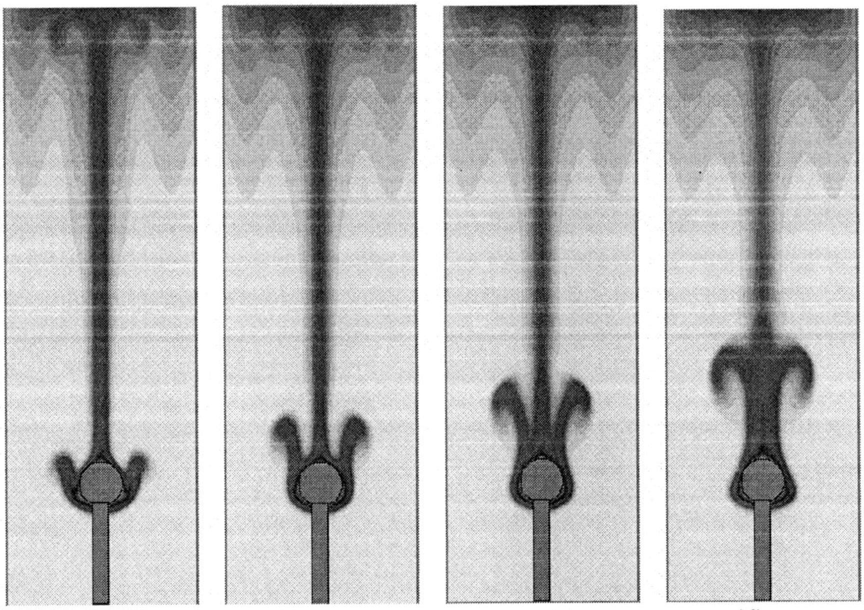

(a) (b) (c) (d)

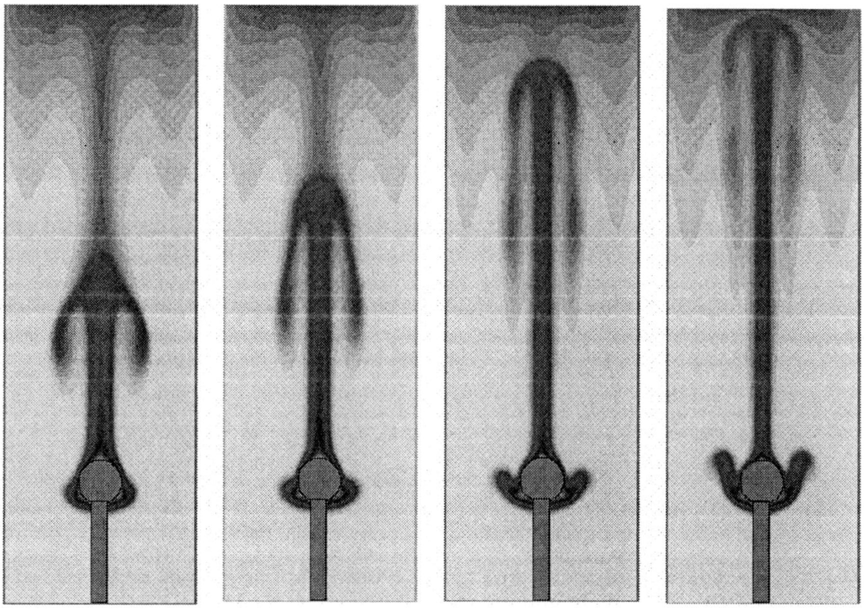

(e) (f) (g) (h)

Figure 10. Computed evolution of the unstable solutal plume: concentration and flow-field patterns for a 5 [μl] droplet of methanol dissolving into a matrix of cyclohexane with $\partial T/\partial x$=1.5 [°C cm^{-1}] ($T_{average}$=27 [°C]): one period of oscillation is shown by means of subsequent snapshots.

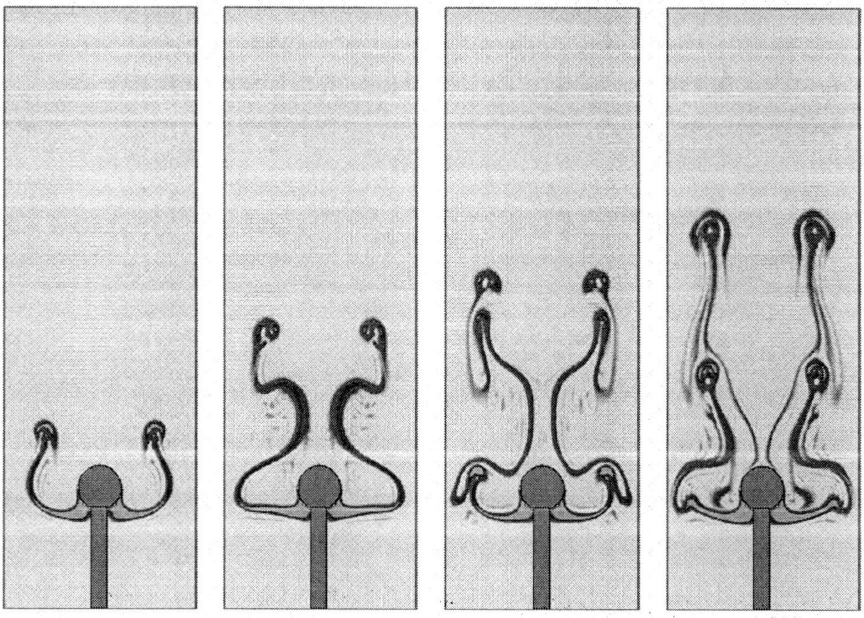

(a) (b) (c) (d)

Figure 11. (Continued).

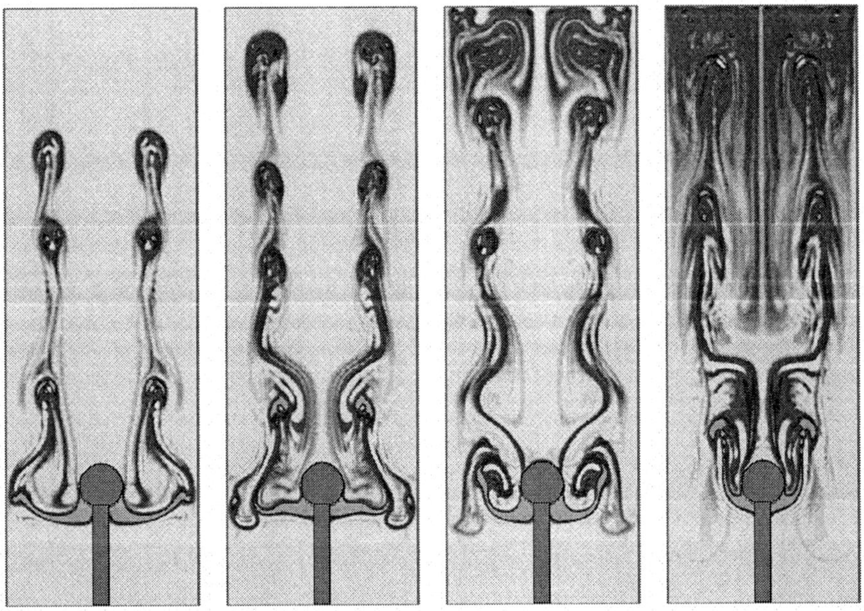

(e) (f) (g) (h)

Figure 11. Computed evolution of the unstable solutal plume: concentration and flow-field patterns for a 5 [μl] droplet of methanol dissolving into a matrix of cyclohexane with $\partial T/\partial x$=5 [°C cm^{-1}] ($T_{average}$=27 [°C]): (a) t= 17.9 [s]; (b) t= 27.7 [s]; (c) t= 34.2 [s]; (d) t= 42.4 [s]; (e) t= 52.2 [s]; (f) t= 68.5 [s]; (g) t= 88.0 [s]; (h) t= 117.4 [s].

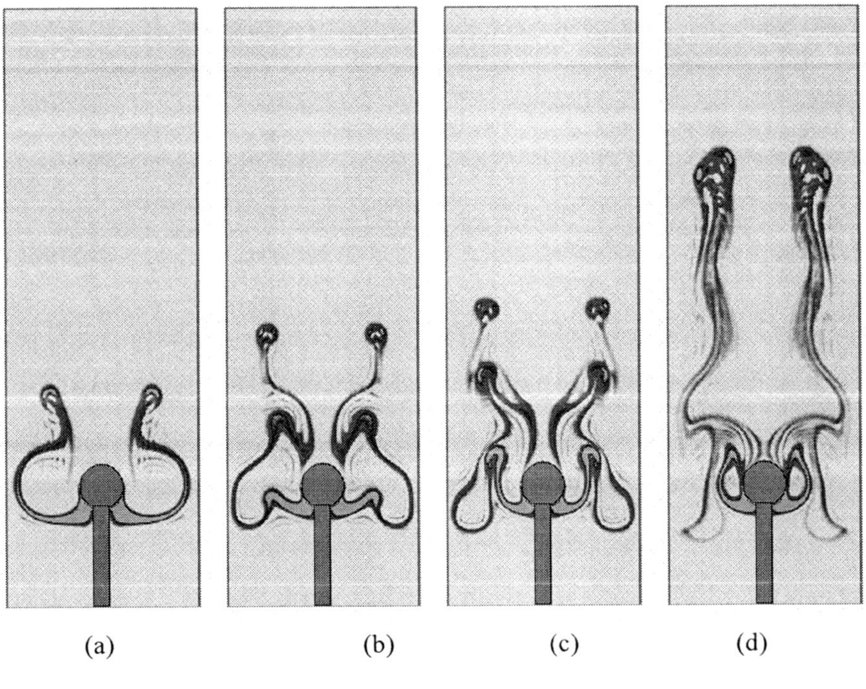

(a) (b) (c) (d)

Figure 12. (Continued).

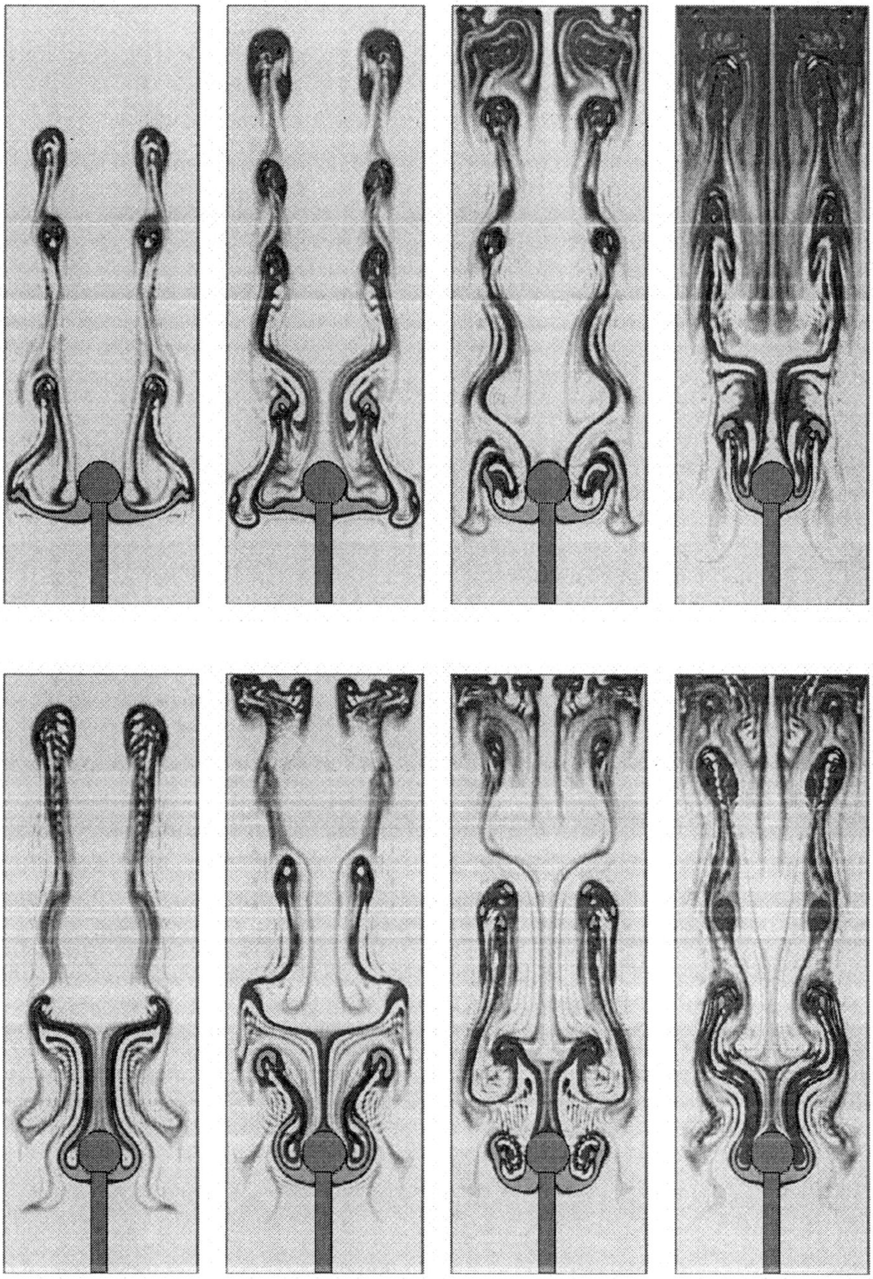

(e) (f) (g) (h)

Figure 12. Computed evolution of the unstable solutal plume: concentration and flow-field patterns for a 5 [µl] droplet of methanol dissolving into a matrix of cyclohexane with $\partial T/\partial x$=8 [°C cm^{-1}] ($T_{average}$=27 [°C]): (a) t= 17.9 [s]; (b) t= 24.5 [s]; (c) t= 27.7 [s]; (d) t= 44.0 [s]; (e) t= 52.2 [s]; (f) t= 65.2 [s]; (g) t= 76.6 [s]; (h) t= 86.4 [s].

5.4. Some Scaling Arguments

This section is devoted to the discussion of the intrinsic nature of the curious instabilities described in the earlier sections for the single-droplet case.

Along these lines, computations obtained "switching off" surface-tension effects, can be used to prove that the presence of the Marangoni effect is crucial in determining them. These simulations show, in fact, that in the absence of Marangoni effects the solutal jet is not broken into discrete events and the aforementioned characteristic curls do not appear.

It is well known that Marangoni forces drive liquid from hot regions towards cold regions. Accordingly, on the surface of the drop, Marangoni effects tend to drive liquid towards the bottom. On the contrary, solutal buoyancy effects act in the opposite direction. Therefore, when the test cell is heated from above and cooled from below, thermal Marangoni and solutal buoyancy flow counteract.

In view of these arguments and due to the fact that the instability does not appear if the surface-tension forces are artificially neglected in the simulations, a theory based on the delicate balance of the aforementioned counteracting phenomena could be proposed to explain the oscillatory behavior.

A particular procedure can be used to get insights into these aspects. The basic approach proposed herein for this purpose relies on the combination of the available experimental and numerical results with theoretical models. First the trends in available data are used to postulate qualitatively mechanical principles for explaining the instability (for the case under investigation the aforementioned delicate balance between solutal buoyancy forces and Marangoni thermal stresses). Then, on the basis of the foregoing principles, explicit laws are proposed and tested for their ability to predict the experimentally determined patterns of the instability threshold.

In practice, this approach is based on a parametric analysis about the effect of the initial volume of the droplet. A scaling analysis, in fact, is carried out to discern the theoretical relationship between the critical temperature gradient and the initial radius of the droplet.

If the instability was the result of a delicate counteracting behavior between solutal buoyancy and thermal surface driven effects, it would be possible to draw information about the critical threshold simply imposing that

Intensity of solutal buoyancy effects/intensity of Marangoni thermal effects = O(1)

The relative importance of these effects can be estimated by comparing the relevant characteristic velocities.

The solutal buoyancy velocity reads:

$$V_g^C = \frac{g\beta_{c2}\Delta C(2R)^2}{\nu} \tag{15}$$

where ΔC is the initial concentration difference between the drop surface and the top of the test cell (for the present case $\Delta C \cong C_{2(o)}$)

The thermal Marangoni velocity reads:

$$V_{Ma}^{T} = \frac{\sigma_T \frac{\partial T}{\partial x} 2R}{\mu} \tag{16}$$

According to the foregoing arguments the conditions for the onset of the oscillatory instability should correspond simply to

$$\frac{V_{Ma}^{T}}{V_{g}^{C}} = O(1) \tag{17a}$$

$$\rightarrow \frac{\partial T}{\partial x} \propto \frac{g\beta_c \rho \Delta C}{\sigma_T} 2R = \frac{g\beta_{c2}\rho\Delta C}{\sigma_T} 2\sqrt[3]{\frac{3}{4\pi}Vol} \tag{17b}$$

Equation (17b) is the aforementioned analytical relationship between the critical gradient and the droplet diameter provided by the theoretical scaling analysis carried out under the assumption that the instability is caused by the opposition of the surface Marangoni stress to the natural direction in which the fluid wants to move (upward, because of buoyancy). This explicit law shows that the critical ∂T/∂x is an increasing function of the initial volume of the droplet.

To provide physical validation to these arguments, eq. (17b) is finally compared with numerical and experimental results. Such a comparison shows that the relationship presented herein fits the dependencies provided by the experimental and numerical results (this is clearly shown in Figure 13). Finally, this good agreement can be regarded as a validation of the proposed theory.

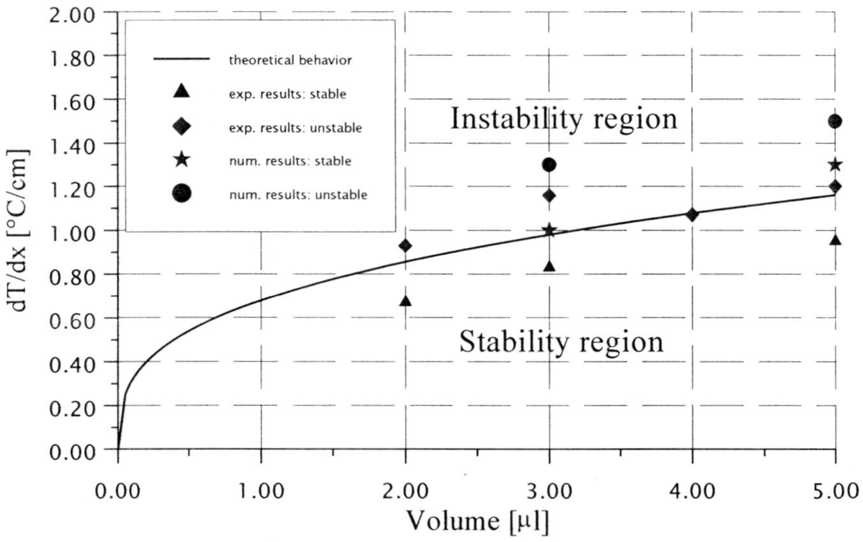

Figure 13. Critical temperature gradient for the onset of oscillations as a function of the initial volume of the dissolving droplet (Lappa et al., 2004a).

The consistency of model predictions with experimental data suggests that rate-controlling steps have been taken into account, and that simplifications do not distort actual behavior. According to Figure 13, the proposed theory about the counteracting interplay of surface-tension and buoyancy forces exhibits quite exhaustive capabilities to predict and elucidate experimental observations (in particular with regard to the behavior of the system as a function of the initial droplet volume) and to identify cause-and-effect relationships, i.e. to give insights into the mechanisms driving the phenomena under investigation.

To recapitulate, when the temperature gradient is sufficiently small, solutal buoyancy effects prevail and the velocity along the drop surface is upward (i.e. directed from the capillary towards the point at the peak of the drop). In this case the release of solute occurs in the form of a continuous rising jet of lighter liquid starting at the peak of the methanol drop.

If the temperature gradient is increased, correspondingly, the relative importance of the surface Marangoni stress with respect to local buoyancy forces increases.

The surface-tension-induced flow tends to carry the hot fluid from the peak of the drop into the lower regions of the domain just below the drop base. The solutal buoyancy force tends to oppose the thermocapillary effect.

The $\partial T/\partial x$ at which the aforementioned counteracting effects can be considered competitive in determining the direction of the surface flow, corresponds to the critical conditions for the onset of the instability. For temperature differences larger than the critical one, a "back and forth" mechanism occurs, i.e. an instability of the mixed buoyancy-Marangoni type.

A clear evidence of the counteracting behavior responsible for the onset of the instability is given by the presence along the drop surface of a “detachment point” (Figure 14). This point corresponds to the detachment of the aforementioned layer of liquid transported close to the interface by surface tension forces. Owing to the solutal buoyancy forces that oppose to downward motion, the surface layer has not enough momentum to reach the rear of the drop (this phenomenon exhibits an outstanding theoretical kinship with the corresponding phenomenon, well known by aerospace engineers, by which air boundary layers “separate” from thin wing airfoils).

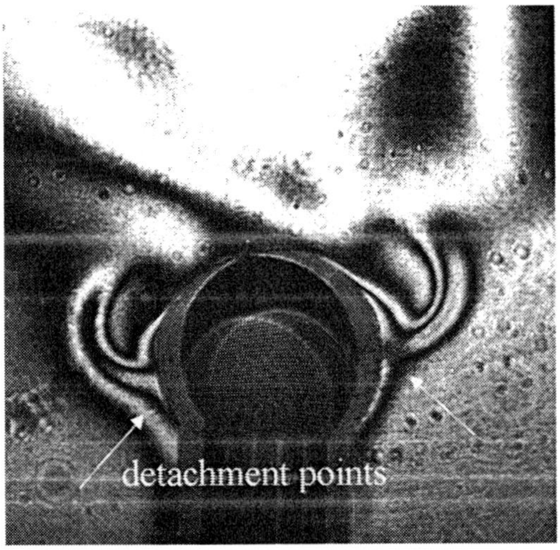

Figure 14: Experimental evidence of “detachment points” along the drop surface.

For this reason return flow is initiated prior to reach the drop base. Since the detached layer is lighter than the liquid matrix, not far away from the drop, the buoyancy forces tend to bring this fluid towards the top. Thus, the restoring upward force exerted by the stabilizing solutal density gradients on the downward flowing hot fluid, restricts the flow (downward motion and return rising flow) to an area close to the drop.

In other words, the surface transfers warm liquid from the peak of the drop to the lower regions until at a certain stage (the detachment point) solutal buoyancy and continuity bring the fluid up again resulting in a closed toroidal vortex (this also means that the generation of curls close to the drop surface is strictly related to the counteracting behavior of buoyancy and surface-tension forces; it is the direct opposition of these forces that leads to the production of vorticity). Of course, the detachment point does not hold a fixed positions, but continuously undergoes rising and descending motion along the drop surface according to the alternate direction of the surface flow. Such a behavior is strictly associated with the aforementioned periodic release of toroidal convection rolls in the region around/above the dissolving drop (i.e. it is responsible for a "continuous" production of vorticity).

The complex interplay among transport of heat, solute and momentum results in a pumping action first pushing down and then pulling up with a time-dependent mechanism the fluid close to the drop surface. This results in a complex flow structure that affects the entire enclosure as depicted in the sequence in Figures 7 and 8.

All these consideration illustrate why the instability occurs only if a certain initial value of the temperature gradient is exceeded, whereas the flow exhibits quite stable solutal rising plumes in the case of small initial ΔT or isothermal conditions.

For supercritical regimes, the surface tension contribution becomes the dominating effect (surface tension forces, in fact, are strongly temperature-dependent whereas solutal buoyancy forces tend to stay constant). This explains why the possible positions of the detachment point tend to stay confined close to the rear of the drop when the applied temperature gradient is increased as shown in Figure 15 (see also the computations in Figures 11 and 12).

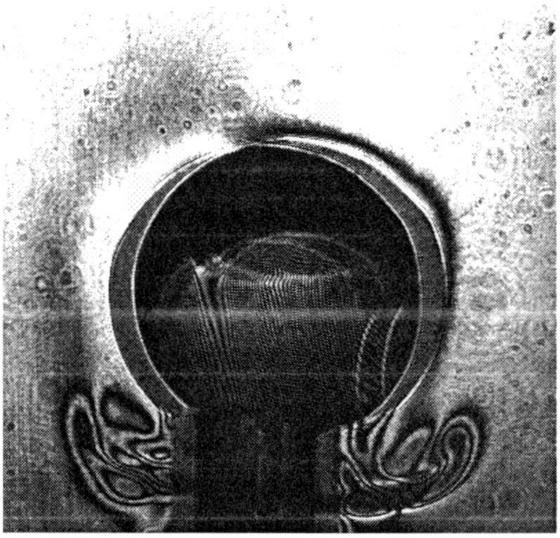

Figure 15. Surface flow for a large temperature gradient ($\partial T/\partial x>3$ [°C cm^{-1}]).

5.5. Comparison with other Systems in Nature

To conclude Sect. 5, it is worth pointing out that the oscillatory phenomena elucidated in the foregoing exhibit some remarkable analogies with other systems in nature.

For instance, typical instabilities of rising buoyant jets were investigated by Cetegen and coworkers for a number of different cases and fluid-fluid systems (e.g., candle flames and buoyant plumes).

Cetegen (1997) carried out experimental studies dealing with the instabilities and flow transitions of buoyant plume/jets of gas mixtures. This author investigated the case of a buoyant plume of helium or helium/air mixture originating from a large axisymmetric nozzle with low velocity. He observed toroidal-vortex formation as a result of rapid buoyant acceleration of light plume fluid in heavier more or less quiescent surroundings.

It was found that these plumes may undergo periodic oscillations: as the buoyant fluid exits the nozzle, the plume boundary contracts towards the plume centerline as a result of buoyant acceleration due to the hydrostatic pressure field and the condition imposed by the conservation of mass; the plumes undergo periodic oscillations of the plume boundary starting in the immediate vicinity of the nozzle lip. This behavior might resemble that discussed before about the periodic release of curls close to the drop surface.

Moreover, in analogy with the present case of solutal plumes, it was disclosed experimentally by Cetegen (1997) that not all buoyant plumes exhibit periodic oscillations. The onset of oscillations is, in general, a function of nozzle diameter, nozzle exit velocity and the density ratio between the plume fluid and its surroundings (Cetegen, 1997) as for the present case, instability occurs only if a critical temperature gradient is exceeded.

Cetegen and Kasper (1996) provided a very interesting explanation of the observations: according to such explanation, the mechanism leading to the periodic oscillatory state of the flow field is connected with the highly unstable (Rayleigh-Taylor) density stratification in the sharply contracting region of the flow just above the nozzle exit.

Important similarities can also be found about the features of pool fires. It is well known (Cetegen and Ahmed, 1993) that axisymmetric pool fires exhibit a periodic oscillatory motion close to their origin, often referred to as puffing. In a fire, these periodic oscillations result in the formation of large-scale (of the order of the burner diameter) flaming vortical structures at a short distance from the burner surface. These structures significantly modify the downstream flame behavior as they rise through the flame and finally burnout near the flame top.

Again the behavior is very similar to that observed in the case of the periodic release of packets of fluid above the dissolving drop. The perturbation in the concentration field originated at the drop surface rises along the core of the jet and dies at the top.

It is crucial to point out how, despite the macroscopic analogies and apparent common features, the cause-and-effect relationships exhibit notable differences with respect to the present case of dissolving drops in nonisothermal conditions.

In the case of isothermal helium plumes and flames the oscillatory condition follows from an instability of the pure buoyant flow. In particular, for helium plumes the force driving the instability is the pressure gradient between the helium reservoir and the ambient. In the case of flames and fires, the driving force is the release of heat in the core of the flame due to the combustion process; puffing in pool fires, in fact, is typically stronger than buoyant nonreacting plumes due to the local heat release and the consequently maintained buoyancy.

These instabilities are hydrodynamic in nature i.e. bifurcation to oscillatory flow occurs when the velocity in the core of the rising plume exceeds a critical value. Despite the similarities on the macroscopic scale the underlying mechanisms for the case of a dissolving drop are very different. The related transition to oscillatory flow is not induced by the effect of gravity alone (like the instabilities investigated by Cetegen and coworkers); moreover, it does not follow from an instability of the rising jet; it is due to a delicate balance between two counteracting effects (buoyancy and surface-tension-driven stresses) at the surface of the dissolving drop (as proven by the numerical computations carried out "switching off" the Marangoni effects).

These special features make this instability more similar to other categories of phenomena where free fluid-fluid interfaces supporting Marangoni stresses are involved. Along these lines, in particular, the oscillatory instability that occurs in the case of air bubbles surrounded by liquid and trapped by gravity below heated walls should be mentioned.

Experiments with transparent liquids indicate that, for a given radius, the flow and temperature fields generated by the bubble become oscillatory beyond a critical Marangoni number. From a kinematic point of view, two distinct classes of oscillatory flows were identified by Kassemi and Rashidnia (1999).

The initial or first class of oscillatory patterns are in the transverse direction and are mainly associated with azimuthally traveling waves in the flow field. In this case the phenomenon is fully three-dimensional and both symmetric and asymmetric oscillatory temperature and flow fields are possible in the generic meridian plane; they can be related to the number (even or odd) of disturbance lobes formed by the hydrothermal-wave disturbance on the annular ring-shaped Marangoni roll around the bubble.

This first class of instabilities falls into the same category of instabilities characterizing the case of liquid bridges with Pr>>1 which have instigated much research in the literature (see, e.g., Lappa et al., 2000 and 2001).

The second class of oscillatory pattern has a higher frequency. It is symmetric in the axial direction, and is associated with an axially traveling wave generating a mainly up-and-down axial movement in the flow and temperature fields.

This second class, not present in the case of liquid bridges, seems to resemble the behavior previously described in the case of a dissolving drop.

The analogy between the two instabilities (bubble and drop), however, also goes beyond the observed macroscopic oscillatory pattern. Some similarities can also be highlighted with respect to the underlying forces responsible for the time-dependent flow.

The available results for the bubble case clearly indicate that the axially-driven oscillatory state that occurs at high Marangoni numbers is induced by the dynamic interaction between the buoyancy and surface-tension effects and the resulting combined gravitational convection and Marangoni flow.

The mechanism is the following: At higher Marangoni numbers, the Marangoni convection is stronger and brings about a significant alteration of the temperature field around the bubble. An interesting feature of this temperature field is the existence of a distinct thermal structure in the form of a cold return flow intruding near the top of the bubble that is not present at lower Ma (hereafter referred to as "cold finger").

This cold finger plays a critical role in instigating the oscillatory response of the flow since the origin of the convective instability is closely related to an intricate temporal coupling between the large-scale thermal structures that develop in the fluid in the form of the cold finger and the temperature-sensitive surface of the bubble.

The cold finger can periodically reverse the local thermal gradient on the free surface leading to the onset of oscillations. Gravity enters the dynamics since it plays the important role of keeping the main Marangoni vortex near the bubble surface where all interaction occurs. In the absence of gravity the Marangoni vortex tends to grow unopposed until it nearly fills the entire enclosure; in such a situation the mechanism with which the cold finger affects the bubble surface is removed and the instability is suppressed.

In the light of these aspects and as also demonstrated in the analysis of Kassemi and Rashidnia (1999) by means of numerical simulation, the bubble-case instability is basically hydrothermal in nature, like that described in the landmark analysis of Smith and Davis (1983) for the case of liquid layers).

Therefore, despite all the analogies highlighted above, this type of instability is also not relevant to the present case related to dissolution of drops. In fact, visualization of the computed thermal field corresponding to Figure 10 does not show the presence of any cold finger impinging on the fluid-fluid interface (Figure 16). There is no reversion of the local thermal gradient on the free surface (Figure 17).

Moreover, the axisymmetric behavior does not belong to a secondary class of instabilities. It is a "primary" bifurcation of the flow.

As mentioned before, the instability investigated by Kassemi and Rashidnia (1999) is basically an instability of the Marangoni toroidal convection roll wrapped around the bubble very similar to that arising in the case of liquid bridges with Pr>>1 (see Lappa et al., 2001). Gravity plays only a marginal role in the related mechanisms: as previously explained it makes possible the existence of a cold finger that can periodically perturb the surface-temperature distribution leading to a periodic reversion of the thermal gradient. For instance, the presence of gravity is not necessary in the case of liquid bridges since even in zero-g conditions a cold finger is always present near the hot disk, which is created by the return flow that brings cool fluid away from the cold wall along the symmetry axis of the liquid bridge (the cold liquid is carried toward the hot surface at a rather high position, see the discussion in Lappa et al., 2000).

The instability related to the dissolving drop is different as it follows from the direct opposition of counteracting forces (a correct denomination would be "thermocapillary-solutogravitational instability of a mixed type"). As a consequence of such a nature, in fact, the drop is not surrounded by a Marangoni vortex with fixed position and undergoing minute accelerations and decelerations in the axial direction (like in the bubble case). Rather, the drop continuously generates vorticity with the periodic (or quasi-periodic) release of rising toroidal rolls (as mentioned before, the generation of "curls" close to the drop surface is strictly related to the counteracting behavior of buoyancy and surface-tension forces that leads to the continuous production of vorticity).

A critical comparison of the results with other worthy and important studies concerning similar (from a macroscopic point of view) phenomena leads to the conclusion that the curious flow instability treated in first part of this chapter, following from the delicate balance between counteracting thermal Marangoni forces and solutal-buoyancy effects, is neither a hydrodynamic transition of the pure buoyant flow nor a hydrothermal bifurcation of the Marangoni convection, but represents a new and relevant class of instabilities driven by lack of thermodynamic equilibrium that can be present when processing liquid-liquid systems with a miscibility gap in normal gravity conditions.

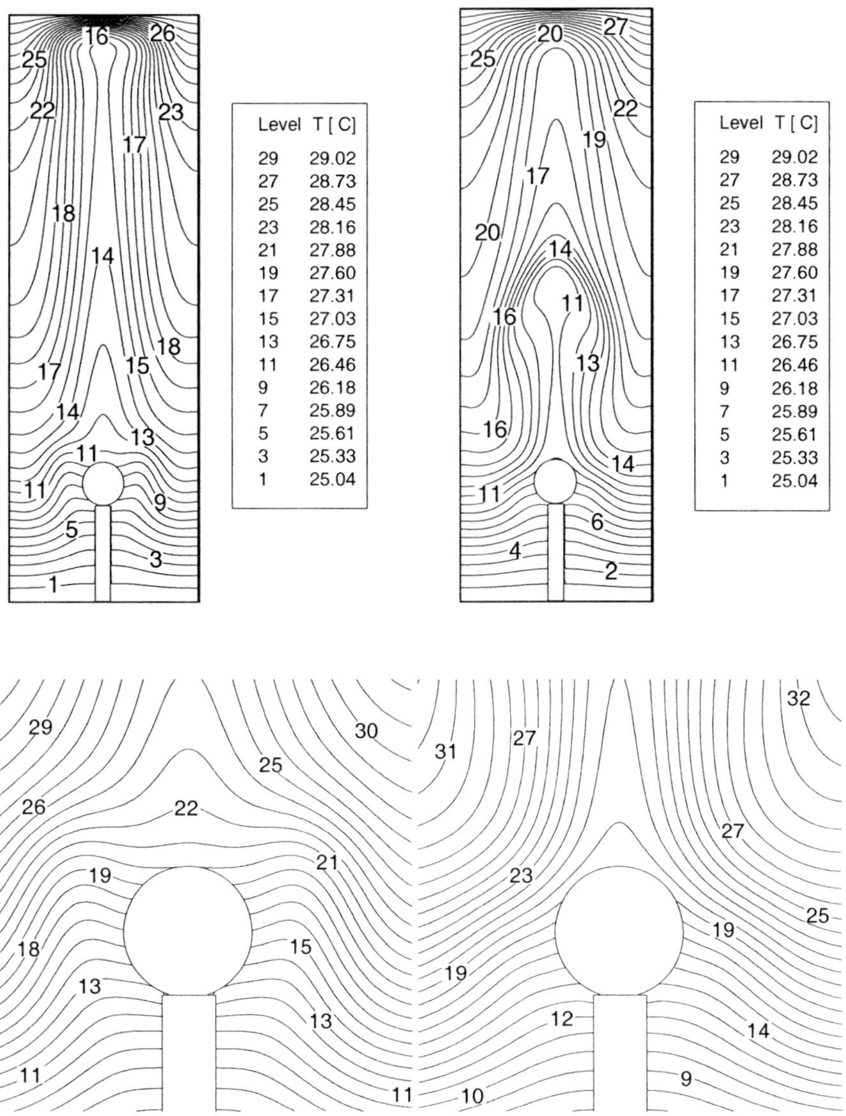

Figure 16. Temperature distribution for a 5 [μl] droplet of methanol dissolving into a matrix of cyclohexane with $\partial T/\partial x=1.5$ [°C cm^{-1}] ($T_{average}=27$ [°C]): two snapshots are shown corresponding to t=to and t=to+τ/2 where τ is the period of oscillation.

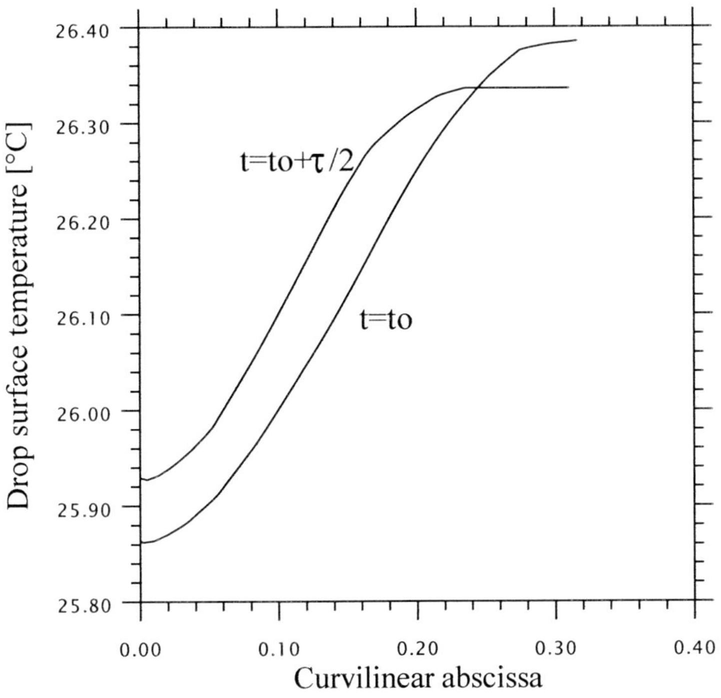

Figure 17. Temperature distribution on the drop surface for the same conditions of Figs. 16 ($\partial T/\partial x$=1.5 [°C cm^{-1}]).

6. MULTIDROPLET CONFIGURATION IN NON-EQUILIBRIUM CONDITIONS

The aim of the present section is to extend the results treated in Sect. 5, to the more realistic situation in which a relatively large number of droplets are uniformly distributed on the bottom of the container (a discrete layer of N drops). It is expected, in fact, that convective structures originated from different droplets can interact in unexpected ways making, therefore, "ensemble behaviors" with many droplets an important aspect of the problem.

As a relevant reference case, a container with aspect ratio $A=L_x/L_y=1$ is considered, x and y being the vertical and the horizontal directions, respectively (i.e. a liquid matrix confined between two parallel and vertical walls with a periodic array of evenly spaced spherical liquid bodies in the interior).

Other relevant non-dimensional geometrical parameters are the non-dimensional diameter of the droplets A=d/L=0.067 (L being the characteristic length of the container, L=3 [cm] and d=2 [mm]) and the parameter $\kappa=d/\ell$ (ratio of drop diameter to the distance ℓ between the centers of two consecutive droplets, Figure 18).

Like in Sect. 5, the droplets are supposed to be attached at the tip of the needles used to inject them into the matrix from the bottom of the container. However, for the present case, it is supposed that the needles protrude in the text cell only for few microns (Figure 18). The number of droplets is changed in order to analyze the sensitivity of the overall system to this parameter (N=4→κ=0.33 and N=5→κ=0.4). As already explained in Sect. 5 for the case of a single droplet, as droplets dissolve in the external matrix, each droplet enriches the surrounding fluid of methanol producing a light concentration-enriched zone around it.

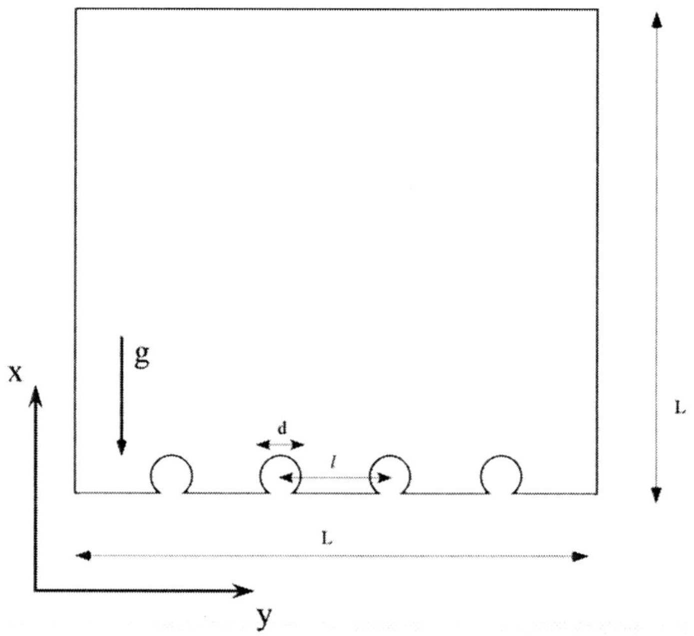

Figure 18. Sketch of the multidroplet configuration and related geometrical parameters.

In this region the solute concentration changes continuously from the local equilibrium concentration at the droplet surface to the concentration in the bulk of the solution. The related concentration profile varies with time and is controlled by the balance between diffusion of solute and the tendency of the fluid to rise owing to solutal buoyancy effects.

If a non-isothermal process is considered (system heated from above), the tendency of the fluid to be driven from the hot upper regions towards the bottom cold regions due to surface tension (Marangoni) effects also plays a crucial role in the dynamics mentioned above. For sufficiently large values of the temperature gradient, in fact, it can force the layer of liquid surrounding a drop to move downward even if it is lighter than the matrix (as already clearly shown in Figure 15).

In addition to these arguments, as outlined before, one must keep in mind that there is an effect related to the fact that, when many droplets dissolve at close distance some mutual influence occurs due to superposition and intersection of the related "enriched zones" and/or due to viscous interaction of convective structures originated from their surfaces (rising currents driven by solutal buoyancy effects, or descending currents driven by thermal Marangoni forces). The coupling between these convective "pluming phenomena" and the overlapping of the enriched zones can make the concentration pattern very complex. The next sub-sections consider the simultaneous presence of many droplets and the related possible mutual interplay.

6.1. Evenly Spaced Droplets in Isothermal Conditions

For relatively small values of the $\partial T/\partial x$ (<1 [°C cm^{-1}]), or for isothermal conditions, the flow around each droplet is dominated by solutal buoyancy effects. According to the results discussed in Sect. 5.1, in particular, it is known that in the case of a single droplet, after an initial transient time, a stable buoyant straight rising jet continuously carrying methanol towards the top of the test cell is created.

The present studies prove that this is not the case when an array of droplets is considered (see the detailed discussion below).

For N=4, at the beginning each droplet is characterized by its own rising solutal plume. In this early phase of the dissolution process, the dynamics of the different droplets can be regarded as independent. Two convective cells are generated around each droplet and then rise attached to the mushroom-shaped head (the cap) of the related solutal jet (hereafter the droplets will be referred to as droplet "i" with i=1→N, respectively, starting from the left side of the container). However, after a short initial transient behavior the two central rising jets i=2 and i=3 tend to be featured by a reduced rising velocity with respect to those at the edges (i=1 and i=4, see Figure 19a). In practice, they experience the return flow of the lateral jets and for this reason tend to be retarded.

After 46 s (see Figure 19c), the lateral solutal plumes (1 and 2) merge giving rise to a single rising current. The same behavior holds for the plumes 3 and 4 since the system exhibits mirror symmetry with respect to the mid-section y=L/2. For this reason, after an initial stage, the lighter fluid is transported upwards by two solutal rising jets only. They are located approximately at y=L/5, y=4L/5, respectively (instead of the four solutal plumes that characterize the system behavior in the early phase located at y=iL/5, i=1→4).

Evolution of the surviving solutal plumes results in a shape that bulges out below x=L/2 and necks in above it. This leads to an amphora-like pattern (Figure 19d).

For a further increase of time, two new plumes are originated from the central droplets 2 and 3. As they rise, the related combined solute paths result in a shape that necks below x=L/2 and bulges out above it (Figure 19f). This phenomenon is then followed in time by the generation of two new plumes above the lateral droplets 1 and 4 (Figure 19g).

During these stages of evolution, the droplets seem to "work" in mirror groups of two. Rising solutal jets are initially created above the couple (1,4), then above the couple (2,3). Then the couple (1,4) is again involved in the generation of rising jets and so on. Therefore a certain degree of periodicity in time can be highlighted with mechanisms exhibiting a somewhat "repetitive" behavior. As time passes new regimes evolve which exhibit a periodic alternance of central or lateral rising plumes.

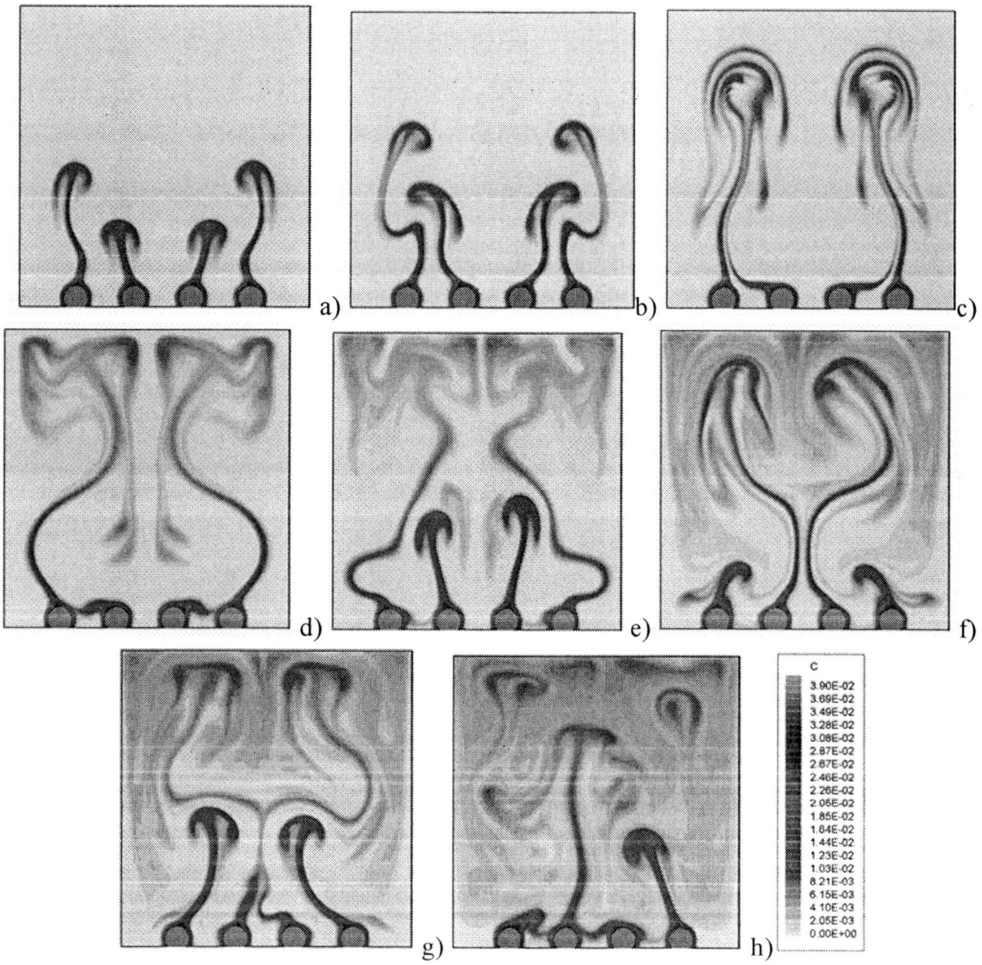

Figure 19. Computed pattern formation and evolution for the case N=4, uniform temperature distribution (T=23 [°C]), subsequent snapshots corresponding to (a) t= 35.6 [s], (b) t=39.2 [s], (c) t=46.3 [s], (d) t=57 [s], (e) t=71.3 [s], (f) t=82 [s], (g) t=89.1 [s], (h) t=114.2 [s].

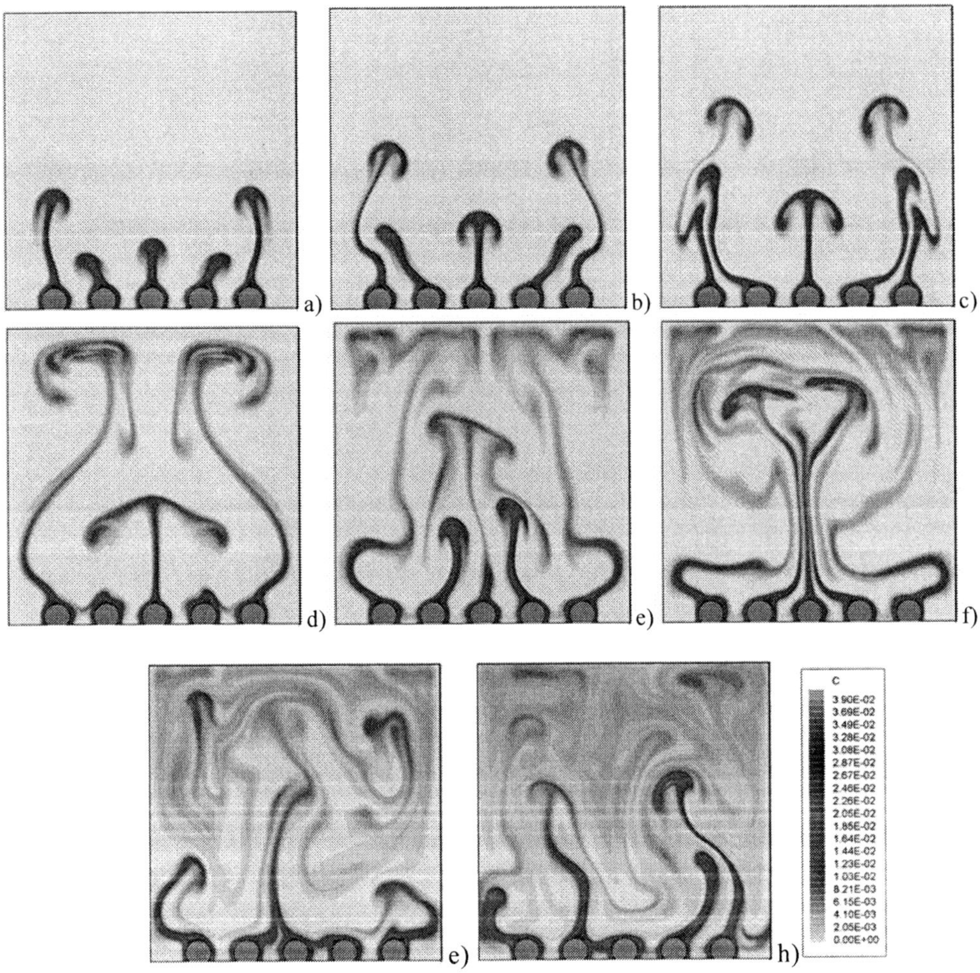

Figure 20. Computed pattern formation and evolution for the case N=5, uniform temperature distribution (T=23 [°C]), subsequent snapshots corresponding to (a) t= 35.6 [s], (b) t=39.2 [s], (c) t=42.75 [s], (d) t=53.4 [s], (e) t=64.1 [s], (f) t=71.2 [s], (g) t=81.9 [s], (h) t=99.75 [s].

These phenomena are also susceptible of an interesting interpretation in terms of relative temporal phase shifts. The phase shift is herein defined as $\pi\Delta L/L$, ΔL being the difference between the height (along x) of two plumes at a given instant (this simply means that two plumes having the same height are featured by a zero phase shift, whereas the phase shift is approximately π if two plumes, one extended throughout the container and the other confined to the bottom are considered).

Periodic convective structures originated from the droplets 1 and 4 do not exhibit phase shift ($\Delta\varphi_{14} = 0$). Convective structures originated from the droplets 2 and 3 also do not exhibit phase shift ($\Delta\varphi_{23} = 0$). A phase shift $\Delta\varphi = \pi$ occurs between droplets 1 and 2, or 3 and 4 (e.g., a new jet is generated above the second drop when the plume above the first one dies at the top of the cell). This also means that only two values of $\Delta\varphi_{ij}$ are possible (0 or π).

For a further increase of time however, the spectrum of possible phase shifts becomes more complex. The initial regular pulsating mechanism described above, with a rhythmic alternance of central and lateral couples of rising jets (featured by mirror symmetry with respect to the midsection y=L/2) is taken over by a new disordered regime in which the release of plumes seems to be quite erratic or chaotic (see Figure 19h). In such a regime, the pattern becomes asymmetric and such a symmetry breaking is associated with the presence of a variety of possible phase shifts $\Delta\varphi_{ij}$.

These trends are partially confirmed by the simulations carried out for N=5 (Figure 20).

In such a case, the major stages of the system evolution are quite similar to those discussed before for the case N=4: at an early stage, the flow undergoes transition to an oscillatory state featured by a mirror symmetric regular alternance of plumes released by lateral and internal droplets ((1,5) and (2,4), respectively). In terms of possible phase shift, however, the central drop i=3, seems to work independently. The related plume, in fact, is extended along half of the height of the domain when the others have already reached the top of the domain and/or are still confined to a region close to the surface of the droplets (see Figure 20d, where plumes 1 and 5 are extended throughout the domain, plumes 2 and 4 are still in very embryonic conditions and plume 3 affects the lower half of the container). Accordingly it exhibits a phase shift $\Delta\varphi_{3j} \cong \pi/2$ with respect to the pulsating phenomena pertaining to the other companion droplets.

As time passes, however, as for the case already discussed for the array of 4 evenly spaced droplets, these regular behaviors are replaced by a quite erratic generation/release of plumes and related possible time shifts (e.g., Figure 20h); this can be regarded as a significant result as it highlights the spontaneous tendency of multidroplet configurations to evolve towards final erratic pulsations regardless of the even or odd number of droplets.

6.2. Evenly Spaced Droplets in Non-Isothermal Conditions

As illustrated in Sect. 5.2 if a sufficiently large temperature gradient is applied along the drop surface, even for a single droplet (for a drop with a diameter d = 2 [mm], the critical $\partial T/\partial x$ is about 1.2 [C cm^{-1}], see Figure 13), the system can undergo transition to oscillatory behavior and such an instability is driven by the opposition of solutal buoyancy to thermal Marangoni stresses (Sect. 5.3).

A clear evidence of the counteracting interplay of solutal buoyancy and thermal Marangoni forces is given by the presence along the drop surface of the "detachment point" already shown in Figure 14.

When the temperature difference applied to the container is sufficiently high and/or the drop is very close to the bottom wall (present case) where a thermal boundary layer with steep temperature gradients is present, the detachment point moves to the rear of the droplets (Figure 15). For this reason return flow is initiated close to the drop base.

For the array of considered evenly spaced droplets the above-mentioned return flows give rise to liquid currents that meet each other at about half of the distance between consecutive drops. Thus rising solutal jets are created there (the positions are y=L/2(N+1)+ iL/(N+1)) whereas the onset of pluming phenomena in the regions above the droplets is prevented (see Figures 21a and 22a). Thus there is a significant difference with respect to the cases discussed before (Sect. 6.1) in which rising jets are originated from the peak of the droplets (i.e. at y=iL/(N+1)).

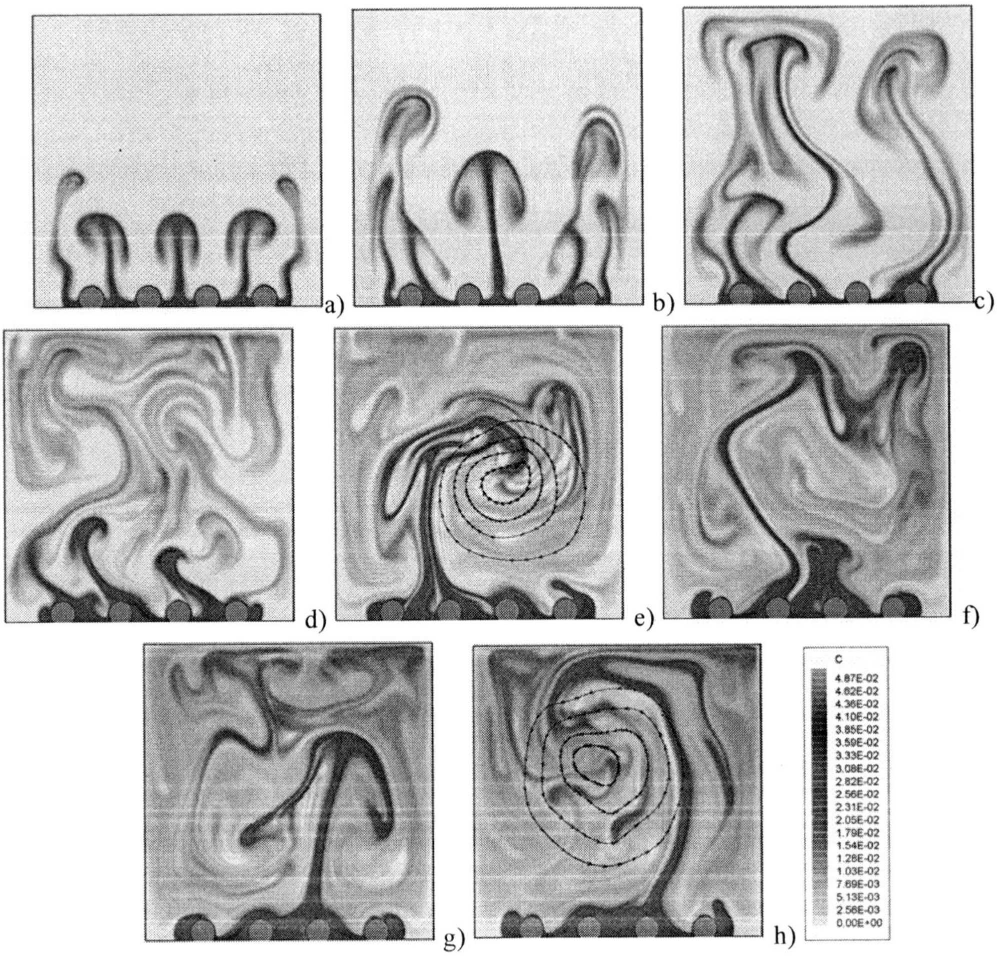

Figure 21. Computed pattern formation and evolution for the case N=4, with $\partial T/\partial x=2$ [°C cm^{-1}] (T_{bottom}=23 [°C], T_{Top}=[29 °C]), subsequent snapshots corresponding to (a) t= 26.9 [s], (b) t=32.9 [s], (c) t=44.9 [s], (d) t=55.3 [s], (e) t=65 [s], (f) t=73.1 [s], (g) t=78.6 [s], (h) t=84.1 [s].

A very complex multicellular structure is created above the discrete distribution of droplets. This convective field undergoes interesting subsequent transitions to different regimes as time passes. It is also worthwhile to stress how, unlike the case of pure buoyancy flow discussed before, the major stages of this spatio-temporal evolution are significantly affected by the value of N (N=4 in Figure 21 and N=5 in Figure 22).

For the case N=4, at the beginning 5 distinct rising convective jets occur (Figure 21a).

After about 30 s (see Figure 21b), the lateral solutal plumes (1 and 2) merge giving rise to a single rising current. The same behavior holds for the plumes 4 and 5 since the system exhibits mirror symmetry with respect to the mid-section y=L/2. For this reason, after an initial stage, the lighter fluid is transported upwards by three solutal rising jets only. For a further increase of the time, the simulations show that, the system exhibits a pronounced tendency to lose its initial mirror symmetry with respect to the midsection y=L/2 (Figure 21c,d).

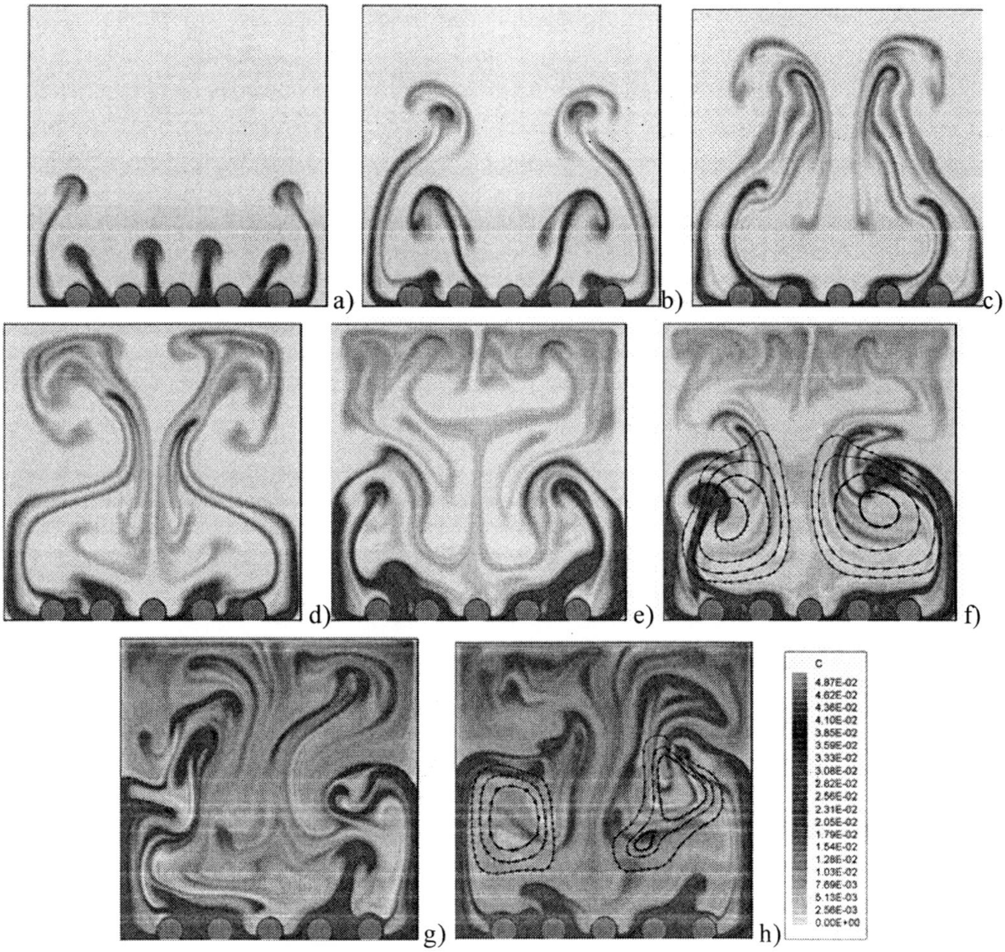

Figure 22. Computed pattern formation and evolution for the case N=5, with $\partial T/\partial x=2$ [°C cm^{-1}] (T_{bottom}=23 [°C], T_{Top}=29 [°C]), subsequent snapshots corresponding to (a) t= 26.9 [s], (b) t=32.9 [s], (c) t=38.9 [s], (d) t=48.3 [s], (e) t=55.8 [s], (f) t=58.9 [s], (g) t=82.7 [s], (h) t=91.8 [s].

The lateral plumes intersect the central one. This is associated with a significant time-dependent mixing of the solute throughout the container. A single vortex expanded throughout the domain is created, in fact, above the droplets (see, e.g., Figure 21e).

This vortex is initially clockwise oriented, but as time passes it is taken over by a counter-clockwise new vortex (Figure 21g,h). The system then comes back to a clockwise circulation and so on, with a periodic alternance of clockwise and counter-clockwise convective structures. Correspondingly, all the rising solutal jets tend to coalesce into a single elongated finger. Such a finger is bended by the action of the aforementioned vortex. It is originated from droplet 2 when the domain is featured by the presence of the clockwise vortex (Figure 21e,f) and from droplet 3 (Figure 21g,h) when the system undergoes transition to the counter-clockwise one, respectively (the frequency of this phenomenon is given in Table III).

As anticipated, for N=5 the roll-pattern formation and evolution are quite different. At the beginning (Figure 22a) 6 rising plumes appear above the seeds ($i=1\rightarrow 6$). For t≅30 s (see Figure 22b), the plumes 1 and 2, and 5 and 6 merge so that the lighter fluid is transported upwards by four solutal rising jets only. The surviving solutal plumes are located approximately above the first two drops (1 and 2) and the last two ones (4 and 5) whereas pluming phenomena at y=L/2 are prevented. For t>40 s further coalescence of adjacent plumes occurs and a new regime arises with only two lateral rising main convective structures and a return fluid current descending from the top of the container towards the central drop at y=L/2; correspondingly, the space above the droplets is characterized by the presence of two counter-rotating vortices, see, e.g., Figures 22f and 22h. This regime is strongly oscillatory since the lateral solute fingers are periodically fed with new "packets" of solute.

At this stage of the discussion, a critical comparison with the results illustrated in Sect. 6.1 can lead to other significant insights into the physics of the problem.

Along these lines, it is worth noting that the time-dependent dynamics of the considered system seems to be significantly influenced for both cases N=4 and N=5 by the fact that the presence of descending currents along the surface of the drops (induced by the surface tension forces) changes the number and the position of the rising solutal jets with respect to isothermal conditions. As illustrated before, owing to the Marangoni surface descending currents, the plumes are originated from the middle space between consecutive droplets rather than from their surface. Also, for the same reason, the number of rising plumes generated by a given number of interacting droplets is larger in the case of non-isothermal conditions. Such a feature, owing to the delicate interplay among a larger number of plumes, affects the possible transitional states of the multidroplet system. Moreover, the final flow spatio-temporal organization with a single vortex above the droplets in the case N=4 and two mirror (approximately) vortices for N=5 tends to be more regular than the final erratic behavior occurring in the absence of Marangoni effects; in other words the presence of thermal surface tension effects favors the establishment of coherent well-defined repetitive structures from a global point of view.

Such "standing" convective phenomena, however, are replaced by the propagation of traveling waves along the horizontal extension of the domain when periodic conditions are assumed at the side boundaries, as illustrated in the next sub-section.

6.3. Periodic Distributions with Infinite Number of Droplets

In this sub-section, the results with no-slip lateral walls treated in Sect. 6.2 are compared with corresponding simulations obtained assuming periodic conditions (hereafter referred to as PBC) at the lateral boundaries. This case is considered to indirectly assess the role played by lateral geometrical constraints in determining edge effects and the pattern selection process. Also, such results provide a variety of interesting insights into the "extreme" behavior, which the system tends to, when the number of interacting droplets is increased to infinite ($N \rightarrow \infty$) at fixed values of the interspace parameter κ (i.e. infinite layer of liquid with evenly spaced droplets on the bottom).

As a first step, comparison between the computations for κ=0.33 and N=4 (PBC, shown in Figure. 23) and those with solid walls (SW), and same value of κ (Figure 21) highlights that in the absence of side solid constraints, the delicate evolutionary equilibrium of the velocity field is featured by a different time history (compare, e.g., Figure 23a with Figure 21b and Figure 23b with Figure 21c). Therefore, as expected, the absence of lateral solid constraints can be thought of as altering in a significant way the mode selection process and the pattern symmetries.

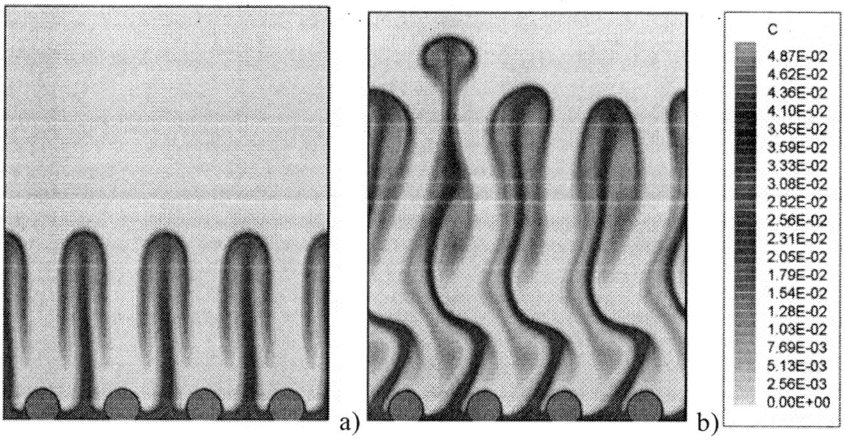

Figure 23. Computed traveling wave for the case N=4 with PBC for the same thermal conditions corresponding to Figure 21, subsequent snapshots: (a) t=35.56 [s], (b) t=47.53 [s].

The major outcome of these studies (specially conceived to allow for lateral freedom), however, is that the final oscillatory state (already depicted in detail in Sect. 6.2 for the SW cases with a well-defined "standing" single cell or a well-defined "standing" couple of vortices above the droplets for N=4 and N=5, respectively), is replaced for PBC by the occurrence of a "traveling wave" migrating along the horizontal direction. This behavior is clearly shown in Figure 24 for the case N=4→κ=0.33 (the phenomena for N=5→κ=0.4 are qualitatively similar to that for κ=0.33, therefore they are not discussed in detail).

The evolutionary scenario is driven by the continuous propagation along y of the position where the shooting of the "packet" of lighter fluid in the surrounding liquid occurs. In other words the front of the aforementioned wave corresponds at any time to the onset of a new rising plume. For instance, in Figure 24a the main rising plume is originated from the space between droplets 1 and 2 whereas in Figures 24c, 24e and 24g, such a position has moved to the interspace between droplets 2 and 3, 3 and 4, and 4 and 1, respectively. This also means that the spectrum of possible plume phase shifts is not discrete but "continuous" along the horizontal direction y, i.e. $\Delta\varphi(y_1-y_2)=(y_1-y_2)/\varpi$, ϖ being the wave propagation velocity (ϖ=f·L, f being the frequency given in Table III). Due to the combined presence of a disturbance traveling towards the top (i.e. a perturbation of the solutal and flow field rising at a given position y) and the simultaneous propagation of the related onset along the horizontal extension, the traveling wave can be regarded as a superposition of two perpendicular components (along y and x, respectively). Such a feature is also evidenced by the initial oblique inclination of the newly formed plumes.

Additional simulations carried out for isothermal conditions have revealed that in the absence of the Marangoni effect such waves do not occur even if PBC are employed. This proves that their origin is not of a pure gravitational type. Therefore, they do not have a solid kinship with the traveling states that have been occasionally reported in the literature in the case of thermosolutal convection in fluid layers of wide horizontal extent (see below).

Deane et al. (1987) revealed, by means of two-dimensional numerical simulations of thermosolutal gravitational convection, the presence of oscillations in the form of traveling, standing, modulated, and chaotic waves (see also Cross, 1986). Similarly, Walden et al. (1985) during studies of Rayleigh-Bénard convection in binary alcohol-water mixtures, observed the convective rolls to move continuously as traveling waves with either periodic or chaotic nature according to the temperature difference across the fluid layer (i.e., the Rayleigh number, Ra); similar results have been obtained by Moses and Steinberg (1986), Barten et al. (1989), Niemela et al. (1990), etc.

As a subsequent natural step to investigate the intrinsic nature of the present waves, some computations have been repeated "uncoupling" the temperature field (i.e. assuming a fixed linear temperature distribution along x with the same temperature gradient considered for the results in Figure 24).

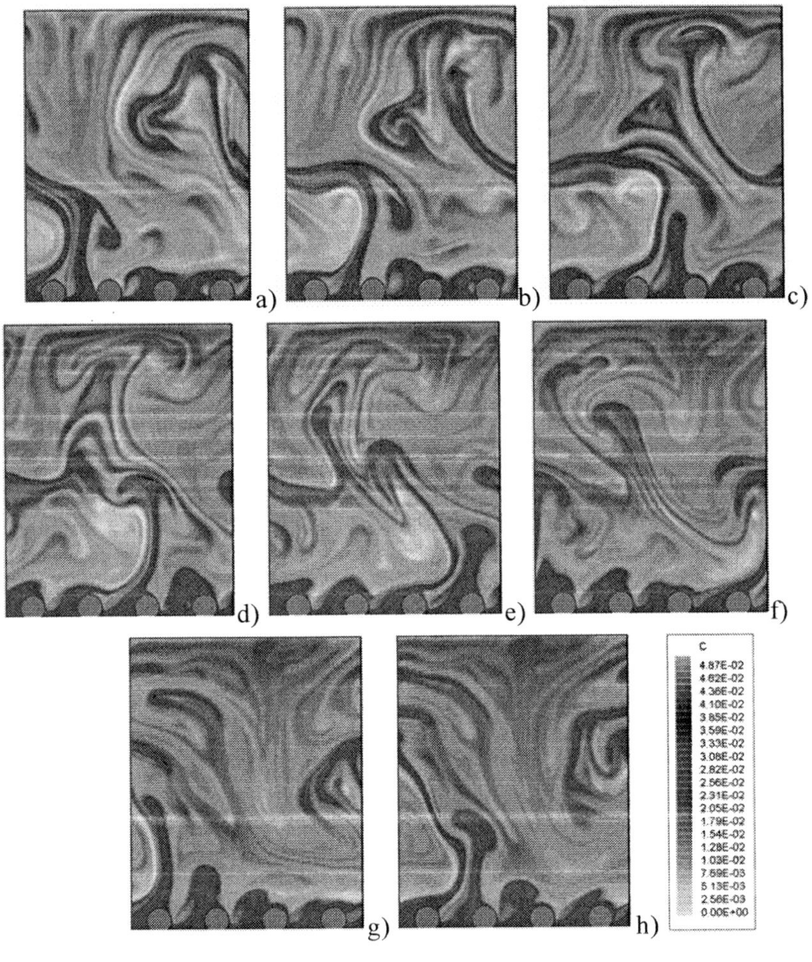

Figure 24. Computed traveling wave for the case N=4 with PBC for the same thermal conditions corresponding to Figs. 21, subsequent snapshots: (a) t=79.9 [s], (b) t=82.3 [s], (c) t=84.8 [s], (d) t=87.5 [s], (e) t=90.1 [s], (f) t=92.8 [s], (g) t=95.5 [s], (h) t=98.3 [s] (the period is about 18 [s]).

Such an artifice has shown that the presence of "static" Marangoni forces is not sufficient to explain the presence of the waves (they are suppressed). In practice, these results can be regarded as an indirect evidence of the fact that the dynamical interplay among solutal buoyancy and thermal surface tension forces and/or the dynamical interaction among velocity, solutal and temperature fields are a necessary requisite for the onset of this type of instability. This conclusion is also supported by the numerical simulations carried out for the case of gravitational effects switched off and droplets subjected to the action of Marangoni forces alone. The simulations show that in the absence of gravity, unlike buoyancy convection, Marangoni convection originated from a multidroplet configuration does not affect the entire enclosure. Rather the flow is restricted to an area close to the drop surfaces (see Figure 25). Continuity brings the hot fluid (carried from the peak of the drop into the lower regions of the domain just below the drop base) up again resulting in a set of closed vortices. Above this region, convection is very weak and solute is simply transported by diffusion. Accordingly no solutal fingers exist and the rich variety of scenarios with interacting plumes occurring under the effect of gravity is suppressed (similar results have been obtained by Morris and Parviz, (2006) during their experimental studies dealing with "Marangoni actuators" on micro-patterned substrates).

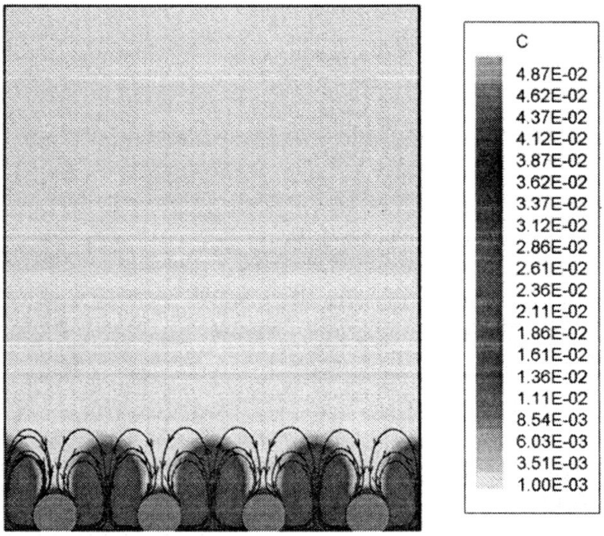

Figure 25. Concentration and velocity fields for the case N=4 with PBC for the same thermal conditions corresponding to Figs. 21 under zero-g conditions (buoyancy absent), t=85.32 [s].

Table III. Computed oscillatory regimes in non-isothermal conditions ($\partial T/\partial x = 2$ °[C cm^{-1}]) and related frequencies as a function of N and of the boundary conditions type

N	Boundary conditions	Regime	f [Hz]
4	SW	Single oscillatory standing vortex	$2.63\ 10^{-2}$
5	SW	Two oscillatory standing vortices	$4.34\ 10^{-2}$
4	PBC	Traveling wave	$5.43\ 10^{-2}$
5	PBC	Traveling wave	$3.52\ 10^{-2}$

7. Erratic Pulsations and Kicking Behaviors

7.1. Rippling and Erratic Pulsations

The results in Sect. 6.1 could be used to shed some light on other (still unexplained) similar phenomena that have been reported in the literature in the absence of imposed thermal gradients. For instance, many authors (see, e.g., the pioneering analysis of Lewis and Pratt, 1953) observed rippling and erratic pulsations at the interface of drops of organic solvents dispersed in various solvents. According to the present analysis, they could be explained in the light of the phenomena depicted in Sect. 6.1, i.e. on the basis of the pure buoyant oscillatory instability that can occur (even in the absence of Marangoni effects) when many close droplets "interact".

The simulations have proven, in fact, that, even in the absence of Marangoni forces, time-dependent "pulsating" phenomena can occur if a multi-droplet configuration is considered. In such a case, in fact, the generation of jets results modulated in time with an intriguing temporal alternation of buoyant plumes originated from different locations within the container.

After a short initial transient time in which the system exhibits mirror symmetry with respect to the midsection, progression from order to disorder occurs and the system undergoes transition to a condition in which the time-dependent release of solutal plumes becomes erratic from both spatial and temporal points of view.

7.2. Kicking

Many recent experimental studies have pointed out that surfactants can also play a crucial role in creating interfacial instabilities in binary liquid-liquid systems and oscillations (referred to as "kicking" by these authors) of drops (see, e.g., Perez de Ortiz and Sawistowski, 1973a,b; Liang and Slater, 1990; Bekki et al., 1990, 1992; Kovalchuk and Vollhardt, 2006) very similar to those discussed in Sects. 6.2. and 6.3.

The effect of surfactants on interfacial mass transfer in binary liquid-liquid systems (dissolving drops) has been experimentally investigated by Agble and Mendes Tatsis (2000) using the Schlieren and the Mach-Zehnder optical techniques. The experiments were focused on the analysis of interfacial stability by means of direct examination of the interface of a drop of an organic phase (saturated with water) that was immersed in an aqueous phase of pure water. The onset of Marangoni convection and of possible fluid-dynamic instabilities at the interface was clearly observed. The role played by the surfactants (like temperature gradients, these substance can alter the surface tension) in inducing possible surface instabilities was investigated in detail in a subsequent analysis (Agble and Mendes Tatsis, 2001). In this analysis much room was devoted to compare the available experimental results with canonical and/or semi-empirical surface stability criteria introduced over the years for the case of various binary liquid-liquid interfaces in the presence of a transferring surfactant; i.e. the Sternling and Scriven (1959) criteria for ternary systems, the Hennenberg et al. (1979, 1980) criteria for surfactant transfer, the semi-empirical criteria of Nakache et al. (1983). It was found that these criteria often fail in predicting the observed phenomena. Within this context, it is opinion of the author of this chapter that all of the phenomena mentioned above in the case of added surfactants could be explained according to the results discussed in Sects. 6.2 and 6.3, the thermal surface tension gradient being simply replaced by the effect of the concentration gradient of surfactant for the cases above (see also the very recent linear stability analysis of Slavtchev et al., 2006).

Experimental conditions somewhat more similar to those considered in the present chapter were investigated by Zander and Schneider (1992), who studied the liquid-liquid phase decomposition induced in a horizontal fluid layer of cyclohexane and methanol being exposed to a vertical temperature gradient. They reported the existence of some convective instabilities in the form of "dissipative structures" (regular convective patterns) inside the layer that indirectly manifested themselves by the organization of the floating precipitate. Some attempts were given to explain the appearance of these instabilities according to the model of a two-component Rayleigh-Bénard instability (see also the focused discussion in Sect. 8) driven by the temperature and concentration differences between the lower and the upper boundaries of the fluid (the concentration difference resulting from the dependence of the local equilibrium composition on the local temperature). Such a criterion, however, was found not to properly model the experimental results; this was attributed to the influence of the Marangoni driven flows near the deposited droplets.

In the light of the present results, the "dissipative structures" indirectly observed by Zander and Schneider (1992), could be regarded as the indirect effect of the standing convective structures described in Sect. 6.2.

7.3. Chaotic, Discrete and Continuous Phase Shifts

To summarize, all the results presented and discussed in Sect. 6 prove that the simultaneous presence of several dispersed droplets in non-equilibrium conditions can be responsible for the onset of a variety of additional instabilities with respect to the case of an isolated drop (which make such a configuration a pattern-forming dynamical system of exceptional complexity). It exhibits sensitivity to the number of droplets and, in turn, each droplet and the related surrounding convective structure undergo different evolution according to the position within the container.

In terms of possible plume time shifts, all the described phenomena can be categorized as follows: chaotic (lateral SW and isothermal conditions), discrete (lateral SW and non isothermal conditions) and continuous (lateral PBC and non isothermal conditions) phase-shifts distributions, respectively. In terms of macroscopic features of the unsteady evolution, a corresponding categorization would read: erratic pulsation, standing pulsation and traveling pulsation, respectively.

8. Comparison with other Multi-Plume Configurations: Analogies and Differences

8.1. Rayleigh-Bénard Systems

In Sect. 5.5 the oscillatory instability affecting the solutal jet released by a single droplet in non-isothermal conditions has been compared with the instability of line-source single plumes (jets of gas mixtures) investigated by Cetegen and co-workers.

It is noteworthy how, similarly, from a fluid-dynamic point of view, the case of a multi-droplet configuration exhibits quite an interesting theoretical kinship with the classical Rayleigh-Bénard (RB) system in which convection is induced in a fluid by uniform heating from below. Therefore, it is opportune to open a rich discussion about the related analogies and differences. Such a critical comparison can be used as an additional theoretical artifice by which additional insights into the physics and the results described in the earlier sections can be obtained.

Thermal Rayleigh–Bénard convection is a canonical example of a pattern forming system. A fascinating question is raised by the pattern selection process upon which it evolves through time to a final state. The number of rising plumes (also referred to as "wave number" in many circumstances) selection has been well studied with many different mechanisms now well characterized.

Most of the initial studies considered infinite systems, i.e. the stability of the quiescent state of nonconfined fluid layers heated from below. Currently, however, it is well known that if systems confined laterally by rigid sidewalls are considered, even in containers of large horizontal dimensions, the lateral walls can have a significant influence on the flow pattern that develops when the thermal Rayleigh number ($Ra=g\beta_T(\Delta T)L^3/\nu\alpha$ where ν and α are the kinematic viscosity and thermal diffusivity, respectively, and their ratio $Pr=\nu/\alpha$ is the Prandtl number) exceeds its critical value (for a given Ra, the wave length basically depends on the aspect ratio of the container). The flow structure, however, becomes very complex when turbulent states are reached (very high values of the Rayleigh number, see Taylor et al., 1956; Turner, 1969; Chu and Goldstein, 1973; Castaing et al., 1989; Solomon and Gollub, 1990; Siggia, 1994). It is known that the flow tends to develop into time-dependent convection with a strong asymmetry and highly convoluted thermal plumes delineating a large-scale circulation. In such conditions smaller thermal plumes can detach from thermal boundary layers and extend over the entire cell, coalescing and creating a local inversion of the temperature gradient adjacent to the boundary layers (see the relevant studies of Werne, 1993; Vincent and Yuen, 1999; Childress, 2000, and Xi et al., 2004 and Parodi et al., 2004).

According to the aforementioned "analogy", these features may provide a somewhat relevant theoretical background to explain the trends observed for the present case of square domain with dissolving droplets on the bottom; among them: the time-dependent plume release and adjustment process and the related coalescence phenomena.

Section 8.2 explains in detail the features of the classical turbulent RB convection (much room is devoted to illustrate the state-of-the-art about this problem that can be regarded as a very classical subject of investigation in the literature concerning the onset of fluid-dynamic instabilities and the ensuing evolution towards turbulent states); then some comparison with the present problem is considered in Sect. 8.4.

8.2. Classical Turbulent Rayleigh-Bénard Convection and the Flywheel Mechanism

As anticipated, the flow structure related to a fluid uniformly heated from below, becomes very complex when turbulent states are reached (very high values of the Rayleigh number).

Turbulent Rayleigh-Bénard convection produces fields of intense updrafts and downdrafts (rising and descending plumes) that are responsible for much of the vertical heat transport. These structures (sometimes also referred to in the literature as "thermals"), have horizontal scales comparable to the thicknesses of the boundary layers in which they arise (see, e.g., Parodi et al., 2004). Figure 26 shows an example of such structures in a square domain. As time passes they organize themselves into clusters with horizontal scales that grow with time and reach the width of the computational domain.

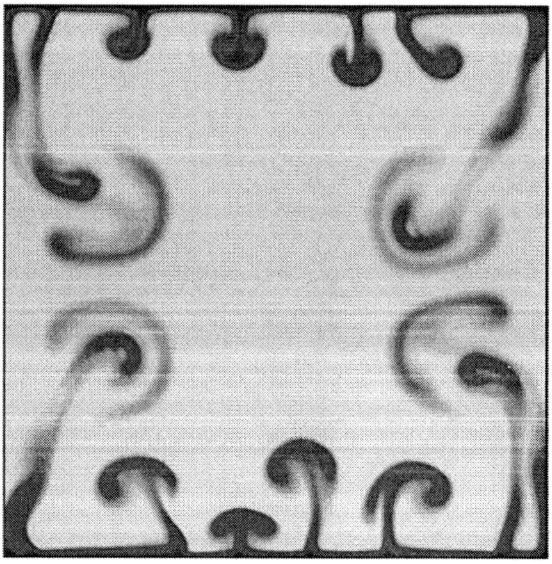

Figure 26. Thermal Rayleigh-Bénard convection (Pr=15, Ra=10^8): early stage of convection with multiple plume-detachment phenomena along the uniformly heated (cooled) bottom (top) wall.

Figure 27 shows the typical flow structure when a fully-developed regime is established. Hot plumes congregate in an upwelling jet of fluid near the right-hand wall of the container. A similar, downward jet formed from cold plumes occurs on the left-hand wall. Large numbers of hot plumes are also found in left-to-right motion in a *mixing zone*, or viscous boundary layer, near the bottom of the container. A similar layer on the top contains cold plumes, moving from right to left. The central region contains a few plumes, hot and cold, in a partially random motion.

These plumes wander chaotically, but also participate in an overall large-scale rotatory flow (fluid circulates as a loop with the plumes rising and falling on opposite sides of the loop). Since, as explained before, the cold and warm plumes are separated laterally in two opposing sidewall regions and exert buoyancy forces to the bulk fluid, an alternating eruption of the thermal plumes, therefore, gives rise to a periodic impulsive torque, which drives a large-scale circulation continuously (see Figure 28).

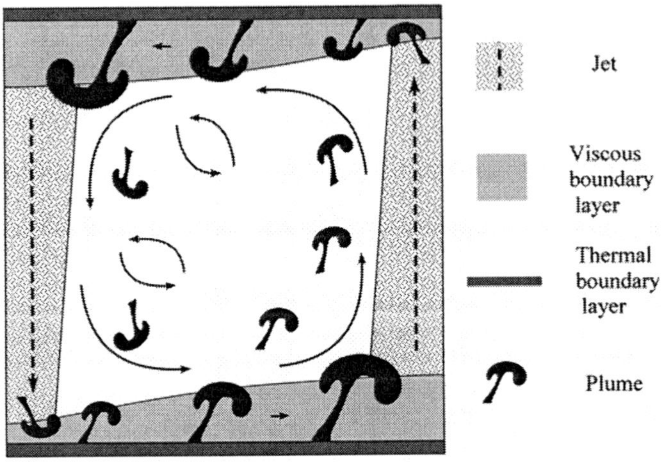

Figure 27. The spatial organization of the thermal plumes produces a unique flow structure, which undergoes a coherent rotation. Plumes evolve from the boundary layers above (below) the bottom (top) plate. First they move laterally toward the sidewall, driven by the prevailing circulation (the horizontal "wind"). They then travel vertically in the region near the sidewall, leaving the central vertical section of the cell relatively free.

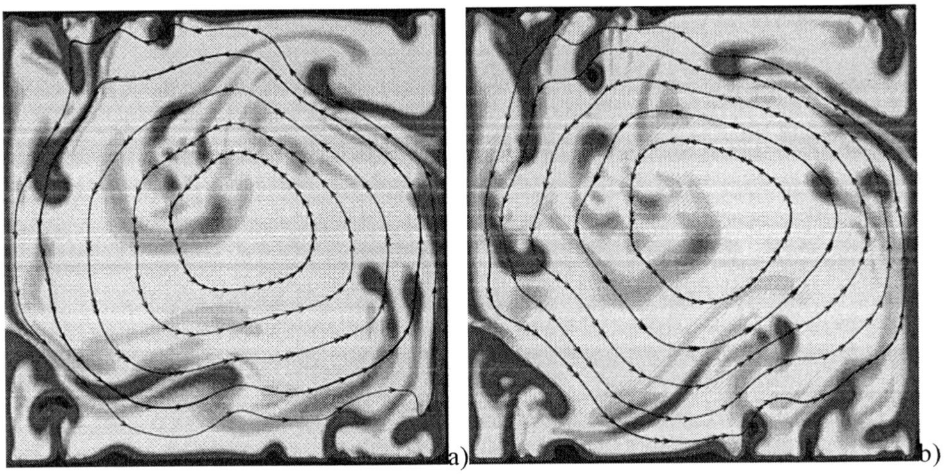

Figure 28. Thermal turbulent Rayleigh-Bénard convection (Numerical simulations, Pr=15, Ra=10^8). Frames (a) and (b) show the temperature and velocity fields at two close instants during the fully developed regime in which a large-scale coherent circulation (the wind) has been established.

This complex mechanism may be regarded as a “machine” (Kadanoff, 2001) containing many different working parts: boundary layers, mixing zones, jets, and a central region.

As outlined above and as experimentally illustrated by Qiu et al. (2000) and Qiu and Tong (2001), despite the large velocity fluctuations, the time-averaged flow field maintains a large-scale quasi-two-dimensional structure, which rotates in a coherent manner. It is usually referred to in the literature as "wind of turbulence" (Grossman and Lohse, 2000), simply "wind", "mean wind" (Niemela and Sreenivasan, 2003) or "flywheel" (Kadanoff, 2001), accordingly, hereafter these terms will be used as synonyms together with "large-scale circulation" or motion.

As outlined before, the self-organized and coherent large-scale structure can be divided into three regions in the rotation plane: (1) a thin viscous boundary layer, (2) a fully mixed central core region with a constant mean velocity gradient, and (3) an intermediate plume-dominated buffer region. In such a context (all the plume activity is carried along in the overall rotatory motion), the plumes play a very important role as they act as heat pumps; they, in fact, shuttle heat flux directly between the top and bottom boundaries (the hot plumes rising with their caps on top, cold plumes falling cap-down).

Because of contributions of a number of investigators, a lot of progress has been made in the understanding of the large-scale circulation, its periodicity, the boundary layers, plume dynamics, and the plume interaction with the large-scale wind.

The large-scale coherent circulation is responsible for the aforementioned horizontal motion of the plumes along the top and bottom walls (the *horizontal* motion of the plumes is slaved to the wind velocity, since the only independent force acting on the plumes is the buoyancy force in the *vertical* direction). The large scale motion, in turn, is a result of plume-interaction phenomena as theoretically proven by Childress (2000) and experimentally observed by Xi et al. (2004).

Childress (2000) has considered the dynamical origin of large-scale flows in RB systems. In this analysis a two-dimensional model of plumes, such as those observed in thermal convection at large Rayleigh and Prandtl numbers, has been introduced. From this model, the author deduces the onset of mean flow as an instability of a convective state consisting of parallel vertical flow supported by buoyancy forces. Along these lines, Xi et al. (2004) have illustrated by direct experimental analysis that the interaction and merging of the vortices from neighboring plumes lead to groupings and/or merging of plumes, which in turn generate vortices of even larger scale; as a result of these interactions, the convective flow evolves into the coherent rotatory motion (the flywheel) consisting of mainly the plumes themselves and spanning the whole convection box. This study has clearly demonstrated that it is the thermal plumes that initiate the horizontal large-scale flow across the top and bottom conducting plates.

Some authors have also shown that the phenomenon related to the transport of plumes along the top and bottom boundary layers (due to the action of the horizontal wind) is somewhat linked to the propagation of horizontal waves (Gluckmann et al., 1993; Villermaux, 1995; Kadanoff, 2001). The Villermaux's theory, in particular, proposes that the temperature oscillation is caused by a thermal boundary layer instability triggered by the incoming thermal plumes from the opposite conducting surface.

These waves form on the thermal boundary layers in regions of high horizontal velocity gradient. The plumes, in turn, form on top of the waves, ride with the wind, and are squeezed into the side walls.

The waves appear as bulges in the height of the thin boundary layer at the bottom of the cell. As they move, the waves throw up a hot spray that tends to rise, forming sheets of hot, upward motion. These sheets break up into columnar structures that form the plumes shown in Figure 28.

As they move across the bottom of the cell, these structures grow larger, as shown in Figure 27. A few plumes come loose and move into the central region. Most hit the right-hand wall and move upward as a jet aimed toward the top of the cell. As a plume hits the upper wall, it makes a splash. The splash makes a wave. Each wave moves along the top, producing a cold spray and thence cold plumes. The plumes form into a downward jet at the left-hand side, splash on the bottom, and produce hot waves, thus keeping the motion going indefinitely: these mechanisms lead to the alternating emission of cold and warm plumes between the upper and lower thermal boundary layers and the side-jets resulting from plume-clustering phenomena pull the fluid around the container, forming the liquid flywheel.

As anticipated, this classical turbulent RB convection (container uniformly heated from below and cooled from above) shares many common features with the solutal convection pertaining to the multi-droplet configuration treated in Sect. 6.1 and 6.2 (among them the time-dependent release of plumes and the formation of large-scale structures). The flywheel mechanism, however, does not occur in the latter case.

This behavior may be explained by the absence of an upper solutal boundary layer.

The absence of such a layer does not make possible the onset of the splash/wave-propagation phenomenon discussed before. In practice the system with plumes generated only from one side (the bottom) lacks the necessary feedback mechanism able to keep a large-scale rotatory motion going indefinitely.

8.3. The Rayleigh-Taylor Instability

It should be also mentioned that possible analogies are not limited to RB systems, the present problem, in fact, also exhibits quite an interesting theoretical link with the Rayleigh-Taylor (RT) instability. Within the present context it is worth considering for comparison/discussion the work of Bizon et al. (1997).

They considered turbulent convection driven by an imposed concentration gradient in an isothermal vertical thin square Hele-Shaw cell (24 [cm] x 24 [cm] x 0.1 [cm]) at very high Rayleigh numbers (up to $7\ 10^4$ times the value for convective onset). The vertical density gradient was, in practice, imposed by these researchers using a heavy fluid at the top horizontal boundary and a light fluid (like in the present work) at the bottom (both imposed fluids being solutions of iodine in ethyl alcohol).

They found the concentration field at such high values of the Rayleigh number to be characterized by sharp concentration gradients at the top and bottom boundaries (the concentration gradients corresponding to the presence of boundary layers) and a central core region with mean concentration not dependent on height; the velocity field was shown to be featured by the release of solutal plumes growing up of the aforementioned boundary layers (the large density difference between the boundary fluid and the central core driving a Rayleigh-Taylor instability be responsible for plume formation).

They observed that the above mentioned instability of the boundary layer tends to excite the formation of many, closely spaced plumes whose number always exceeds that predicted in the case of simple RB convection. They also reported coalescence of these boundary-layer plumes resulting from 1) sweeping motions caused when fluid moving from the interior strikes the boundary and then spreads out horizontally, and/or 2) mutual interactions between neighboring plumes already known in the case of canonical thermal RB (Moses et al., 1991, Xi et al., 2004).

Conglomerates of plumes leaving the boundary were observed to form "conduits" that organize the vertical transport and "channel" subsequent plumes (the channeled fluid, upon impacting and spreading along the far boundary, provides the sweeping motion that organizes the boundary-layer plumes there into return vertical channels).

8.4. Analogies and Differences

Despite the macroscopic qualitative similarities with turbulent RB and RT systems, however, the dynamics and the mechanisms underlying the evolution of the present multi-droplet system are quite different.

In the case of RB and RT configurations the driving force for the onset of convection is the lighter fluid on the bottom of the container, this "force" being replaced by the presence of a discrete distribution of sources of solute for the case of a discrete distribution of droplets. However, there is no doubt that the geometries that arise in immiscible alloys processing, are more complicated than a simple classical RB or RT system. For such a case the distribution of light fluid at the bottom is not uniform. The light solute-enriched zones do not exhibit uniform behavior. Moreover, during the initial stages of convection fluid motion also arises along the lateral boundaries of the droplets (this effect is not present in the classical Rayleigh-Bénard system; within the framework of the proposed analogy, it would correspond to a "heating-from-the-side" thermal condition).

For many interacting droplets, the adjustment in the number of plumes involves the overlapping mechanisms of the enriched zones that merge as time passes; there is an undeniable effect of the number of droplets and their distribution.

Therefore, for the case of many aligned evenly spaced dissolving droplets, after the early stages (with an alternating array of clockwise and counter-clockwise rotating cells), the process by which the instantaneous mode is selected is very complicated (possibly involving the effects of multiple incommensurate selection mechanisms). There are at least five independent parameters influencing the aforementioned evolution of the multicellular and multi-plume structure, i.e. the aspect ratio of the container and/or the type of lateral boundary conditions SW or PBC), the time-evolution of the phenomenon of release of solute (this would correspond to a quasi-steady variation of the Rayleigh number for the case of the layer of fluid heated from below), the coalescence process of the enriched zones originated from different drops, the number of interacting droplets (as previously highlighted) and finally the possible presence of surface Marangoni effects.

In practice, such a system would "ideally" tend to the behavior of the classical infinite RB or RT problem only if PBC conditions and a zero separation distance among the droplets (i.e. an uniform horizontal layer of liquid formed by the coalescence process of very close droplets) were considered.

8.5. Other Multi-Plume Systems

Many investigations have also appeared in the literature for the case of multiple localized sources of buoyancy (instead of uniform conditions at the bottom). To the author's knowledge, however, most of these studies has been focused on flows of different types, but all dealing with large-scale scenarios, e.g., geological flows (Farnetani et al., 2002) or environmental problems and (see, e.g., Lavelle, 1997 and the rich reference list therein). Hydrothermal plumes originated from deep-sea floor are one of many examples in the natural environment where a localized source of buoyancy causes materials to rise into an overlying fluid and forced circulation to develop. Plumes from terrestrial volcanoes, forest fires, urban mass fires and oceanic thermal and waste water discharges (e.g., jet groups issuing from ocean outfalls) provide instances of buoyant plumes or buoyant jets occurring in a cross flow. A related problem in the industrial realm involves the dispersion of power plant plumes into the atmosphere and associated chemical reactions, interaction and spreading of cooling tower plumes, burning of oil spills on water, etc.

Studies about smoke dispersion from multiple fire plumes have shown that multiple-plume interactions can push particulate up to altitudes exceeding that of an equivalent single plume (Trelles et al., 1999). These interactions can have unexpected results on the containment of airborne material. The mixing promoted by the large-scale plume vortex structures can transport combustion products to areas that would not have been covered by a single plume produced by a single fire.

The investigation of König and Mokhtarzadeh-Dehghan (2002) was aimed to verify the question of how closely the overall characteristics of merged plumes from a multi-flue chimney match those of an equivalent single plume. The results for multi-flue plumes were compared with those for a single plume under the same release conditions for volume flow rate, momentum and temperature. The differences in the velocity, temperature and turbulence energy fields of a single plume and multiple plumes were found to be mainly significant in the early stages of rise and spreading.

Bornoff and Mokhtarzadeh-Dehghan (2001), presented the results of a numerical investigation into the interaction of two adjacent plumes in a cross-flow. The computations were performed for three-dimensional, turbulent, buoyant and interacting plumes, and for a single plume for comparison. Two double-source arrangements, namely, tandem and side-by-side, with respect to the oncoming atmospheric boundary layer were considered. The results showed that the interaction of side-by-side plumes is dominated by the interaction of the rotating vortex pairs within the plumes. A tandem source arrangement leads to early merging and efficient rise enhancement.

Davies et al. (2000), carried out experiments to measure concentration fluctuations downwind of two tracer sources in the atmospheric surface layer. They showed that the correlation between the concentrations from the two sources varies with downwind distance and source separation.

These studies about non confined flows (free plumes) and the emission and dispersion of pollutants or the rise of oceanic or geophysical currents, though very interesting, however are not very relevant to the present case of interacting plumes in a closed geometry and direct comparison is very difficult or impossible. They have been cited here for the sake of completeness and are of little help when trying to get additional insights into the physics of the problem treated in this chapter.

More relevant analyses available in the literature are due to Wang and Jaluria (2002), Lee et al., (2004), Icoz and Jaluria (2005), Bazylak et al., (2006), Lappa, (2003) and (2005e).

Lappa (2005e), in particular has considered the growth of several macromolecular seeds uniformly distributed on the bottom of a protein reactor under microgravity conditions (i.e. a discrete layer of N crystals embedded within a horizontal layer of liquid with no-slip boundaries, Figure 29).

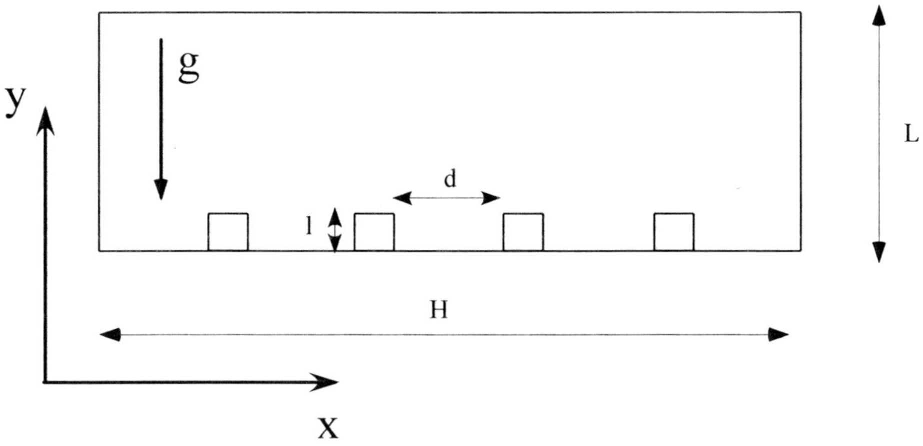

Figure 29. Sketch of the growth reactor with the protein seeds and of the relative direction of the residual g.

Such a problem exhibits significant analogies with the multi-droplet configuration as when crystal growth sets in, the solute concentration around a crystal decreases. With this depletion of the heavier solute, the solution around a crystal becomes lighter and, thus, rises forming a plume above each crystal.

Figure 30 shows that at the beginning of the growth process two convective cells are generated around each crystal. As explained above, the formation of halos due to the lowered solute concentration around growing seeds, in fact, has the effect of producing density gradients in these areas. These in turn (under the effects of gravity) result in the onset of convection.

After this initial transient behavior, however, the depletion zones for two growing consecutive seeds intersect and overlap (Figure 30b). At this stage mutual interference occurs. Crystals, in fact, absorb protein from common regions.

It has been shown by Lappa (2005e) that the delicate evolutionary equilibrium among crystal growth (absorption process), solution-depletion phenomena and competition among several seeds is coupled to significant and intriguing adjustments in the roll pattern inside the reactor.

In fact, a very complex multicellular structure is created above the interacting crystals and this convective field undergoes interesting subsequent transitions to different regimes as time passes as shown in Figure 30.

It has also been found that the prominent features of this evolution exhibit large sensitivity to the number of crystals simultaneously growing in the protein reactor if this number is not sufficiently large (this case corresponds to typical experiments carried out by organic crystal growers interested in studying morphological and kinetic details). It, in fact, plays a crucial role in determining the mode of the convective field (i.e. the number of vortex cells), the number and positions of rising solutal jets and the related time-evolution.

Such arguments provide additional evidence for the finding that in systems with multiple localized sources of convection, the number of these sources and their distribution must be always regarded as significant parameters potentially affecting the system response and evolution.

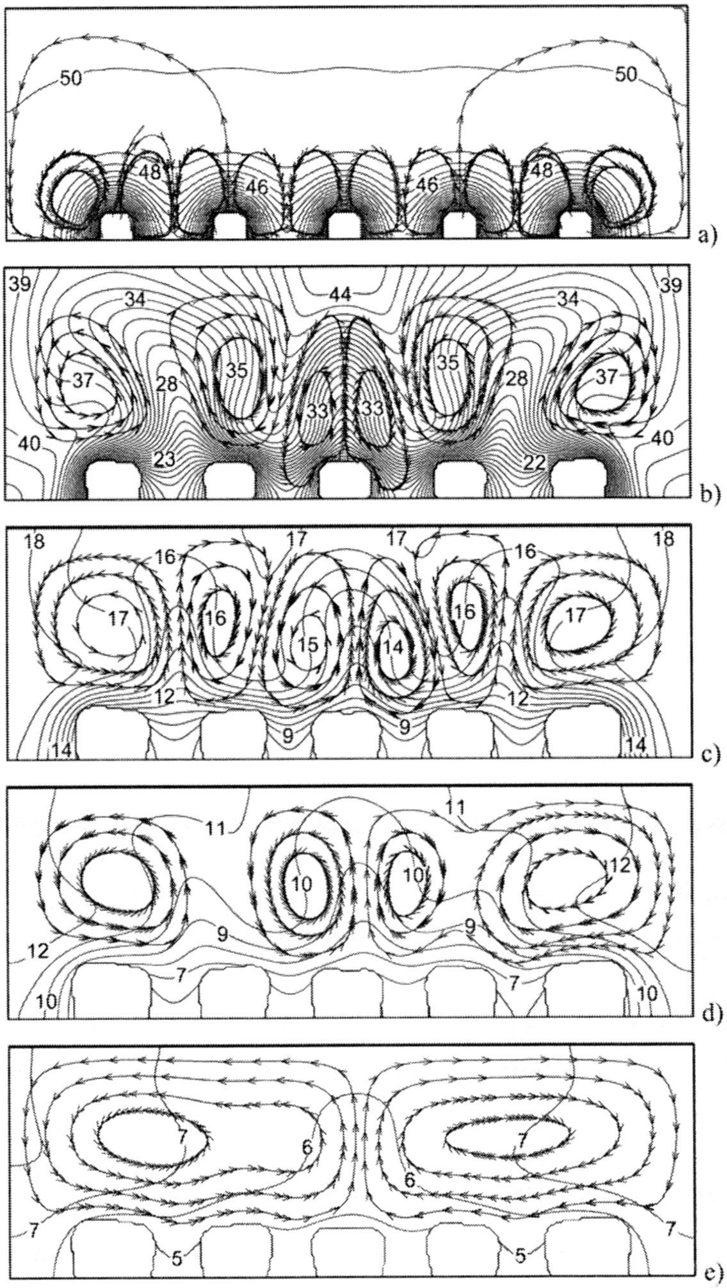

Figure 30. Snapshots of growing lysozyme crystals (A=H/L=3, N=5, κ=l/d=0.24), concentration distribution and velocity field reduced-gravity conditions ($g=10^{-4}\ g_o$): (a) t= 1.38×10^{4} [s], (b) t= 1.38×10^{5} [s], (c) t=3.59×10^{5} [s], (d) t=4.97×10^{5} [s], (e) t=7.18×10^{5} [s] (level 1→2.8×10^{-2} [g cm^{-3}], level 50→ 6×10^{-2} [g cm^{-3}], ΔC=6.4×10^{-4} [g cm^{-3}]).

REFERENCES

Agble D., Mendes Tatsis M.A., (2000), "The effect of surfactants on interfacial mass transfer in binary liquid-liquid systems", *Int. J. Heat and Mass Transfer*, 43: 1025-1034.

Agble D., Mendes-Tatsis M.A., (2001), "The prediction of Marangoni convection in binary liquid-liquid systems with added surfactants", *Int. J. heat Mass Transfer*, 44: 1439-1449.

Armfield S. and Street R., (2002), "An analysis and comparison of the time accuracy of fractional-step methods for the Navier-Stokes equations on staggered grids", Int. *J. Numer. Meth. Fluids*, 38(3): 255–282.

Barten W., Lücke M., Hort W., and Kamps M. (1989), "Fully developed traveling-wave convection in binary fluid mixtures", *Phys. Rev. Lett.*, 63, 376–379.

Bassano E., (2003), "Numerical simulation of thermo-solutal-capillary migration of a dissolving drop in a cavity", *Int. J. Numer. Meth. Fluids*, 41: 765-788.

Batchelor G. K., (1954), "Heat convection and buoyancy effects in fluids",. *Q. J. R. Met. Soc.* 80: 339–358.

Bazylak A., Djilali N., Sinton D., (2006), "Natural Convection in an Enclosure with Distributed Heat Sources", *Numerical Heat Transfer: Part A: Applications*, 49(7): 655 - 667.

Bekki S., Vignes-Adler M., Nakache E., (1992), "Solutal Marangoni effect: II Dissolution", *J. Colloid Interface Sci.,* 152(2): 314-324.

Bekki S., Vignes-Adler M., Nakache E., Adler P.M., (1990), "Solutal Marangoni effect: I Pure interfacial transfer", *J. Colloid Interface Sci.*, 140(2): 492-506.

Bizon C., Werne J., Predtechensky A. A., Julien K., McCormick W. D., Swift J. B., and Swinney H. L., (1997), "Plume dynamics in quasi-2D turbulent convection", *Chaos* 7 (1): 107-124.

Bornoff R.B., Mokhtarzadeh-Dehghan M.R., (2001), "A numerical study of interacting buoyant cooling-tower plumes", *Atmospheric Environment*, 35(3): 589-598.

Brand R. and Lahey F., (1967), "The heated laminar vertical jet", *J. Fluid Mech.*, 29: 305–315.

Braun B., Ikier C., Klein H. and Woermann D., (1993), "Thermocapillary Migration of Droplets in a Binary Mixture with Miscibility Gap during Liquid/Liquid Phase Separation under Reduced Gravity", *J. Colloid Interface Sci.*, 159(2): 515-516.

Brown D.L., Cortez R. and Minion M.L., (2001), "Accurate projecion methods for the incompressible Navier-Stokes equations", *J. Comput. Phys.*, 168 (2): 464–499.

Castaing B., Gunaratne G., Heslot F., Kadanoff L., Libchaber A., Thomae S., Wu X., Zaleski S., and Zanetti G., (1989), "Scaling of hard thermal turbulence in Rayleigh-Bénard convection", *J. Fluid Mech.*, 204: 1-30.

Cetegen B.M. , (1997), "Behaviour of naturally oscillating and periodically forced axisymmetric buoyant plumes of helium and helium-air mixtures", Phys. Fluids, 9 (12): 3742-3753.

Cetegen B.M., Ahmed T. A. , (1993), "Experiments on the periodic instability of buoyant plumes and pool fires", *Combustion and flame*, 93: 157-184.

Cetegen B.M., Kasper K.D. , (1996), "Experiments on the oscillatory behaviour of buoyant plumes of helium and helium-air", *Phys. Fluids*, 8 (11): 2974-2984.

Childress S., (2000), "Eulerian mean flow from an instability of convective plumes", *Chaos*, 10(1): 28-38.

Chu T. Y. and Goldstein R. J., (1973), "Turbulent convection in a horizontal layer of water", *J. Fluid Mech.*, 60: 141-159.

Cross M.C., (1986), "Traveling and Standing Waves in Binary-Fluid Convection in Finite Geometries", *Phys. Rev. Lett.* 57: 2935–2938.

Davies B.M., Jones C.D., Manning A.J., Thomson D.J., (2000), "Field experiments on the interaction of plumes from two sources", *Quarterly Journal of the Royal Meteorological Society,* 126(565), Part B: 1343-1366.

Deane A. E., Knobloch E., Toomre J., (1987), "Traveling waves and chaos in thermosolutal convection" , *Phys. Rev. A*, 36: 2862-2869.

Esmaeeli A., (2005), "Phase Distribution of Bubbly Flows under Terrestrial and Microgravity Conditions", *Fluid Dyn. Mater. Process. FDMP* 1(1): 63-80.

Farnetani C., Legras B., and Tackley P., (2002), "Mixing and deformations in mantle plumes", *Earth Planet. Sci. Lett.*, 196: 1-15.

Fujii T., (1963), "Theory of the steady laminar natural convection above a horizontal line source and a point heat source", *J. Heat Mass Transfer*, 16: 597–606.

Gluckman B. J., Willaime H., and Gollub J. P., (1993), " Geometry of isothermal and isoconcentration surfaces in thermal turbulence", *Phys. Fluids*, 5(3): 647-661.

Grossman S. and Lohse D., (2000), "Scaling in thermal convection: a unifying theory", *J. Fluid Mech.*, 407: 27-56.

Guermond J.L., Minev P., Shen J., (2006), "An Overview of Projection Methods for Incompressible Flows. Comput. Methods", *Comput. Methods Appl. Mech. Eng.*, 195: 6011-6045.

Haj-Hariri H., Shi Q., Borhan A., (1997),"Thermocapillary motion of deformable drops at finite Reynolds and Marangoni numbers", *Phys. Fluids* 9 (4):845-855.

Hennenberg M., Bisch P.M., Vignes-Adler M., Sanfeld A., (1979), "Mass Transfer Marangoni and Instability of interfacial longitudinal waves I. Diffusional exchanges", *J. Colloid Interfacial Sci.*, 69(1): 128-137.

Hennenberg M., Bisch P.M., Vignes-Adler M., Sanfeld A., (1980), "Mass Transfer Marangoni and Instability of interfacial longitudinal waves II. Diffusional exchanges and adsorption-desorption processes ", *J. Colloid Interface Sci.*, 74(2): 495-508.

Icoz T. and Jaluria Y., (2005), *"Numerical Simulation of Boundary Conditions and the Onset of Instability in Natural Convection due to Protruding Thermal Sources in an Open Rectangular Channel"*, 48(9): 831-847.

Kadanoff L. P., (2001), "Turbulent Heat Flow: Structures and Scaling", *Physics Today*, 54 (8): 34-39.

Kaminski E. and Jaupart C., (2003), "Laminar starting plumes in high-Prandtl-number fluids", *J. Fluid Mech.*, 478: 287-298.

Kassemi M., Rashidnia N., (1999), "Steady and oscillatory thermocapillary convection generated by a bubble", *Phys. Fluids*, 12 (12): 3133-3146.

König C.S. and Mokhtarzadeh-Dehghan M.R., (2002), "Numerical study of buoyant plumes from a multi-flue chimney released into an atmospheric boundary layer", *Atmospheric Environment,* 36(22): 3951-3962.

Kovalchuk N.M., Vollhardt D., (2006), "Marangoni instability and spontaneous non-linear oscillations produced at liquid interfaces by surfactant transfer", *Adv. Colloid Interface Sci.*, 120 (1-3): 1-31.

Langbein D. and Heide W., (1984), "The separation of liquids due to Marangoni convection", *Advances in Space Research*, 4(5): 27-36.

Langbein D. and Heide W., (1987), "Mixing and separation of transparent liquids under microgravity", [Mischen und Entmischen transparenter Fluessigkeiten unter Mikrogravitation] *Zeitschrift fuer Flugwissenschaften und Weltraumforschung.* 11: 29-36.

Lappa M., (2003) "Growth and Mutual Interference of Protein Seeds under reduced gravity conditions", *Phys. Fluids*, 15 (4), 1046-1057.

Lappa M., (2005a) "Assessment of VOF Strategies for the analysis of Marangoni Migration, Collisional Coagulation of Droplets and Thermal wake effects in Metal Alloys under Microgravity conditions ", *Computers, Materials and Continua CMC*, 2(1): 51-64.

Lappa M., (2005b), "Coalescence and non-coalescence phenomena in multi-material problems and dispersed multiphase flows: Part 1, a critical review of theories ", *Fluid Dynamics and Materials Processing,* 1(3): 201-212.

Lappa M., (2005c), "Coalescence and non-coalescence phenomena in multi-material problems and dispersed multiphase flows: Part 2, a critical review of CFD approaches", *Fluid Dynamics and Materials Processing*, 1(3): 213-234.

Lappa M., (2005d), "Numerical strategies for the simulation of dissolution kinetics and fluid-dynamic instabilities in liquid couples with a miscibility gap", *Journal of Non-Equilibrium Thermodynamics,* 30 (3): 243-259.

Lappa M., (2005e), "Discrete layers of interacting growing protein seeds: convective and morphological stages of evolution", *Phys. Rev.* E 71 (3): 031904.

Lappa M., (2006), "Oscillatory convective structures and solutal jets originated from discrete distributions of droplets in organic alloys with a miscibility gap", *Phys. Fluids*,18 (4) 042105 (14 pages).

Lappa M., Piccolo C., (2004),"Higher modes of Mixed Buoyant-Marangoni unstable convection originated from a droplet dissolving in a liquid/liquid system with miscibility gap", *Phys. Fluids*, 16(12) : 4262-4272.

Lappa M., Piccolo C., Carotenuto L., (2004a)," Mixed buoyant-Marangoni convection due to dissolution of a droplet in a liquid-liquid system with miscibility gap ", *Eur. J. Mech./B Fluids* 23/5: 781-794.

Lappa M., Piccolo C., Carotenuto L., (2004b), "Fluiddynamics and dissolution kinetics in immiscible organic systems with dispersed droplets", *Colloids and Surfaces A*, 261/1-3: 177-186.

Lappa M., Savino R. and Monti R., (2000), "Influence of buoyancy forces on Marangoni flow instabilities in liquid bridges"; *Int. J. Num. Meth. Heat Fluid Flow*, 10, (7): 721-749.

Lappa M., Savino R. and Monti R., (2001), "Three-dimensional numerical simulation of Marangoni instabilities in liquid bridges: influence of geometrical aspect ratio"; *Int. J. Num. Meth. Fluids,* 36 (1): 53-90.

Lavelle J.W., (1997), "Buoyancy-driven plumes in rotating, stratified cross flows: Plume dependence on rotation, turbulent mixing, and cross-flow strength", *Journal of Geophysical Research,* 102(C2): 3405-3420.

Lee J.R., Ha M.Y, Balachandar S., Yoon H.S., Lee S.S, (2004), "Natural convection in a horizontal layer of fluid with aperiodic array of square cylinders in the interior", *Phys. Fluids*, 16(4): 1097-1117.

Lewis J.B., Pratt H.R.C, (1953), "Oscillating Droplets", *Nature*, 171: 1155-1156.

Liang T.B., Slater M.J., (1990), "Liquid-liquid extraction drop formation: mass transfer and the influence of surfactant", *Chem. Eng. Sci.*, 45(1): 97-105.

Mehdi-Nejad V., Mostaghimi J., Chandra S., (2003),"Two-fluid Heat Transfer Based on a Volume Tracking advection Algorithm", *Proceedings of 9th International Conference on Liquid Atomization and Spray Systems* (ILASS-2003), 13-17 July, 2003, Sorrento, Italy.

Morris C.J. and Parviz B.A., (2006), "Self-assembly and characterization of Marangoni microfluidic actuators", *J. Micromech. Microeng.* 16: 972–980

Moses E. and Steinberg V., (1986), "Flow patterns and nonlinear behavior of traveling waves in a convective binary fluid", *Phys. Rev. A* 34: 693–696.

Moses E., Zocchi G., and Libchaber A., (1993), "An experimental study of laminar plumes", *J. Fluid Mech.*, 251: 581-601.

Moses E., Zocchi G., Procaccia I., and Libchaber A., (1991), "The dynamics and interaction of laminar thermal plumes",' *Europhys. Lett.*, 14: 55-60.

Nakache E., Dupeyrat M., Vignes-Adler M., (1983), "Experimental and theoretical study of an interfacial instability at some oil-water interfaces involving a surface-active agent; I Physicochemical description and outlines for a theoretical approach", *J. Colloid Interface Sci.*, 94(1): 120-127.

Niemela J. J. and Sreenivasan K. R., (2003), "Rayleigh-number evolution of large-scale coherent motion in turbulent convection ", *Europhys. Lett.*, 62 (6): 829-833.

Niemela J.J., Ahlers G., and Cannell D. S., (1990), "Localized traveling-wave states in binary-fluid convection", *Phys. Rev. Lett.* 64: 1365–1368.

Parodi A, von Hardenberg J., Passoni G., and Spiegel E. A., (2004), "Clustering of Plumes in Turbulent Convection", *Phys. Rev. Lett.* 92, 194503

Perez de Ortiz S. E., Sawistowski H. , (1973a), "Interfacial Instability of binary liquid-liquid systems, I. Stability Analysis, *Chem. Eng. J.*, 28: 2051-2061.

Perez de Ortiz S. E., Sawistowski H. , (1973b), "Interfacial Instability of binary liquid-liquid systems, II. Stability behaviour of selected systems, *Chem. Eng. J.*, 28: 2063-2069.

Qiu X.-L. and Tong P., (2001), "Large-scale velocity structures in turbulent thermal convection", *Phys. Rev.* E 64(3), 036304 [13 pages]

Qiu X.-L., Yao S. H., and Tong P., (2000), "Large-scale coherent rotation and oscillation in turbulent thermal convection", *Phys. Rev.* E 61(6), R6075–R6078.

Ratke L.,"Decomposition, Phase separation, Solidification of immiscible liquid alloys", in *Immiscible liquid metals and organics, Proceedings of an international workshop organized by the ESA expert working group*, "Immiscible alloys" held in the Physikzentrum, Bad Honnef 1992, edited by L. Ratke, Germany, (1992), pp.3-34.

Rider W. J., Kothe D. B., (1998), "Reconstructing Volume Tracking", *J. Comput. Phys.*, 141: 112-152.

Siggia E., (1994), "High Rayleigh number convection", Annu. Rev. Fluid Mech., 26: 137-168.

Slavtchev S., Kalitzova-Kurteva P., Mendes M.A., (2006), "Marangoni instability of liquid-liquid systems with a surface-active solute", *Colloids and Surfaces* A, 282: 37-49.

Smith M.K., Davis S.H. (1983), "Instabilities of dynamic thermocapillary liquid layers. Part 1 : convective instabilities ", *J.Fluid Mech* 132: 119-144.

Solomon T.H. and Gollub J.P., (1990), "Sheared boundary layers in turbulent Rayleigh-Bénard convection", *Phys. Rev. Lett.*, 64: 2382-2385.

Sternling C.V., Scriven L.E., (1959), "Interfacial Turbulence: hydrodynamic instability and the Marangoni effect", *AIChE J.*, 5(4): 514-523.

Taylor B. R., Taylor G., and Turner J. S., (1956), "Turbulent gravitational convection from maintained and instantaneous sources", *Proc. R. Soc. London, Ser.* A 234: 1-23.

Trelles J., McGrattan K. B. and Baum H. R., (1999), "Smoke Dispersion from Multiple Fire Plumes", AIAA Journal, 37(12): 1588-1601.

Turner J. S., (1969), "Buoyant plumes and thermals" *Annu. Rev. Fluid Mech.* 1: 29-44.

Villermaux E., (1995), "Memory-induced low frequency oscillations in closed convection boxes", *Phys. Rev. Lett.* 75: 4618–4621.

Vincent A. P. and Yuen D. A., (1999), "Plumes and waves in two-dimensional turbulent thermal convection", *Phys. Rev.* E 60(3): 2957-2963.

Walden R. W., Kolodner P., Passner A., and Surko C. M., (1985), "Traveling waves and chaos in convection in binary fluid mixtures, *Phys. Rev. Lett.* 55(5): 496–499.

Wang Q. and Jaluria Y., (2002), "Instability and Heat Transfer in Mixed-Convection Flow in a Horizontal Duct with Discrete Heat Sources", *Numerical Heat Transfer: Part A: Applications,* 42(5): 445 - 463

Warren C., Webb W.W., , (1969), "Interfacial Tension of Near-Critical Cyclohexane-Methanol Mixtures", *J. Chem. Phys.* 50 (9): 3694-3700.

Werne J., (1993), "Structure of hard-turbulent convection in two dimensions: Numerical evidence", *Phys. Rev. E*, 48: 1020-1035.

Worster M. G., (1986), "The axisymmetric laminar plume: asymptotic solution for large Prandtl number", *Stud. Appl. Maths*, 75: 139–152.

Xi H.D., Lam S. and Xia K.Q., (2004), "From laminar plumes to organized flows: the onset of large-scale circulation in turbulent thermal convection", *J. Fluid Mech.*, 503: 47-56.

Zander R., Schneider G.M., "Appearance of convective instabilities caused by liquid-liquid phase separation under small buoyancy forces", Proceedings of the VIIIth European Symposium on Materials and Fluid Sciences in Microgravity, Bruxelles, Belgium, 12-16 April 1992, *ESA SP 333, Vol II*, (1992), pp. 755-759.

In: New Colloid and Surface Science Research
Editor: Lawrence V. Schwartz, pp. 61-129
ISBN: 978-1-60021-594-0

Chapter 2

DIFFUSIOOSMOSIS OF ELECTROLYTE SOLUTIONS

Hsien Chen Ma and Huan J. Keh*
Department of Chemical Engineering, National Taiwan University,
Taipei 10617, Taiwan, ROC

ABSTRACT

In general, driving forces for the fluid transport through micropores include dynamic pressure differences between the two ends of a capillary pore (convection), concentration differences of an impermeable solute between the two bulk solutions outside the pores (osmosis), and tangential electric fields that interact with the electric double layer adjacent to a charged pore wall (electroosmosis). Another driving force for the flow of liquid solutions in a capillary pore, which has commanded less attention, involves concentration gradients of a permeable solute along the capillary that interacts with the pore wall. The fluid motion associated with this mechanism is termed diffusioosmosis.

The steady diffusioosmotic flows of an electrolyte solution along a charged plane wall, in a capillary slit, and in a capillary tube generated by an imposed tangential concentration gradient are theoretically examined in this chapter. The charged walls may have either a constant surface potential or a constant surface charge density of an arbitrary quantity. The electric double layers adjacent to the charged walls may have an arbitrary thickness and their electrostatic potential distributions are determined by the Poisson-Boltzmann equation. Solving a modified Navier-Stokes equation with the constraint of no net electric current arising from the cocurrent diffusion, electric migration, and diffusioosmotic convection of the electrolyte ions, the macroscopic electric field and the fluid velocity in the tangential direction induced by the imposed electrolyte concentration gradient are obtained as functions of the lateral position for various cases. The results indicate that the effect of the deviation of the local induced tangential electric field inside the double layer from its bulk-phase quantity and the relaxation effect on the diffusioosmotic velocity of the fluid are significant in most practical situations, even for the case of very thin double layer.

* E-mail: huan@ntu.edu.tw

1. Introduction

1.1. Diffusiophoresis

A colloidal particle can be driven to move by the application of a nonuniform solute concentration field that interacts with the surface of the particle. This phenomenon, known as diffusiophoresis [1–3], has been demonstrated experimentally for both charged [4] and uncharged [5] solutes. Diffusiophoresis is of practical importance in some applications to particle analysis or separation and in certain latex-particle coating processes [1,6]. In a solution of nonionic solute, the solute molecules interact with the particle through the van der Waals/dipole forces. For charged particles in an electrolyte solution, the particle-solute interaction is electrostatic in nature and its range is the Debye screening length κ^{-1} (defined right after Eq. (3)). Particles with zeta potentials of order kT/e (~ 25 mV; e is the elementary electric charge, k is Boltzmann's constant, and T is the absolute temperature) in electrolyte gradients of order 1 M/cm will move by diffusiophoresis at speeds of several micrometers per second.

Using the classical model of the diffuse electric double layer, Prieve [7] derived a formula for the diffusiophoretic velocity of a nonconducting particle of arbitrary shape in a solution of a symmetrically charged electrolyte with a constant concentration gradient for arbitrary zeta potential (ζ) of the particle in the limit of thin double layer ($\kappa a \to \infty$, where a is the linear dimension of the particle). This formula agrees with that previously deduced by a more intuitive method [1]. Using a method of matched asymptotic expansions with κa^{-1} as the small parameter, Prieve et al. [2] evaluated the effect of particle curvature on the diffusiophoretic mobility of a spherical particle accurate to $O(\kappa a^{-1})$. When the double-layer distortion from equilibrium was taken as a small perturbation, Prieve and Roman [8] obtained a numerical solution for the diffusiophoresis of a dielectric sphere in concentration gradients of 1:1 electrolytes (KCl or NaCl), which was applicable to a broad range of ζ and κa Later, analytical expressions for the diffusiophoretic velocity of a dielectric sphere [9-15], cylinder [10,16], and spheroid [17] with a "thin but polarized" double layer were derived. Recently, analytical formulas in close forms for the diffusiophoretic mobility of a colloidal sphere [18-20] and cylinder [21,22] in symmetric electrolytes at low surface charge density σ (valid for ζ up to 50 mV) and arbitrary κa have also been obtained. All results of the above investigations show that the diffusiophoretic mobility of the particle in general decreases with an increase in κa^{-1}. Also, the particle can reverse the direction of its migration when the value of $|\zeta|$, κa^{-1}, or the valence of the electrolyte increases.

1.2. Diffusioosmosis

The electrokinetic flows of an electrolyte solution in a small pore with a charged wall are of much fundamental and practical interest in various areas of science and engineering. Perhaps the most familiar example of electrokinetic flows is electroosmosis, which results from the interaction between an external tangential electric field and the electric double layer adjacent to the charged wall. Problems of fluid flow caused by this well-known mechanism were studied extensively in the past [23-34].

Another example of electrokinetic flows in a capillary pore, which is termed diffusioosmosis (conjugated with diffusiophoresis) or capillary osmosis and has caught less attention, involves a concentration gradient of the electrolyte along the capillary that interacts with the charged wall. Same as in the case of electroosmosis, the electrolyte-wall interaction in diffusioosmosis is electrostatic in nature and its range is the Debye screening length κ^{-1}. The fluid motion caused by diffusioosmosis has been analytically examined for flows near a plane wall [35-37] and inside a capillary pore [38-44]. Some experimental results and interesting applications concerning diffusioosmosis are also available in the literature [45].

A tangential gradient of a dissociating electrolyte produces fluid flow along a charged solid surface by two mechanisms. The first involves the stresses developed by the tangential gradient of the excess pressure within the electric double layer (chemiosmotic effect), and the second is based on the macroscopic electric field that is generated because the tangential diffusive and convective fluxes of the two electrolyte ions are not equal (electroosmotic effect). Both mechanisms were considered to some extent in previous investigations for the diffusioosmotic flow [16,21,22,35-44]. In these studies, however, either the effect of lateral distributions of the counterions and co-ions (or of the electrostatic potential) on the local electric field induced by the imposed electrolyte concentration gradient in the tangential direction inside the double layer or the effect of the ionic convection on it caused by the diffusioosmotic flow was neglected. Moreover, the analyses concerning the diffusioosmotic flow in capillary slits or tubes [41-44] are subject to the severe restriction that the zeta potential is sufficiently low (less than about 25-50 mV) for the Debye-Huckel approximation to be acceptable. In practical applications, however, zeta potentials as high as 100-200 mV are frequently encountered.

In this chapter we present a comprehensive analysis of the diffusioosmosis of an electrolyte solution with a constant prescribed concentration gradient in the tangential direction along three different geometric circumstances, including a plane, a slit, and a tube. The zeta potential or surface charge density of the charged walls is assumed to be uniform, but no assumption is made concerning the magnitude of the zeta potential or the thickness of the double layer, and both the lateral distribution of the induced tangential electric field and the effect of the ionic convection on it are allowed. The Poisson-Boltzmann equation governing the electrostatic potential near the walls is solved analytically. Semi-analytical results for the diffusioosmotic velocity profile are obtained for various cases. These results show that the effect of the deviation of the induced tangential electric field in the electric double layer from its bulk-phase quantity and the effect of the ionic convection on the diffusioosmotic velocity of the fluid are dominantly significant in most practical situations, even for the case of a very thin double layer.

2. Diffusioosmosis along a Charged Plane Wall

In this section, the diffusioosmotic flow of an infinitely thick solution of a symmetrically charged binary electrolyte of valence Z (where Z is a positive integer) tangential to a dielectric plane wall of length L, as illustrated in Figure 1, at the steady state is considered. The discrete nature of the surface charges, which are uniformly distributed over the wall, is ignored. The applied electrolyte concentration gradient ∇n^{∞} is a constant along the tangential (z) direction, where $n^{\infty}(z)$ is the linear concentration (number density) distribution of the electrolyte in the bulk solution phase far from the wall (with $y \to \infty$ or beyond the electrostatic influence of the charged wall). The end effects are neglected. It is assumed that n^{∞} is only slightly non-uniform such that $L|\nabla n^{\infty}|/n^{\infty}(0) << 1$, where $z = 0$ is set at the midpoint along the plane wall. Thus, the variation of the electrostatic potential (excluding the macroscopic electric field induced by the electrolyte gradient, which will be discussed in Section 2.2) and ionic concentrations in the electric double layer adjacent to the wall with the tangential position can be neglected in comparison with their corresponding quantities at $z = 0$.

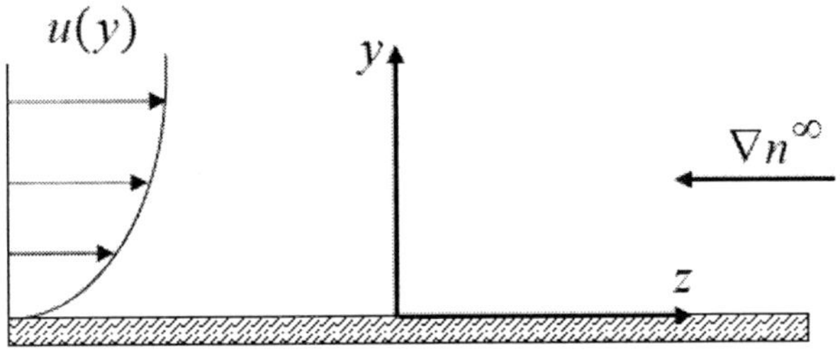

Figure 1. Geometrical sketch for the diffusioosmosis due to an applied concentration gradient of an electrolyte tangential to a plane wall.

2.1. Electrostatic Potential Distribution

We first consider the electrostatic potential distribution in the electrolyte solution near the charged plane wall. Let $\psi(y)$ be the electrostatic potential at a position y from the wall relative to that in the bulk solution and $n_+(y,z)$ and $n_-(y,z)$ be the local concentrations of the cations and anions, respectively. Then the Poisson equation gives

$$\frac{d^2\psi}{dy^2} = -\frac{4\pi Ze}{\varepsilon}[n_+(y,0) - n_-(y,0)], \tag{1}$$

with $\varepsilon = 4\pi\varepsilon_0\varepsilon_r$, where ε_r is the relative permittivity of the electrolyte solution and ε_0 is the permittivity of a vacuum.

The local ionic concentrations can also be related to the electrostatic potential by the Boltzmann equation,

$$n_{\pm} = n^{\infty} \exp(\mp\bar{\psi}), \tag{2}$$

where $\bar{\psi} = Ze\psi / kT$ is the dimensionless potential profile. Substitution of Eq. (2) into Eq. (1) leads to the well-known Poisson-Boltzmann equation,

$$\frac{d^2\bar{\psi}}{dy^2} = \kappa^2 \sinh\bar{\psi}, \tag{3}$$

where $\kappa = [8\pi(Ze)^2 n^{\infty}(0) / \varepsilon kT]^{1/2}$ is the Debye screening parameter.

2.1.1. General Analysis

For the case of constant surface potential, the boundary conditions for $\bar{\psi}$ are

$$y = 0: \bar{\psi} = \bar{\zeta} \equiv \frac{Ze\zeta}{kT}, \tag{4a}$$

$$y \to \infty: \bar{\psi} = 0, \tag{4b}$$

where the constant ζ is the zeta potential at the shear plane of the wall adjacent to the electrolyte solution having a uniform bulk concentration $n^{\infty}(0)$. The effects of the Stern layer between the solid surface and the diffuse part of the electric double layer are neglected. The solution to Eqs. (3) and (4) is the well-known Gouy-Chapman result [1,29],

$$\bar{\psi} = 2\ln[\frac{1+\gamma\exp(-\kappa y)}{1-\gamma\exp(-\kappa y)}] \tag{5}$$

where $\gamma = \tanh(\bar{\zeta}/4)$.

If the constant surface charge density σ, instead of the surface potential ζ, is known at the plane wall, the boundary condition specified by Eq. (4a) should be replaced by the Gauss condition,

$$y = 0: \frac{d\psi}{dy} = -\frac{4\pi\sigma}{\varepsilon}. \tag{6}$$

Now, the solution for $\bar{\psi}$ given by Eq. (5) is still valid for this condition, with the relation between ζ and σ as

$$\bar{\zeta} = 2\sinh^{-1}\left(\frac{2\pi Ze\sigma}{\varepsilon\kappa kT}\right). \tag{7}$$

2.1.2. Case of Low Surface Potential

When the surface potential $\bar{\zeta}$ is small (say, $|\bar{\zeta}| \leq 1$), Eq. (3) can be linearized (known as the Debye-Huckel approximation), and Eqs. (5) and (7) reduce to

$$\bar{\psi} = \bar{\zeta}\exp(-\kappa y) \tag{8}$$

and

$$\bar{\zeta} = \frac{4\pi Ze\sigma}{\varepsilon\kappa kT}. \tag{9}$$

Both Eq. (7) and Eq. (9) indicate that, for a given electrolyte solution, σ increases with an increase in κ or $[n^{\infty}(0)]^{1/2}$ for the case of constant surface potential and ζ decreases with an increase in κ or $[n^{\infty}(0)]^{1/2}$ for the case of constant surface charge density.

2.2. Induced Tangential Electric Field

The ionic concentrations n_+ and n_- in the fluid undergoing diffusioosmosis along the plane wall are not uniform in both tangential (z) and normal (y) directions, and their gradients in the tangential direction can give rise to a "diffusion current" distribution. To prevent a continuous separation of the counterions and co-ions, an electric field distribution along the tangential direction arises spontaneously in the electrolyte solution to produce another electric current distribution which exactly balances the diffusion current [2,7-10]. This induced electric field generates an electroosmotic flow of the fluid parallel to the plane wall, in addition to the chemiosmotic flow caused by the electrolyte gradient directly. Both the chemiosmotic and the electroosmotic flows also generate an electric current distribution by the ionic convection (known as the relaxation effect), and alternately, this secondary "convection current" again needs to be balanced by the electric current contributed from the induced electric field.

2.2.1. General Analysis

The total flux of either ionic species can be expressed as the general form

$$\mathbf{J}_{\pm} = -D_{\pm}[\nabla n_{\pm} \pm \frac{Ze}{kT} n_{\pm}(\nabla \psi - \mathbf{E})] + n_{\pm}\mathbf{u} \tag{10}$$

where $\mathbf{u} = u(y)\mathbf{e}_z$ is the fluid velocity relative to the plane wall in the direction of decreasing electrolyte concentration (i.e., $\mathbf{e}_z$ is the unit vector in the direction of $-\nabla n^{\infty}$), D_+ and D_- are the diffusion coefficients of the cations and anions, respectively, $\mathbf{E} = E(y)\mathbf{e}_z$ is the macroscopic electric field induced by the concentration gradient of the electrolyte, and the principle of superposition for the electric potential is used. To have no net electric current arising from the cocurrent diffusion, electric migration, and diffusioosmotic convection of the cations and anions, one must require that $\mathbf{J}_+ = \mathbf{J}_- = \mathbf{J}$ [obviously, the normal component of $\mathbf{J}$ vanishes and the ionic fluxes induced by $\nabla \psi$ in Eq. (10) are balanced by the normal components of the diffusive ionic fluxes as required by the Boltzmann distribution given by Eq. (2)].

Applying the constraint $\mathbf{J}_+ = \mathbf{J}_-$ to Eq. (10), we obtain [46]

$$\mathbf{E} = \frac{kT}{Ze}\frac{\nabla n^{\infty}}{n^{\infty}(0)}[\frac{(1+\beta)e^{-\bar{\psi}} - (1-\beta)e^{\bar{\psi}}}{(1+\beta)e^{-\bar{\psi}} + (1-\beta)e^{\bar{\psi}}} + \frac{\mathrm{Pe}\sinh\bar{\psi}}{(1+\beta)e^{-\bar{\psi}} + (1-\beta)e^{\bar{\psi}}}\frac{u}{U^*}] \tag{11}$$

where

$$U^* = \frac{\varepsilon |\nabla n^{\infty}|}{4\pi\eta n^{\infty}(0)}(\frac{kT}{Ze})^2 = \frac{2kT}{\eta\kappa^2}|\nabla n^{\infty}| \tag{12}$$

which is a characteristic value of the diffusioosmotic velocity,

$$\beta = \frac{D_+ - D_-}{D_+ + D_-} \tag{13}$$

$$\mathrm{Pe} = \frac{4n^{\infty}(0)U^*}{(D_+ + D_-)|\nabla n^{\infty}|} = \frac{8n^{\infty}(0)kT}{(D_+ + D_-)\eta\kappa^2} \tag{14}$$

and η is the fluid viscosity. As it is defined by Eq. (13), $-1 \leq \beta \leq 1$, with the upper and lower bounds occurring as $D_- / D_+ \to 0$ and ∞, respectively.

Typical values of the physical quantities in Eqs. (11)-(14) are $U^* = 10^{-5}$ m/s, $D_\pm = 10^{-9}$ m^2/s, $n^\infty(0)/|\nabla n^\infty| = 10^{-4}$ m, and Pe of order unity. The induced electric filed $\mathbf{E}$ given by Eq. (11) in a self-consistent way depends on the dimensionless local electrostatic potential $\overline{\psi}$ and fluid velocity u. It indicates that $\mathbf{E}$ is collinear with and proportional to the tangentially imposed electrolyte gradient ∇n^∞.

2.2.2. Bulk-Phase Quantity

If we consider the situation that $\kappa y \to \infty$, then $\overline{\psi} \to 0$ and Eq. (11) for the induced electric field caused by the imposed electrolyte concentration gradient reduces to its bulk-phase quantity, $\mathbf{E}^\infty = \dfrac{kT}{Ze} \dfrac{\beta \nabla n^\infty}{n^\infty(0)}$. (15) For the special case of an uncharged wall ($\overline{\zeta} = 0$), $\mathbf{E}$ at any location y is also identical to this bulk-phase quantity. Note that $\mathbf{E}^\infty$ is linearly proportional to the parameter β, but $\mathbf{E}(y)$ is not necessarily to vanish if $\beta = 0$, even as Pe=0, as shown in Eq. (11).

2.3. Fluid Velocity Distribution

For the steady diffusioosmotic flow along a plane wall, the momentum balances on the symmetric electrolyte solution, which is taken to be incompressible and Newtonian, in the y and z directions give

$$\frac{\partial p}{\partial y} + Ze(n_+ - n_-)\frac{d\psi}{dy} = 0 \tag{16a}$$

$$\eta \frac{d^2 u}{dy^2} = \frac{\partial p}{\partial z} - Ze(n_+ - n_-)E \tag{16b}$$

where $p(y,z)$ is the dynamic pressure distribution. The boundary conditions for u at the no-slip wall and at infinity are

$$y = 0: \ u = 0, \tag{17a}$$

$$y \to \infty: \frac{du}{dy} = 0. \tag{17b}$$

After the substitution of Eq. (2) into Eq. (16a) based on the assumption that the equilibrium ionic distributions are not affected by the net electrolyte flux $\mathbf{J}$, which is warranted if $|\nabla n^{\infty}| / \kappa n^{\infty}(0) << 1$, the pressure distribution can be determined as

$$p = p_{\infty} + 2n^{\infty}(z)kT(\cosh\overline{\psi} - 1). \tag{18}$$

Here, p_{∞} is the pressure far away from the wall, which is a constant in the absence of the applied pressure gradient, and the electric potential distribution $\overline{\psi}(y)$ is given by Eq. (5). Note that the pressure distribution in the y and z directions results from the variation of the electrostatic potential ψ with y and the variation of the bulk concentration n^{∞} with z, respectively.

2.3.1. General Analysis

Substituting the ionic concentration distributions of Eq. (2) and the pressure profile of Eq. (18) into Eq. (16b) and then performing the integration with respect to *y* twice subject to the boundary conditions in Eq. (17), we obtain [46]

$$\frac{u}{U^*} = \int_0^{\kappa y}\int_{\infty}^{\kappa y}[\cosh\overline{\psi} - 1 + \frac{Zen^{\infty}(0)}{kT|\nabla n^{\infty}|}E\sinh\overline{\psi}]\mathrm{d}(\kappa y)\mathrm{d}(\kappa y) \tag{19}$$

After the substitution of Eq. (19) for *u* and Eq. (5) for $\overline{\psi}$ into Eq. (11), the induced electric field *E* can be numerically solved as a function of the dimensionless parameters κy, $\overline{\zeta}$, β, and Pe. With the known results of $\overline{\psi}$ and *E*, the diffusioosmotic velocity distribution of the electrolyte solution as a function of $\overline{\zeta}$, β, and Pe can be determined from Eq. (19) with the numerical integrations. As expected, Eq. (19) yields $u/U^* = 0$ (since $\overline{\psi} = 0$) everywhere if $\zeta = 0$.

It is understood that, for given values of κy and Pe, the quantity u/U^* with specified values $-\overline{\zeta}$ and β is equal to that with the values $\overline{\zeta}$ and $-\beta$. The bulk-phase diffusioosmotic velocity u_{∞} of the electrolyte solution at a large distance from the plane wall as a function of the relevant parameters can be obtained from the distribution *u* by taking $\kappa y \to \infty$.

2.3.2. Result of Previous Work

In previous studies of the diffusioosmosis of electrolyte solutions along a plane wall [35,36,41], the constant bulk-phase quantity of the induced electric field $\mathbf{E}$ given by Eq. (15) taking $\overline{\psi} = 0$ everywhere was used (and thus, the effect of the lateral distribution of the induced electric field and the relaxation effect on the fluid velocity was excluded) in Eq. (16b) and its solution gives

$$\frac{u_\infty}{U^*} = \beta\overline{\zeta} + 4\ln\cosh(\frac{\overline{\zeta}}{4}) \tag{20}$$

In this equation, the electroosmotic contribution represented by the first term on the right-hand side and the chemiosmotic contribution denoted by the second term are decoupled as a result of the assumption of a constant induced electric field.

2.4. Diffusioosmosis without Relaxation Effect

In this section, we consider a situation of the diffusioosmosis of an infinitely thick solution of a symmetrically charged electrolyte tangential to a dielectric plane wall when the effect of the ionic convection on the induced electric field caused by the diffusioosmotic flow itself (relaxation effect) is negligible. Close-form formulas for the local and bulk-phase diffusioosmotic velocities as functions of relevant parameters can be derived in this special case.

2.4.1. Induced Electric Field and Net Electrolyte Diffusivity

When the fluid velocity u is taken to be zero (or Pe=0), Eq. (11) reduces to [37,43]

$$\mathbf{E} = \frac{kT}{Ze}(\frac{G_+ - G_-}{G_+ + G_-})\frac{\nabla n^\infty}{n^\infty(0)} \tag{21}$$

where

$$G_\pm = D_\pm \exp(\mp\frac{Ze\psi}{kT}) \tag{22}$$

The coefficients G_+ and G_- defined by the above equation reflect the fact of an increase in the tangential diffusive flux of the counter-ions and a decrease in the flux of the co-ions inside the electric double layer.

Substitution of Eqs. (21) and (22) into Eq. (10) taking **u**=**0** leads to a net flux distribution of the electrolyte,

$$\mathbf{J} = -D\nabla n^{\infty} \tag{23}$$

where the position-dependent net diffusivity is

$$D = \frac{2G_+G_-}{G_+ + G_-} = \frac{2D_+D_-}{G_+ + G_-} \tag{24}$$

which is always positive. Equation (23) indicates clearly that $\mathbf{J}$ is also collinear with and proportional to the tangentially imposed electrolyte gradient ∇n^{∞}. For the special case of an uncharged wall ($\zeta = 0$), D at any location y is identical to its corresponding bulk-phase quantity.

When $|\overline{\psi}| \to \infty$, the ratio $(G_+ - G_-)/(G_+ + G_-)$ in Eq. (21) approaches the value $-\zeta/|\zeta|$ (whose magnitude is unity) and the induced electric field $\mathbf{E}$ becomes independent of the ionic diffusion coefficients. In this limit, the net diffusivity D given by Eq. (24) will vanish asymptotically as $2G_+$ if $\zeta > 0$ and as $2G_-$ if $\zeta < 0$.

When the magnitude of the electrostatic potential $\overline{\psi}$ is small, a Taylor expansion applies to Eq. (22), and Eqs. (21) and (24) can be expressed as

$$\mathbf{E} = \frac{kT}{Ze}[\beta - (1-\beta^2)\overline{\psi} - \beta(1-\beta^2)\overline{\psi}^2 + O(\overline{\psi}^3)]\frac{\nabla n^{\infty}}{n^{\infty}(0)} \tag{25}$$

and

$$D = \frac{2D_+D_-}{D_+ + D_-}[1 + \beta\overline{\psi} - (\frac{1}{2} - \beta^2)\overline{\psi}^2 + O(\overline{\psi}^3)] \tag{26}$$

where β is defined by Eq. (13). If we consider the situation that $\kappa y \to \infty$, then $\overline{\psi} \to 0$ and Eq. (25) reduces to Eq. (15).

Note that, even if the cation and anion diffusivities are identical (i.e., $\beta = 0$), the $O(\overline{\psi})$ term in Eq. (25) for the induced electric field $\mathbf{E}$ and the $O(\overline{\psi}^2)$ term in Eq. (26) for the net diffusivity D still exist, due to the adsorption of the counter-ions and depletion of the co-ions near the plane wall. For the limiting case of $\beta = \pm 1$ (one of the ionic diffusion coefficients vanishes), the induced electric field at any location is independent of the local electrostatic potential and equals its bulk-phase quantity, as shown by Eq. (21) or (25).

2.4.2. Fluid Velocity

Substituting the ionic concentration distributions of Eq. (2), the electrostatic potential distribution of Eq. (5), the pressure profile of Eq. (18), and the induced electric field profile of Eq. (21) into Eq. (16b) yields [37]

$$\frac{d}{dx}[(x^2-1)\frac{d(u/U^*)}{dx}] = \frac{4(x^2-\omega^4)}{x^4+\omega^4} \tag{27}$$

where

$$x = \frac{1+\gamma\exp(-\kappa y)}{1-\gamma\exp(-\kappa y)} \tag{28}$$

$$\omega = (\frac{1+\beta}{1-\beta})^{1/4} = (\frac{D_+}{D_-})^{1/4} \tag{29}$$

and the diffusioosmotic velocity is normalized by the characteristic value given by Eq. (12).

Integration of Eq. (27) from $y \to \infty$ to $y = y$ using Eq. (17b) leads to

$$\frac{d(u/U^*)}{dx} = \frac{2}{x^2-1}[g(x)-g(1)] \tag{30}$$

where

$$g(x) = \frac{1}{\sqrt{2}\omega}[(1-\omega^2)\tan^{-1}\frac{\sqrt{2}\omega x}{\omega^2-x^2} + \frac{1}{2}(1+\omega^2)\ln\frac{x^2-\sqrt{2}\omega x+\omega^2}{x^2+\sqrt{2}\omega x+\omega^2}]. \tag{31}$$

The diffusioosmotic velocity profile of the electrolyte solution parallel to the plane wall can be solved by integrating Eq. (30) from $y = 0$ to $y = y$ subject to the boundary condition in Eq. (17a), with the result

$$\frac{u}{U^*} = F(x) - F(\frac{1+\gamma}{1-\gamma}) \tag{32}$$

where

$$F(x) = [\frac{1-\omega^2}{\sqrt{2}\omega} m\pi + g(x) - g(1)] \ln \frac{x-1}{x+1}$$

$$+ \frac{1+i}{2\sqrt{2}\omega} \{(\omega^2 - i)[h_{0+}(x) - h_{1+}(x) + h_{1-}(x)]$$

$$+ (1 - i\omega^2)[h_{0-}(x) - h_{2+}(x) + h_{2-}(x)]\} \tag{33}$$

$$h_{0\pm}(x) = \ln \frac{\sqrt{2}\omega + (1 \pm i)}{\sqrt{2}\omega - (1 \pm i)} \ln(x^2 \pm i\omega^2) \tag{34}$$

$$h_{1\pm}(x) = \mathrm{Li}_2[\frac{\sqrt{2}\omega + (1-i)x}{\sqrt{2}\omega \pm (1-i)}] + \mathrm{Li}_2[\frac{\sqrt{2}\omega - (1-i)x}{\sqrt{2}\omega \pm (1-i)}] \tag{35a}$$

$$h_{2\pm}(x) = \mathrm{Li}_2[\frac{\sqrt{2}\omega + (1+i)x}{\sqrt{2}\omega \pm (1+i)}] + \mathrm{Li}_2[\frac{\sqrt{2}\omega - (1+i)x}{\sqrt{2}\omega \pm (1+i)}] \tag{35b}$$

$i = \sqrt{-1}$, and Li_2 is the polylogarithm function of order 2, also known as the Euler dilogarithm function [47]. In Eq. (33), $m = 1$ if $x \geq \omega \geq 1$ or $x \leq \omega \leq 1$, and $m = 0$ otherwise. Equations (32)-(35) predict that the normalized fluid velocity u/U^* is only a function of the dimensionless parameters γ (or $\overline{\zeta}$), ω (or β), and κy.

If we consider the situation that $\kappa y \to \infty$, the fluid velocity at a large distance from the plane wall caused by the imposed electrolyte concentration gradient can be evaluated from Eq. (32). The result of this bulk-phase diffusioosmotic velocity is

$$\frac{u_\infty}{U^*} = F(1) - F(\frac{1+\gamma}{1-\gamma}) \tag{36}$$

regardless of the thickness of the electric double layer. For the diffusioosmosis of electrolyte solutions in a straight capillary tube or slit with $\kappa R \to \infty$, where R is the radius of the tube or half thickness of the slit, the fluid velocity at a large distance from the capillary wall (with $\kappa y >> 1$) can also be evaluated by Eq. (36). The difference between Eq. (20) and Eq. (36) is quite obvious, and their comparison will be graphically displayed in the next section.

When the Debye-Huckel approximation is used, Eqs. (2) and (18) for the concentration and pressure distributions can be expressed by their Taylor expansions in $\overline{\psi}$, and Eqs. (5) and (21) for the electrostatic potential and induced tangential electric field neglecting the relaxation effect reduce to Eqs. (8) and (25). With this simplification, Eqs. (32), (36), and (20) become

$$\frac{u}{U^*} = \beta\overline{\zeta}[1-\exp(-\kappa y)] + [\frac{1}{8} - \frac{1}{4}(1-\beta^2)]\overline{\zeta}^2[1-\exp(-2\kappa y)] + O(\overline{\zeta}^3) \quad (37)$$

$$\frac{u_\infty}{U^*} = \beta\overline{\zeta} + [\frac{1}{8} - \frac{1}{4}(1-\beta^2)]\overline{\zeta}^2 + O(\overline{\zeta}^3) \quad (38)$$

$$\frac{u_\infty}{U^*} = \beta\overline{\zeta} + \frac{1}{8}\overline{\zeta}^2 + O(\overline{\zeta}^3) \quad (39)$$

respectively.

Equations (33)-(39) illustrate that both the effect of the lateral distribution of the induced electric field and the chemiosmotic effect (always directs the fluid toward lower electrolyte concentration) are of the order $\overline{\zeta}^2$. The two effects contribute to the fluid flow in opposite directions, and the former is dominant in most practical situations. Note that the $O(\zeta)$ and $O(\zeta^2)$ terms in Eqs (37) and (38) may act either in concert or in competition to produce the net diffusioosmotic flow, depending on the sign of $\beta\zeta(2\beta^2-1)$.

2.5. Results and Discussion

2.5.1. Induced Electric Field

The distribution of the macroscopic electric field $E(y)$ induced by a concentration gradient of a symmetric electrolyte prescribed parallel to a plane wall can be numerically determined after the substitution of Eqs. (19) and (5) into Eq. (11). A simple method of numerical calculation is to make an initial guess of the fluid velocity distribution $u_1(\kappa y)/U^*$ for a given combination of the dimensionless parameters $\overline{\zeta}$, Pe, and β, and to obtain the resulting induced electric field $E_1(\kappa y)$ from Eq. (11). Then, the next result of the velocity distribution $u_2(\kappa y)/U^*$ can be determined from the double integral involving $E_1(\kappa y)$ in Eq. (19). If the difference between $u_2(\kappa y)$ and $u_1(\kappa y)$ is beyond the tolerable error, the same procedure will be repeated ($E_2(\kappa y)$ is calculated from Eq. (11) using $u_2(\kappa y)/U^*$) until an acceptable result of the velocity distribution is obtained.

The induced electric field normalized by its bulk-phase quantity, $ZeEn^\infty(0)/\beta kT|\nabla n^\infty|$, as a function of the electrokinetic distance κy from the wall is plotted in Figure. 2 for several values of the dimensionless parameters $\bar{\zeta}$, Pe, and β. Note that each curve with specified values of $-\bar{\zeta}$ and β in the figure would be identical to that with the values $\bar{\zeta}$ and $-\beta$.

As expected, the magnitude of the normalized induced electric field in general is a sensitive function of the electrokinetic coordinate κy and becomes infinity for the special case of $\beta = 0$, irrespective of the parameters $\bar{\zeta}$, Pe, and κy. The normalized induced electric field approaches unity (the bulk-phase value) as $\kappa y \geq 5$ and has a typical value $-\zeta/|\zeta|\beta$ at the wall (with $\kappa y = 0$) if $|\bar{\zeta}| >> 1$. For the situation that $\zeta\beta > 0$, the direction of $\mathbf{E}(y)$ can be opposite to that of its bulk-phase quantity $\mathbf{E}^\infty$. When Pe=0, the magnitude of $ZeEn^\infty(0)/\beta kT|\nabla n^\infty| - 1$ (or the deviation of the induced electric field from its bulk-phase quantity) decreases with an increase in κy, increases with an increase in $|\bar{\zeta}|$, and decreases with an increase in the magnitude of β (vanishes in the limit of $\beta = \pm 1$), for an otherwise specified condition. When the value of Pe is finite, the magnitude of $ZeEn^\infty(0)/\beta kT|\nabla n^\infty| - 1$ may not be a monotonic function of κy, and the effect of the electrolyte convection on the local induced electric field in the double layer can be quite significant.

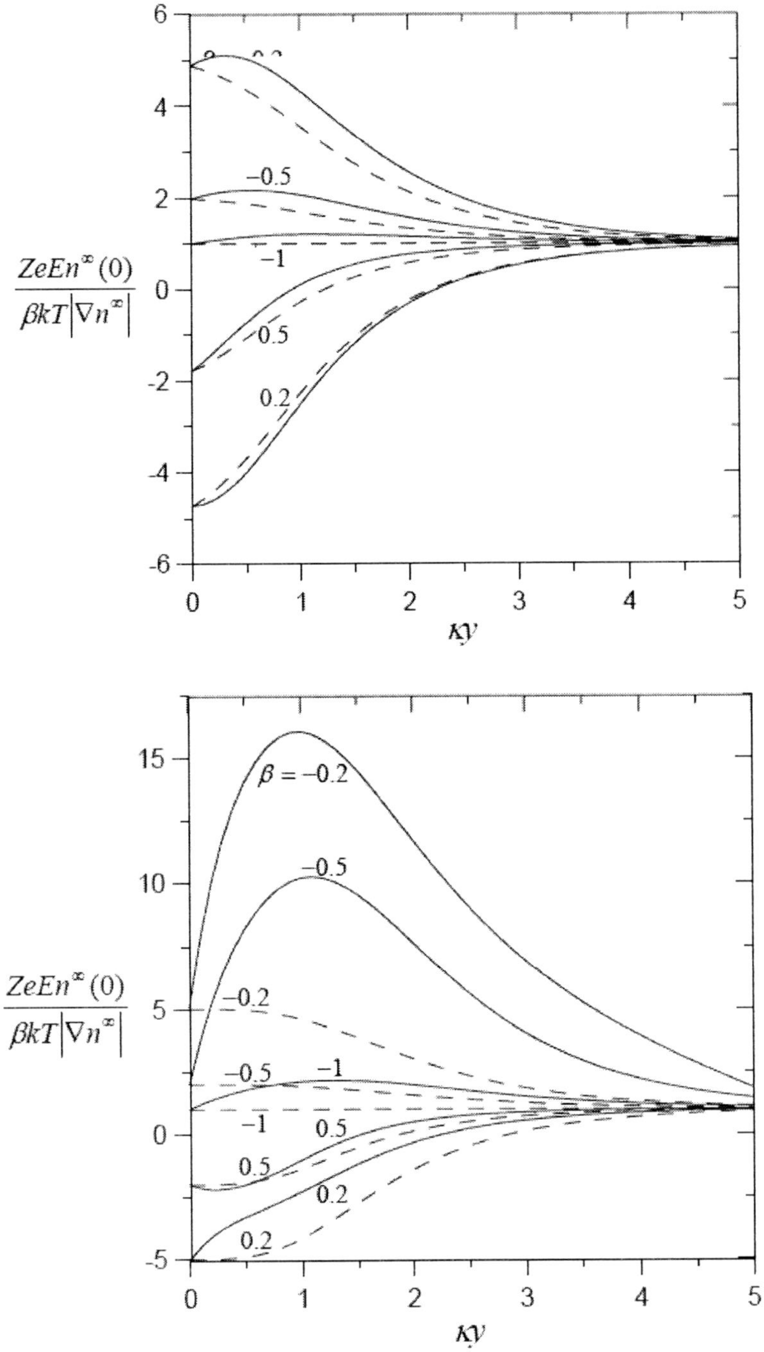

Figure 2. Plots of the normalized electric field induced by an electrolyte gradient along a plane wall versus the electrokinetic coordinate κy for various values of the parameter β: (a) $\bar{\zeta} = 2$; (b) $\bar{\zeta} = 6$. The solid curves represent the case Pe=1 and the dashed curves denote the case Pe=0.

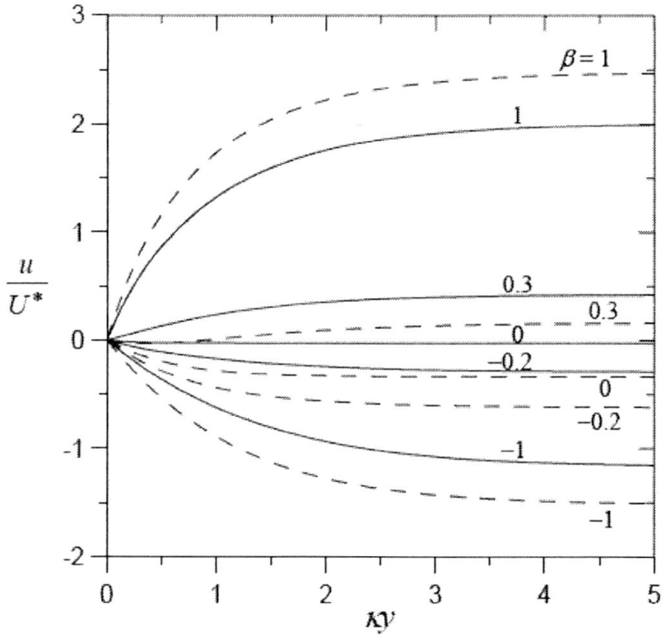

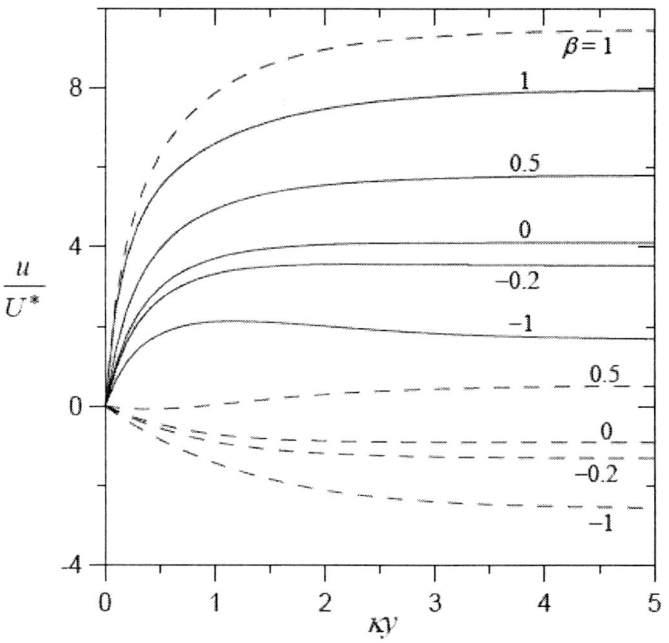

Figure 3. Plots of the normalized diffusioosmotic velocity along a plane wall versus the electrokinetic coordinate κy for various values of the parameter β : (a) $\bar{\zeta} = 2$; (b) $\bar{\zeta} = 6$. The solid curves represent the case Pe=1 and the dashed curves denote the case Pe=0.

2.5.2. Diffusioosmotic Velocity

The dimensionless diffusioosmotic velocity distribution u/U^* of an electrolyte solution along a plane wall can be numerically calculated using Eq. (19) with the known distributions of $\overline{\psi}$ and E. The results are plotted in Figures 3 and 4 for several values of the parameters $\overline{\zeta}$, Pe, and β. For fixed values of $\overline{\zeta}$, Pe, and β, the value of u/U^* approaches a constant (the bulk-phase value u_∞/U^*) as $\kappa y \geq 5$.

When Pe=0 and the product of ζ and β is negative, u is also negative, meaning that the diffusioosmotic flow is in the direction of increasing electrolyte concentration. The magnitude of u/U^* increases monotonically with an increase in the electrokinetic distance κy from the wall, with an increase in $|\overline{\zeta}|$, and with an increase in $|\beta|$, for an otherwise specified condition. When Pe=0 and $\zeta\beta$ is positive, the direction of the diffusioosmotic flow depends on the combination of parameters $\overline{\zeta}$, β, and κy. If the magnitude of β is sufficiently large, the fluid flows against the electrolyte concentration gradient (u is positive) and u/U^* is a monotonic increasing function of κy and of $|\overline{\zeta}|$. If the magnitude of β is sufficiently small for a given value of $\overline{\zeta}$, the fluid flows toward the opposite direction (u is negative) and the magnitude of u/U^* is also a monotonic increasing function of κy and of $|\overline{\zeta}|$. For a specified value of $\overline{\zeta}$ with an intermediate magnitude of β, the diffusioosmotic flow may reverse its direction from along with the concentration gradient to against it as the value of κy increases not much from zero.

When the value of Pe is finite, the dependence of u on κy is similar to that for the case of Pe=0 if the value of $|\overline{\zeta}|$ is small, but u can be positive for any given value of β if the value of $|\overline{\zeta}|$ is large. In general, the value of u/U^* increases monotonically with an increase in the value of Pe for specified values of κy, $\overline{\zeta}$, and β except for the situation that the value of $\beta\zeta/|\zeta|$ is close to unity. In the limit of $\mathrm{Pe} \to \infty$, u/U^* is finite. Note that the cases with $\mathrm{Pe} \geq 10$, which are not likely to exist in practice, is exhibited in Figure 4 for the sake of numerical comparison.

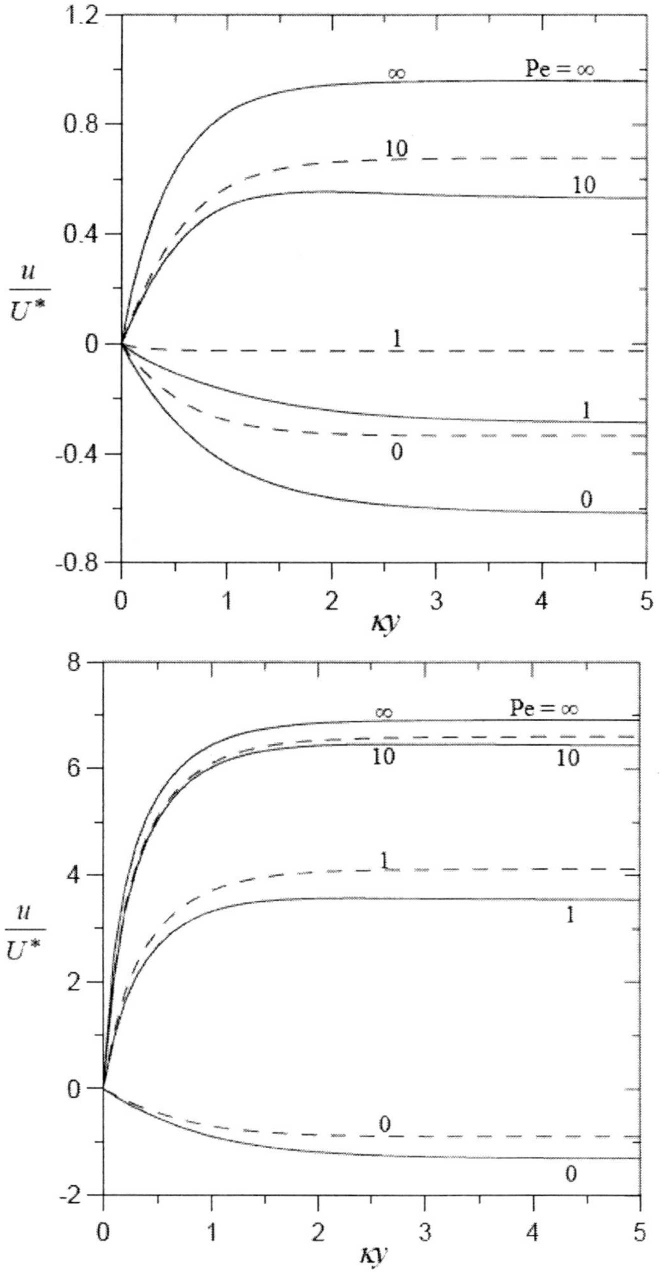

Figure 4. Plots of the normalized diffusioosmotic velocity along a plane wall versus the electrokinetic coordinate κy for various values of the parameter Pe: (a) $\overline{\zeta} = 2$; (b) $\overline{\zeta} = 6$. The solid curves represent the case $\beta = -0.2$ and the dashed curves denote the case $\beta = 0$.

In Figure 5, the normalized diffusioosmotic velocity u_∞ / U^* of the electrolyte solution far away from the plane wall is plotted versus the parameters $\overline{\zeta}$ and β for the cases of Pe=0 and 1. Maps showing the direction of this velocity are also drawn in Figure 6.

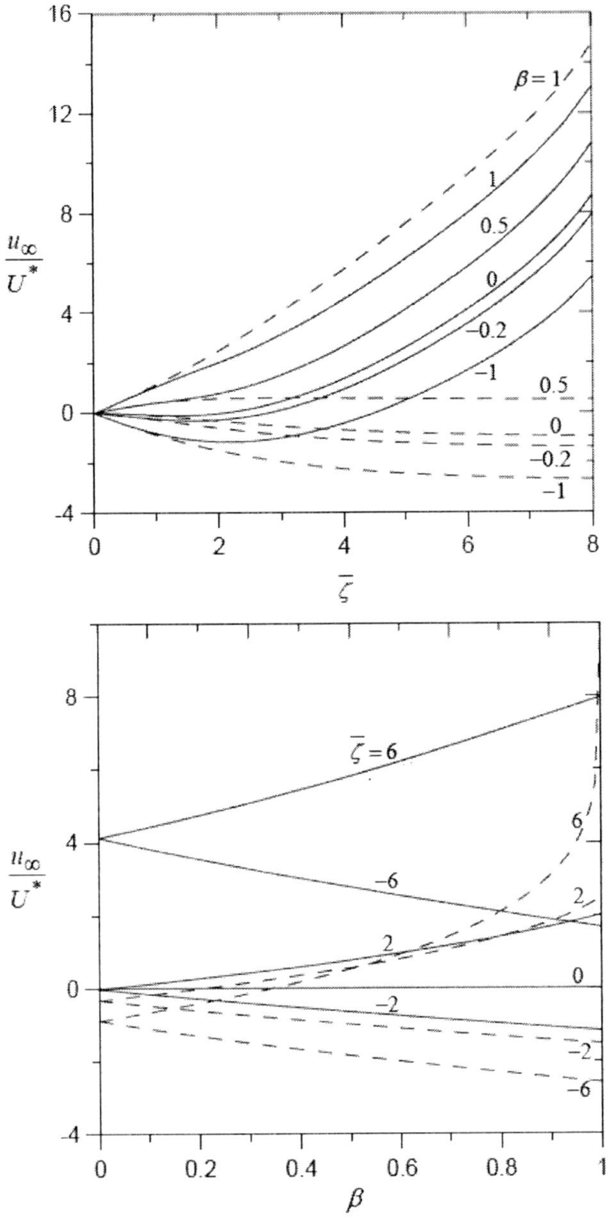

Figure 5. The normalized bulk-phase diffusioosmotic velocity along a plane wall: (a) plots versus the dimensionless surface potential $\bar{\zeta}$ for various values of the parameter β; (b) plots versus β for various values of $\bar{\zeta}$. The solid curves represent the case Pe=1 and the dashed curves denote the case Pe=0.

The dependence of u_∞ / U^* on $\bar{\zeta}$, β, and Pe is quite similar to that of u / U^* for a given value of κy. When Pe=0 and the product of $\bar{\zeta}$ and β is negative (inside the second and fourth quadrants in Figure 6a), u_∞ is negative and the electrolyte solution flows toward higher concentration. When Pe=0 and the product of $\bar{\zeta}$ and β is positive (inside the first and third quadrants in Figure 6a), the bulk-phase diffusioosmotic velocity may reverse its direction from against the concentration gradient to along with it as $|\overline{\zeta}|$ increases not much from zero for the case of a small magnitude of β (in addition to a reversal occurring at $\zeta = 0$), or as $|\beta|$ decreases from 1 to 0 for a given value of $\overline{\zeta}$.

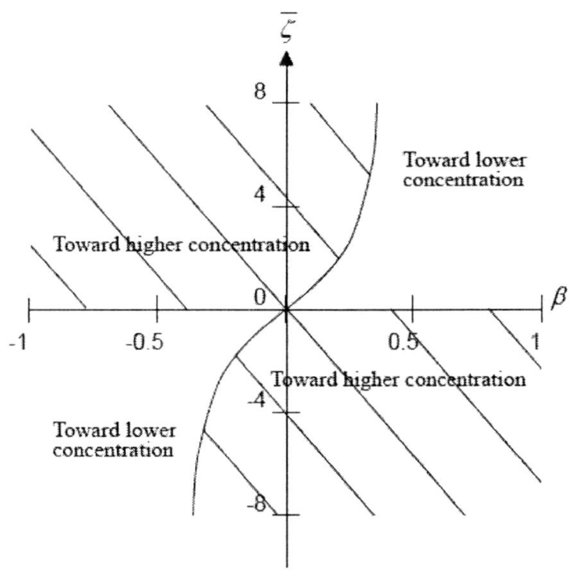

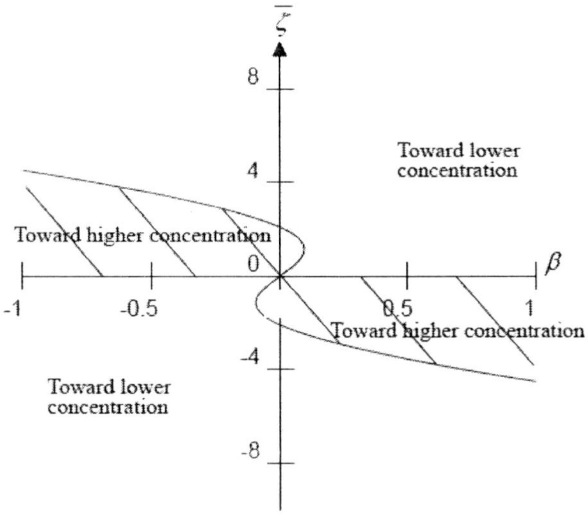

Figure 6. Maps showing the direction of the bulk-phase diffusioosmotic velocity along a plane wall: (a) Pe=0; (b) Pe=1.

When the value of Pe is finite, as indicated in Figure 6b, the probability of the diffusioosmotic flow of the electrolyte solution in the direction toward higher concentration for a combination of β and $\overline{\zeta}$ is greatly reduced, due to the effect of the electrolyte convection. This effect is quite significant, irrespective of the thickness of the electric double layers adjacent to the capillary walls.

2.5.3. Diffusioosmosis without Relaxation Effect

The distribution of the macroscopic electric field $\mathbf{E}(y)$ induced by a concentration gradient of a symmetric electrolyte prescribed parallel to a plane wall excluding the relaxation effect is expressed by Eq. (21). This induced electric field normalized by its bulk-phase quantity, $(G_+ - G_-)/(G_+ + G_-)\beta$ $[=(\omega^4 - x^4)(\omega^4 + 1)/(\omega^4 + x^4)(\omega^4 - 1)]$, as a function of the electrokinetic distance κy from the wall is plotted in Figure 7 for several values of the dimensionless parameters $\bar{\zeta}$ and β. As expected, the value of the normalized induced electric field equals unity in the limit of $\beta = \pm 1$ and its magnitude becomes infinity for the special case of $\beta = 0$, irrespective of the parameters $\bar{\zeta}$ and κy. This normalized electric field approaches unity as $\kappa y \geq 5$, with a typical value at the wall (with $\kappa y = 0$) as $-\beta\zeta/|\zeta|$ if $|\bar{\zeta}| >> 1$. The magnitude of $(G_+ - G_-)/(G_+ + G_-)\beta - 1$ decreases with an increase in κy, increases with an increase in $|\bar{\zeta}|$, and decreases with an increase in $|\beta|$, for an otherwise specified condition. Note that the numerical results presented in Figure 2 for the case of Pe=0 are quite accurate in comparison with the exact solution given in Figure 7. The profile of the net diffusivity $D(y)$ of a symmetric electrolyte in the direction parallel to a plane wall is given by Eq. (24). Figure 8 shows this net diffusivity normalized by its bulk-phase quantity, $(D_+ + D_-)/(G_+ + G_-)$ $[=(\omega^4 + 1)x^2/(\omega^4 + x^4)]$, as a function of the electrokinetic position κy with $\bar{\zeta}$ and β as parameters.

β. The solid curves represent the case $\bar{\zeta} = 6$ and the dashed curves denote the case $\bar{\zeta} = 2$.

When the product of ζ and β is negative, the normalized net diffusivity increases with an increase in κy, decreases with an increase in $|\bar{\zeta}|$, and decreases with an increase in $|\beta|$, for an otherwise fixed condition. When $\zeta\beta$ is positive, the value of the normalized net diffusivity at a position near the wall can be much greater than unity and its dependence on κy is not necessarily monotonic, depending on the combination of $\bar{\zeta}$ and β. For a constant value of κy, the normalized net diffusivity increases with an increase in β for the case of $\zeta > 0$ but decreases with it for the case of $\zeta < 0$. As expected, the value of $(D_+ + D_-)/(G_+ + G_-)$ gets close to unity as $\kappa y \geq 5$ and vanishes at the wall if $Ze|\zeta|/kT \to \infty$ and $\beta\zeta/|\zeta| \neq 1$.

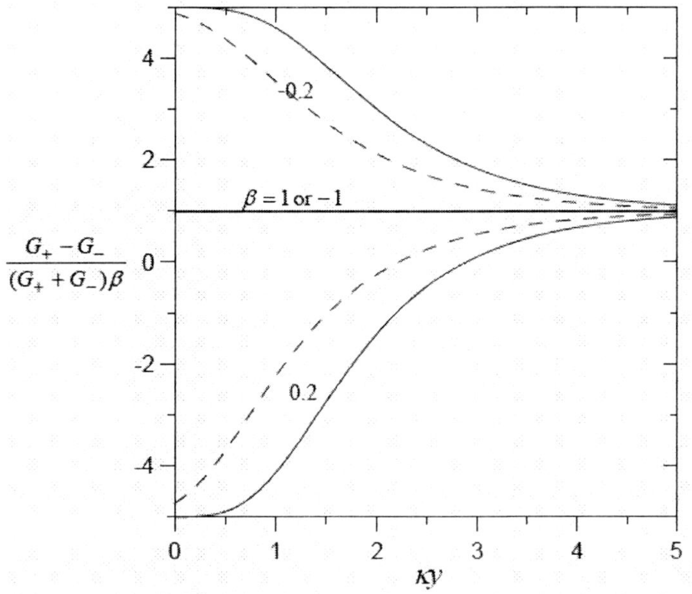

Figure 7. Plots of the normalized electric field induced by an electrolyte gradient along a plane wall as calculated from Eq. (21) versus the electrokinetic coordinate κy for various values of the parameter β. The solid curves represent the case $\bar{\zeta} = 6$ and the dashed curves denote the case $\bar{\zeta} = 2$.

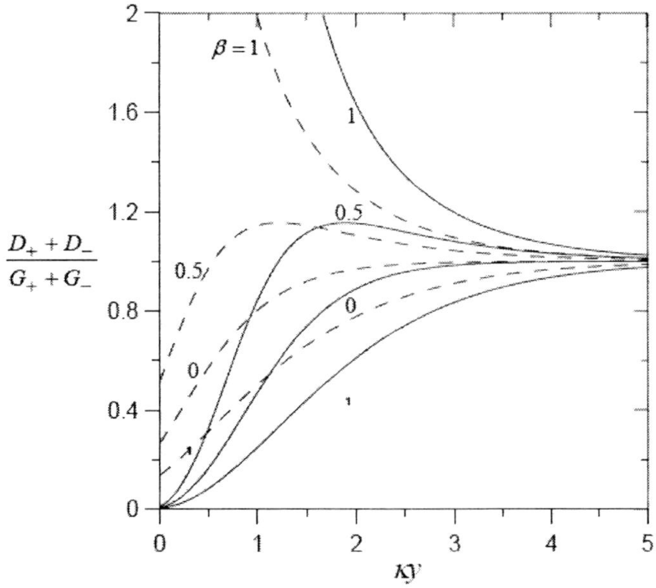

Figure 8. Plots of the normalized net diffusivity of the electrolyte tangential to a plane wall as calculated from Eq. (24) versus the electrokinetic coordinate κy for various values of the parameter

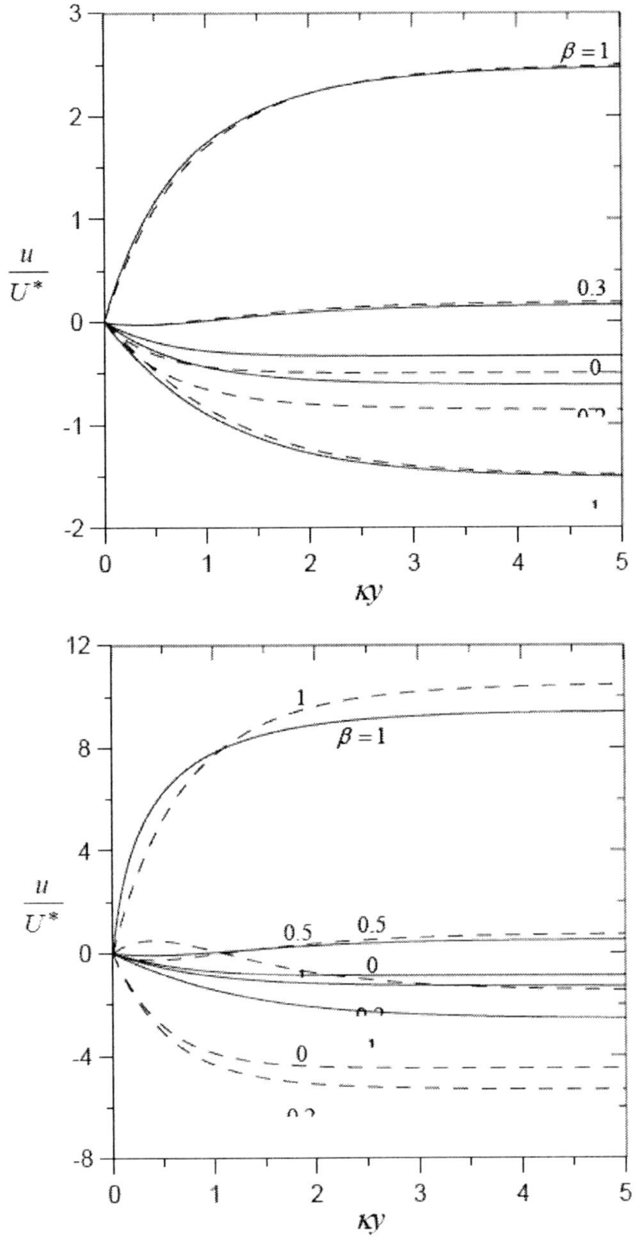

Figure 9. Plots of the normalized diffusioosmotic velocity along a plane wall as calculated from Eqs. (32) (solid curves) and (37) (dashed curves) versus the electrokinetic coordinate κy for various values of the parameter β : (a) $\overline{\zeta} = 2$; (b) $\overline{\zeta} = 6$.

Without the relaxation effect, the normalized diffusioosmotic velocity distribution u / U^* of an electrolyte solution along a plane wall is given by Eq. (32) for the general case and by Eq. (37) for the case of small magnitude of zeta potential up to $O(\zeta^2)$. The results of both cases are plotted in Figure 9 for several values of the parameters $\overline{\zeta}$ and β. When the product of ζ and β is negative, u is also negative, meaning that the diffusioosmotic flow is in the direction of increasing electrolyte concentration. The magnitude of u/U^* increases monotonically with an increase in the electrokinetic distance κy from the wall, with an increase in $|\bar{\zeta}|$, and with an increase in $|\beta|$, for an otherwise specified condition. When $\zeta\beta$ is positive, the direction of the diffusioosmotic flow depends on the combination of parameters $\bar{\zeta}$, β, and κy. If the magnitude of β is sufficiently large, the fluid flows against the electrolyte concentration gradient (u is positive) and u/U^* is a monotonic increasing function of κy and of $|\bar{\zeta}|$.

If the magnitude of β is sufficiently small, the fluid flows toward the opposite direction (u is negative) and the magnitude of u/U^* is also a monotonic increasing function of κy and of $|\bar{\zeta}|$. For a specified value of $\bar{\zeta}$ with an intermediate magnitude of β, the diffusioosmotic flow may reverse its direction from along with the concentration gradient to against it as the value of κy increases not much from zero. For fixed values of $\bar{\zeta}$ and β, the value of u/U^* approaches a constant as $\kappa y \geq 5$. The asymptotic formula of Eq. (37) for u/U^* from the Debye-Huckel approximatin (depicted by dashed curves in Figure 9) agrees reasonably well with the exact result of Eq. (32) for the case of small magnitude of zeta potential (say, $|\bar{\zeta}| \leq 2$), but the accuracy of this approximation deteriorates rapidly, as expected, when the value of $|\bar{\zeta}|$ becomes large. Again, the numerical results presented in Figures 3 and 4 for the case of Pe=0 are quite accurate in comparison with the exact solution given in Figure 9.

In Figure 10, the normalized diffusioosmotic velocity u_∞ / U^* of the electrolyte solution far away from the plane wall given by Eq. (36) is plotted versus the parameters $\bar{\zeta}$ and β. A map showing the direction of this velocity has already been provided in Figure 6a. For comparison, the result of u_∞ / U^* given by Eq. (20) without accounting for the effect of the lateral distribution of the induced tangential electric field in the electric double layer adjacent to the plane wall is also plotted in Figures 10 and 11. Surprisingly, the difference between this previously-obtained bulk-phase diffusioosmotic velocity and the presently-obtained quantity is quite substantial unless the magnitude of β approaches unity. Thus, the effect of the lateral distribution of the induced electric field always exists and can not be neglected even as $\kappa y \to \infty$.

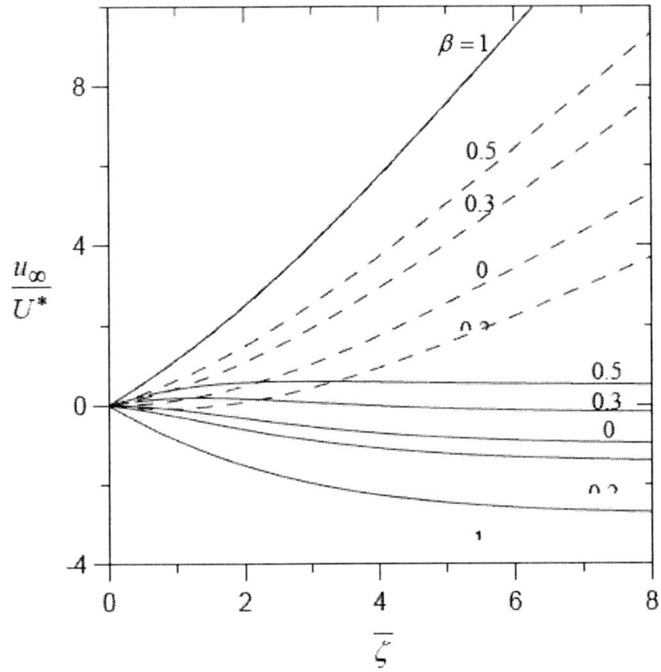

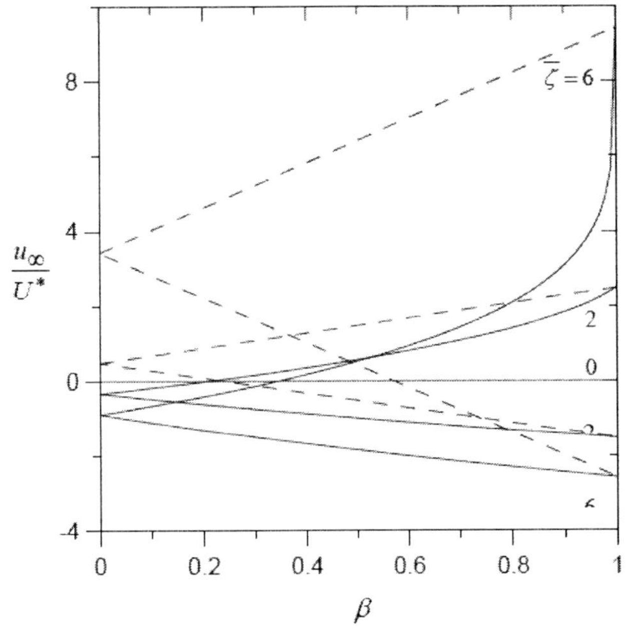

Figure 10. The normalized bulk-phase diffusioosmotic velocity along a plane wall as calculated from Eqs. (36) (solid curves) and (20) (dashed curves): (a) plots versus the dimensionless surface potential $\bar{\zeta}$ for various values of the parameter β; (b) plots versus β for various values of $\bar{\zeta}$.

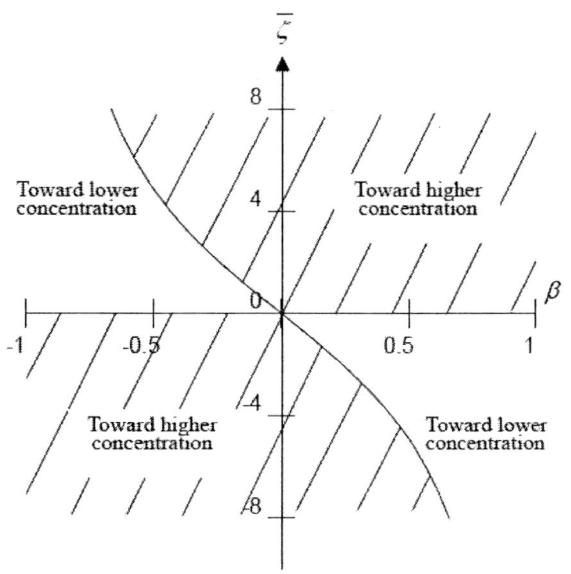

Figure 11. A map showing the direction of the bulk-phase diffusioosmotic velocity along a plane wall as calculated form Eq. (20).

3. Diffusioosmosis in a Capillary Slit

In this section, we deal with the steady diffusioosmotic flow of a solution of a symmetrically charged electrolyte in a capillary channel between two identical parallel charged plates with a separation distance $2h$, as shown in Figure 12. The applied electrolyte concentration gradient ∇n^{∞} is a constant along the tangential (z) direction in the capillary slit, where $n^{\infty}(z)$ is the linear concentration distribution of the electrolyte in the bulk solution phase in equilibrium with the fluid inside the capillary. The electrolyte ions can diffuse freely in the capillary, so there exists no regular osmotic flow of the solvent. The end effects are neglected. Owing to the planar symmetry of the system, we need consider only the half region $0 \leq y \leq h$, where y measures distance from the median plane between the capillary walls in a normal direction. The analysis for this case is similar to that presented in the previous section for the case of diffusioosmosis along a single plane wall.

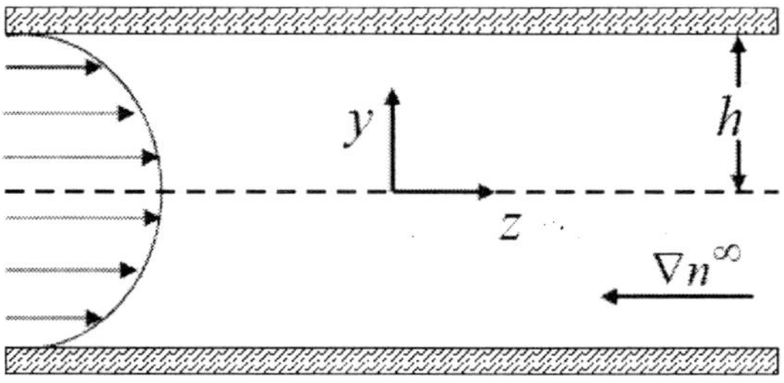

Figure 12. Geometrical sketch for the diffusioosmosis due to an axially applied concentration gradient of an electrolyte in a capillary slit.

3.1. Electrostatic Potential Distribution

3.1.1. General Analysis

For the electrolyte solution in the capillary slit, the electrostatic potential distribution $\psi(y)$ is still governed by Eq. (3), but the boundary conditions become

$$y = h: \ \psi = \zeta \,, \tag{40a}$$

$$y = 0: \ \frac{d\psi}{dy} = 0 \,. \tag{40b}$$

Integration of Eq. (3) from $y = 0$ to $y = y$ using Eq. (40b) leads to

$$\frac{d\overline{\psi}}{dy} = 2\kappa[\cosh^2(\frac{\overline{\psi}}{2}) - \cosh^2(\frac{\overline{\psi}_c}{2})]^{1/2} \tag{41}$$

where $\overline{\psi}_c = \overline{\psi}(0)$ represents the potential at the median plane between the slit walls. Integrating Eq. (41) again from $y = 0$ to $y = y$, one obtains [33]

$$b[F(b,\frac{\pi}{2}) - F(b,\phi)] = \kappa y \tag{42}$$

where

$$\phi = \sin^{-1}\frac{\cosh(\overline{\psi}_c/2)}{\cosh(\overline{\psi}/2)}, \tag{43}$$

$$b = [\cosh(\overline{\psi}_c/2)]^{-1}, \tag{44}$$

and $F(b,\phi)$ is the elliptic integral of the first kind,

$$F(b,\phi) = \int_0^{\phi}(1 - b^2\sin^2\theta)^{-1/2}d\theta \,. \tag{45}$$

Equation (42) is the solution for the potential profile $\overline{\psi}(\kappa y)$ in an implicit form, and the quantity $\overline{\psi}_c / \overline{\zeta}$ can be determined as a function of the parameters $\overline{\zeta}$ and κh by Eq. (42) setting $y = h$ and using Eq. (40a).

If the surface charge density σ, instead of the surface potential ζ, is known at the capillary walls, the boundary condition (40a) should be replaced by the Gauss condition,

$$y = h: \qquad \frac{d\psi}{dy} = \frac{4\pi\sigma}{\varepsilon}. \tag{46}$$

Substitution of this into Eq. (41) yields

$$\cosh\overline{\zeta} - \cosh\overline{\psi}_c = \frac{1}{2}\left(\frac{\overline{\sigma}_1}{\kappa h}\right)^2 \tag{47}$$

where $\overline{\sigma}_1 = 4\pi h Z e \sigma / \varepsilon k T$ is the dimensionless surface charge density. The relation among the parameters $\overline{\sigma}_1$, $\overline{\zeta}$, and κh given by Eq. (47) is displayed in Figure 13, which indicates that, for a given electrolyte solution in a capillary slit with a specified thickness, σ increases (almost linearly) with an increase in κ or $[n^\infty(0)]^{1/2}$ for the case of constant surface potential, and ζ decreases with an increase in κ or $[n^\infty(0)]^{1/2}$ for the case of constant surface charge density.

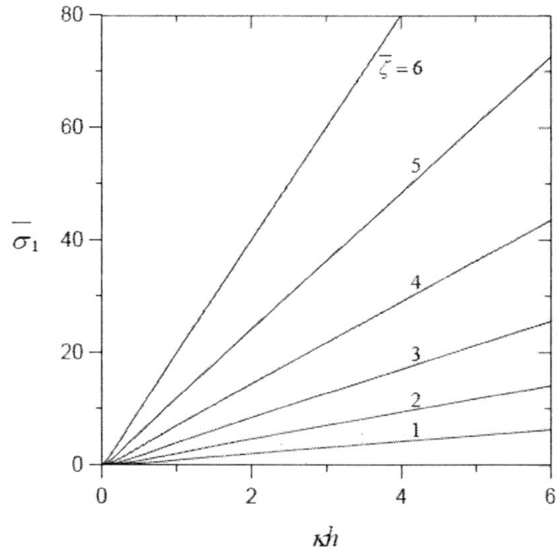

Figure 13. A plot of the relation among the dimensionless parameters $\overline{\sigma}_1$, $\overline{\zeta}$, and κh.

3.1.2. Case of Low Surface Potential

When the surface potential $\bar{\zeta}$ is small, the Debye-Huckel approximation applies. The solution to the linearized form of Eq. (3) subject to Eq. (40) is

$$\bar{\psi} = \bar{\zeta}\frac{\cosh(\kappa y)}{\cosh(\kappa h)} \tag{48}$$

Obviously, $\bar{\psi}/\bar{\zeta} = 1$ at any position in the limit $\kappa h = 0$, and $\bar{\psi}/\bar{\zeta} = 0$ at any position other than $y = h$ in the limit $\kappa h \to \infty$.

If the Gauss condition (46) at the capillary wall is used to replace Eq. (40a), the solution for $\bar{\psi}$ given by Eq. (48) is still valid, with

$$\bar{\zeta} = \frac{4\pi\sigma Ze}{\varepsilon\kappa kT}\frac{\cosh(\kappa h)}{\sinh(\kappa h)}. \tag{49}$$

3.2. Fluid Velocity Distribution

3.2.1. General Analysis

The momentum equations for the steady flow of the electrolyte solution in a capillary slit with an applied concentration gradient of the electrolyte parallel to the slit walls are still given by Eq. (16), and the induced electric field distribution can also be expressed as Eq. (11). The boundary conditions for the fluid velocity u are

$$y = h:\ u = 0, \tag{50a}$$

$$y = 0:\ \frac{du}{dy} = 0. \tag{50b}$$

Substituting Eq. (2) into Eq. (16a), we solve for the pressure distribution to obtain

$$p = p_0 + 2n^{\infty}(z)kT(\cosh\bar{\psi} - \cosh\bar{\psi}_c), \tag{51}$$

where p_0 is the pressure on the mid-plane between the slit walls. Substituting Eqs. (51) and (2) into Eq. (16b) and then performing the integration with respect to y twice subject to the boundary conditions in Eq. (50) yields [46]

$$\frac{u}{U^*} = (\kappa h)^2 \int_0^{y/h} \int_{\infty}^{y/h} [\cosh \overline{\psi} - \cosh \overline{\psi}_c + \frac{Zen^\infty(0)}{kT |\nabla n^\infty|} E \sinh \overline{\psi}]\} \mathrm{d}(y/h) \mathrm{d}(y/h) \tag{52}$$

After the substitution of u expressed by Eq. (52) and $\overline{\psi}$ calculated from Eq. (42) into Eq. (11), the induced electric field distribution E can be numerically solved as a function of the dimensionless parameters κh, $\overline{\zeta}$, β, and Pe. With the results of $\overline{\psi}$ and E, the diffusioosmotic velocity distribution of the electrolyte solution as a function of the relevant parameters can be calculated from Eq. (52) through the numerical integrations. It is understood that, for given values of κh, Pe, and y/h, the quantity u/U^* with specified values $-\overline{\zeta}$ and β is equal to that with the values $\overline{\zeta}$ and $-\beta$.

3.2.2. Case of Low Surface Potential without Relaxation Effect

In this subsection, we consider the case of low zeta potential at the slit walls and Pe=0. Now, the momentum equations for the steady flow of an electrolyte solution in the capillary slit with an applied concentration gradient of the electrolyte parallel to the slit walls can be solved by substituting the linearized form of Eq. (2) into Eq. (16). Thus, the pressure distribution given by Eq. (51) reduces to

$$p = p_0 + n^\infty(z)kT(\overline{\psi}^2 - \overline{\psi}_c^2) + O(\zeta^4). \tag{53}$$

When the effect of the ionic convection on the induced electric field caused by the diffusioosmotic flow itself (relaxation effect) is negligible, the fluid velocity u can be solved from Eqs. (16b) and (50) by the substitution of Eqs. (2), (25), (48), and (53), with the result [41,43]

$$u = -\frac{2Ze}{\eta\kappa^2} |\nabla n^\infty| \{\beta\zeta\Phi_1(\kappa h, y/h) + \frac{Ze}{8kT}\zeta^2 [\Phi_2(\kappa h, y/h)$$

$$+ 2(\beta^2 - 1)\Phi_3(\kappa h, y/h)] + O(\zeta^3)\}, \tag{54}$$

where the functions Φ_1, Φ_2, and Φ_3 are

$$\Phi_1 = 1 - \frac{\cosh(\kappa y)}{\cosh(\kappa h)}, \tag{55a}$$

$$\Phi_2 = \frac{1}{2\cosh^2(\kappa h)}[\cosh(2\kappa h) - \cosh(2\kappa y) - 2\kappa^2(h^2 - y^2)] \tag{55b}$$

$$\Phi_3 = \frac{1}{2\cosh^2(\kappa h)}[\cosh(2\kappa h) - \cosh(2\kappa y) + 2\kappa^2(h^2 - y^2)] \tag{55c}$$

The average fluid velocity $\langle u \rangle$ over a cross section of the slit is

$$\langle u \rangle = -\frac{2Ze}{\eta\kappa^2}|\nabla n^\infty|\{\beta\zeta\langle\Phi_1\rangle + \frac{Ze}{8kT}\zeta^2[\langle\Phi_2\rangle + 2(\beta^2-1)\langle\Phi_3\rangle] + O(\zeta^3)\}, \tag{56}$$

where the definition of the angle brackets is

$$\langle u \rangle = \frac{1}{h}\int_0^h u(y)\,\mathrm{d}y. \tag{57}$$

Substitution of Eq. (55) into Eq. (57) leads to

$$\langle\Phi_1\rangle = 1 - \frac{\sinh(\kappa h)}{\kappa h\cosh(\kappa h)}, \tag{58a}$$

$$\langle\Phi_2\rangle = \frac{1}{2\cosh^2(\kappa h)}[\cosh(2\kappa h) - \frac{1}{2\kappa h}\sinh(2\kappa h) - \frac{4}{3}(\kappa h)^2], \tag{58b}$$

$$\langle\Phi_3\rangle = \frac{1}{2\cosh^2(\kappa h)}[\cosh(2\kappa h) - \frac{1}{2\kappa h}\sinh(2\kappa h) + \frac{4}{3}(\kappa h)^2]. \tag{58c}$$

It can be found from Eq. (55) that all the functions Φ_1, Φ_2, and Φ_3 are always positive. Evidently, $\Phi_1 = \Phi_2 = \Phi_3 = 0$ and $u = 0$ at the capillary walls ($y/h = 1$) due to the no-slip requirement of Eq. (50a). In the limiting situation that $\kappa h = 0$ (there is no interaction between the solute ions and the capillary walls), Eq. (55) gives $\Phi_1 = \Phi_2 = \Phi_3 = 0$ at any position in the slit, which also leads to $\langle\Phi_1\rangle = \langle\Phi_2\rangle = \langle\Phi_3\rangle = 0$. Note that $\Phi_3/(\kappa h)^2 = 4\Phi_1/(\kappa h)^2 = 2[1-(y/h)^2]$ and

$\langle \Phi_3 \rangle/(\kappa h)^2 = 4\langle \Phi_1 \rangle/(\kappa h)^2 = 4/3$, although $\Phi_2/(\kappa h)^2 = \langle \Phi_2 \rangle/(\kappa h)^2 = 0$, as $\kappa h = 0$; thus, the diffusioosmotic flow caused by the induced electric field $\mathbf{E}$ still exists (with a parabolic profile) for this case with a finite half-thickness h of the slit, as shown in Eq. (54). In the other limiting situation that $\kappa h \to \infty$, Eq. (55) yields $\Phi_1 = \Phi_2 = \Phi_3 = 1$ at an arbitrary position with $y/h \neq 1$ (so $\langle \Phi_1 \rangle = \langle \Phi_2 \rangle = \langle \Phi_3 \rangle = 1$, too). It is surprising and interesting that Φ_3, which reflects the effect of the lateral distribution of the induced axial electric field in the electric double layer, does not vanish as $\kappa h \to \infty$, where $\psi = 0$ at any position other than $y/h = 1$ (but the solute-wall interaction is very strong). For all cases of finite κh, Φ_3 and $\langle \Phi_3 \rangle$ are comparable with (especially when κh is large) but always greater than Φ_2 and $\langle \Phi_2 \rangle$, respectively.

The terms in Eqs. (54) and (56) involving the functions Φ_1 and Φ_3 represent the contribution from electroosmosis, while the remainder terms (containing the function Φ_2) are the chemiosmotic component. Note that the direction of the local or average fluid velocity caused by electroosmosis is determined by the combination of the parameters κh, $\bar{\zeta}$, and β, while the fluid flow generated by chemiosmosis for the case of small values of ζ and Pe=0 is always in the direction of decreasing electrolyte concentration. Equation (54) shows that the $O(\zeta^2)$ contributions from chemiosmosis and electroosmosis are always in opposite directions, and the net diffusioosmotic flow is dominated by the electroosmotic effect in most practical situations.

3.3. Results and Discussion

3.3.1. Induced Electric Field

The distribution of the macroscopic electric field $E(y)$ induced by a concentration gradient of a symmetric electrolyte prescribed parallel to the slit walls can be numerically determined after substituting the fluid velocity $u(y)/U^*$ in the form of Eq. (52) and the electric potential $\bar{\psi}(y)$ calculated from Eq. (42) into Eq. (11). This induced electric field normalized by its quantity at the mid-plane between the slit walls, $E(0)$, as a function of the normalized coordinate y/h is plotted in Figures 14 and 15 for some values of the dimensionless parameters $\bar{\zeta}$, β, Pe, and κh. Note that each curve with specified values of $-\bar{\zeta}$ and β in the figures would be identical to that with the values $\bar{\zeta}$ and $-\beta$. As expected, the magnitude of the normalized induced electric field in general is a sensitive function of y/h and can deviate much from its bulk-phase value.

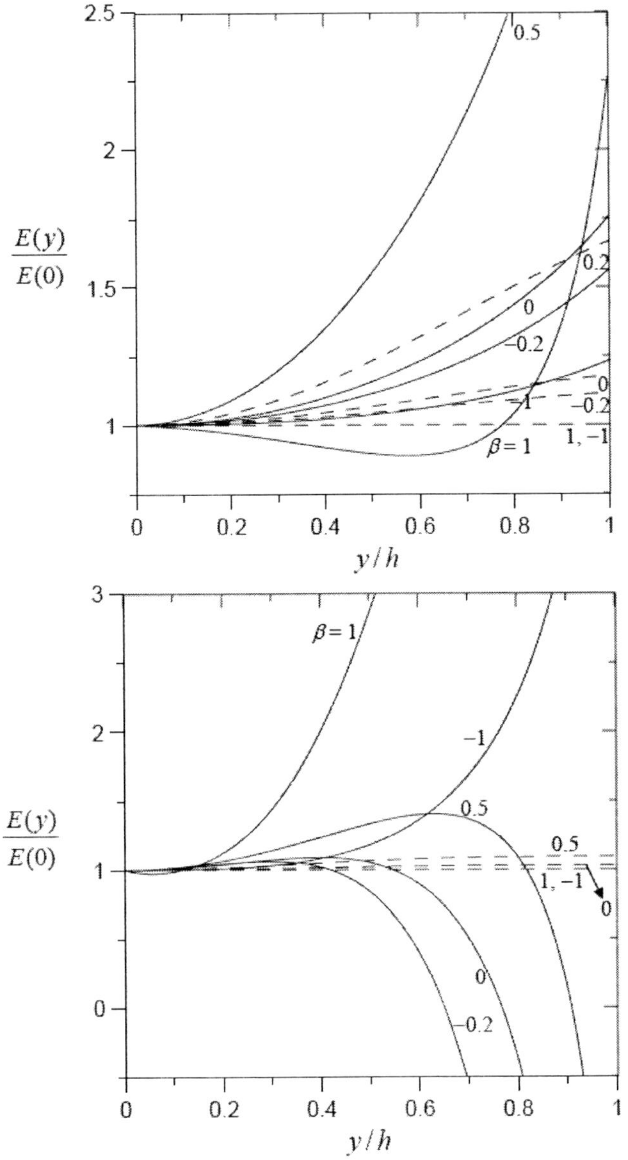

Figure 14. Plots of the normalized electric field induced by an electrolyte gradient parallel to the walls of a slit versus the electrokinetic coordinate κy for the case of $\kappa h = 1$ with various values of the parameter β: (a) $\bar{\zeta} = 2$; (b) $\bar{\zeta} = 6$. The solid curves represent the case Pe=1 and the dashed curves denote the case Pe=0.

When Pe=0, $E(y)/E(0)$ is positive and its value increases with an increase in y/h from unity at the midplane of the slit walls (with $y = 0$) to a maximum at the slit wall (with $y = h$), increases with an increase in κh, decreases with an increase in $|\bar{\zeta}|$, equals unity in the limits $\beta\zeta/|\zeta| = \pm 1$, and increases with an increase in $\beta\zeta/|\zeta|$ if it is not too close to unity (depending on the value of κh), for an otherwise specified condition.

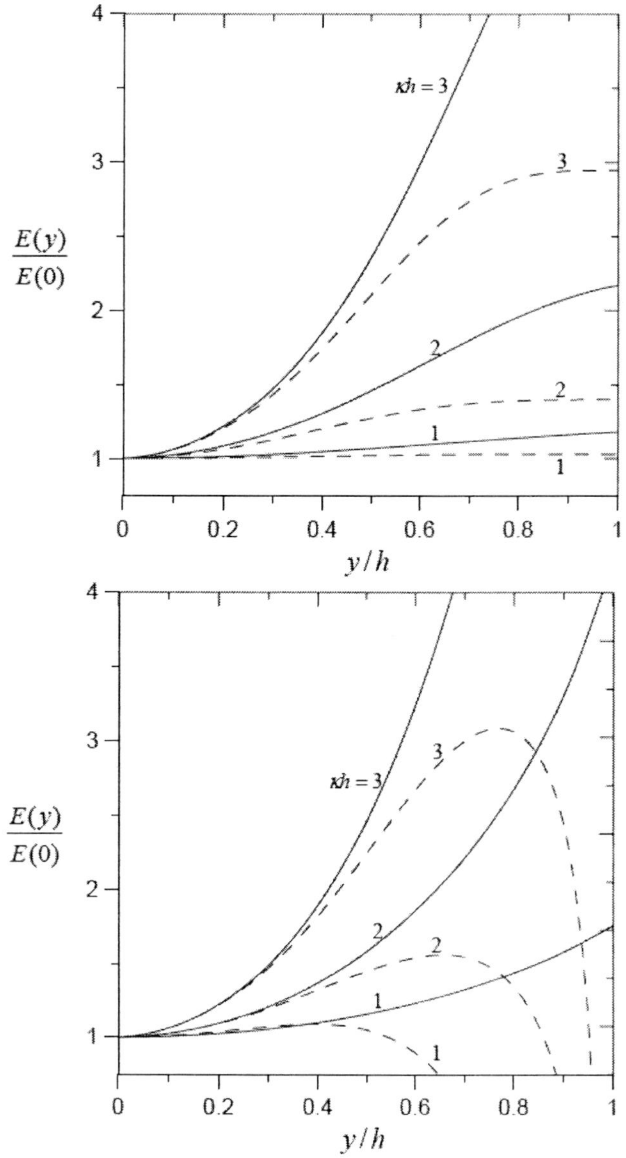

Figure 15. Plots of the normalized electric field induced by an electrolyte gradient parallel to the walls of a slit versus the electrokinetic coordinate κy for the case of $\beta = 0$ with various values of the parameter κh: (a) Pe=0; (b) Pe=1. The solid curves represent the case $\overline{\zeta} = 2$ and the dashed curves denote the case $\overline{\zeta} = 6$.

When the value of Pe is finite, the value of $E(y)/E(0)$ is larger than that for the case of Pe=0 if the magnitude of $\overline{\zeta}$ is small (depending on the values of κh and β), but it may not be a monotonic function of y/h and may become negative if the magnitude of $\overline{\zeta}$ is large. In general, the effect of the electrolyte convection on the local induced electric field in the electric double layers adjacent to the slit walls can be quite significant even for the case of low zeta potential at the walls.

3.3.2. Diffusioosmotic Velocity

The normalized diffusioosmotic velocity distribution $u(y)/U^*$ of an electrolyte solution in a capillary slit can be numerically calculated using Eq. (52) with the known results of $\overline{\psi}$ and E. This distribution is plotted in Figures 16 and 17 for several values of the parameters $\overline{\zeta}$, β, Pe and κh. The diffusioosmotic velocity u/U^* can be either positive or negative and is a monotonic increasing function of $\beta\zeta/|\zeta|$. In general, the magnitude of u/U^* decreases monotonically with an increase in the normalized coordinate y/h (there are exceptions). When Pe=0 and $\beta\zeta/|\zeta|$ is not too close to unity, u is negative, meaning that the diffusioosmotic flow is in the direction of increasing electrolyte concentration. The magnitude of u/U^* increases with an increase in $|\overline{\zeta}|$ and with a decrease in $\beta\zeta/|\zeta|$, for an otherwise specified condition. When Pe=0 and $\beta\zeta/|\zeta|$ approaches unity, the fluid flows against the electrolyte concentration gradient (u is positive) and u/U^* is a monotonic increasing function of $|\overline{\zeta}|$.

When the value of Pe is finite, the dependence of u on y/h is similar to that for the case of Pe=0 if the value of $|\overline{\zeta}|$ is small, but u can be positive for any given value of β if the value of $|\overline{\zeta}|$ is large. In general, the value of u/U^* increases monotonically and remarkably with an increase in the value of Pe (the relaxation effect on the diffusioosmotic flow is quite significant) for specified values of κh, $\overline{\zeta}$, and β except for the situation that the value of $\beta\zeta/|\zeta|$ is close to unity. In the limit of $\mathrm{Pe}\to\infty$, u/U^* is finite.

In Figure 18, the normalized diffusioosmotic velocity $u(0)/U^*$ of the electrolyte solution at the central plane of the slit is plotted versus the parameters κh and $\overline{\zeta}$ at specified values of Pe and β. Maps showing the direction of this velocity for a typical value of κh are also drawn in Figure 19. The dependence of $u(0)/U^*$ on $\overline{\zeta}$, β, κh, and Pe is quite similar to that of u/U^* for a given value of y/h, and it is not necessarily a monotonic function of κh for given values of β and Pe.

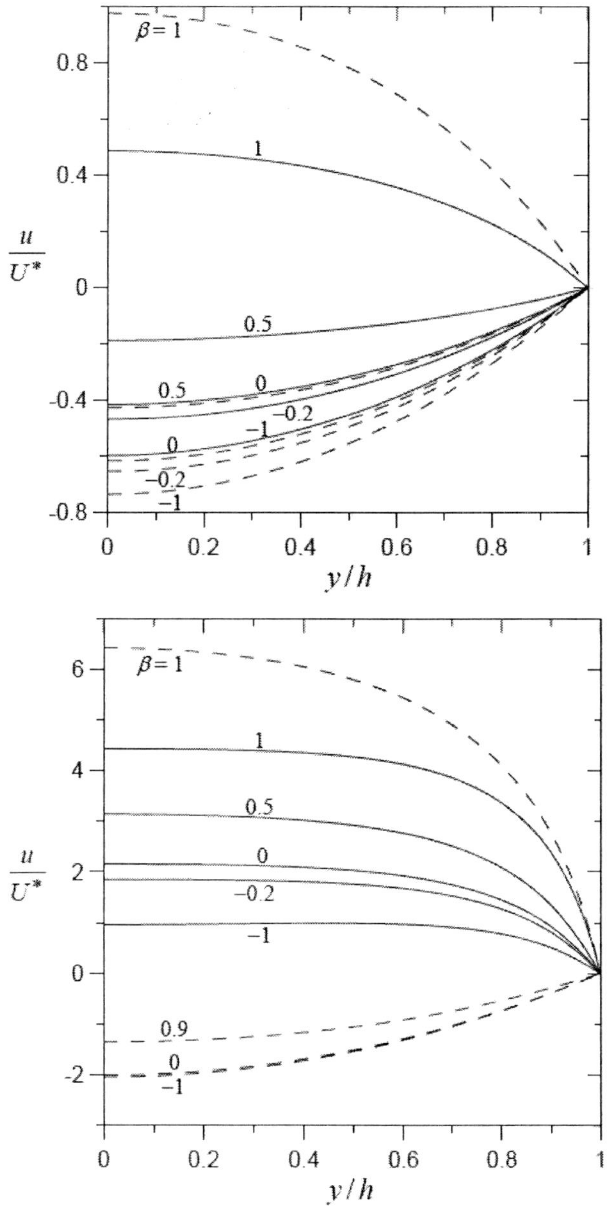

Figure 16. Plots of the normalized diffusioosmotic velocity in a slit versus the electrokinetic coordinate y/h for the case of $\kappa h = 1$ with various values of the parameter β: (a) $\overline{\zeta} = 2$; (b) $\overline{\zeta} = 6$. The solid curves represent the case Pe=1 and the dashed curves denote the case Pe=0.

When Pe=0 and the product of ζ and β is negative (inside the second and fourth quadrants in Figure 19a), $u(0)$ is negative and the electrolyte solution flows toward higher concentration, analogous to the case of diffusioosmosis along a plane wall discussed in the previous section. When Pe=0 and the product of ζ and β is positive (inside the first and third quadrants in Figure 19a), the diffusioosmotic velocity $u(0)$ may reverse its direction from against the concentration gradient to along with it as $|\bar{\zeta}|$ increases not much from zero for all practical cases of β (in addition to a reversal occurred at $\zeta = 0$), or as $|\beta|$ decreases from 1 to 0 for a small magnitude of $\bar{\zeta}$.

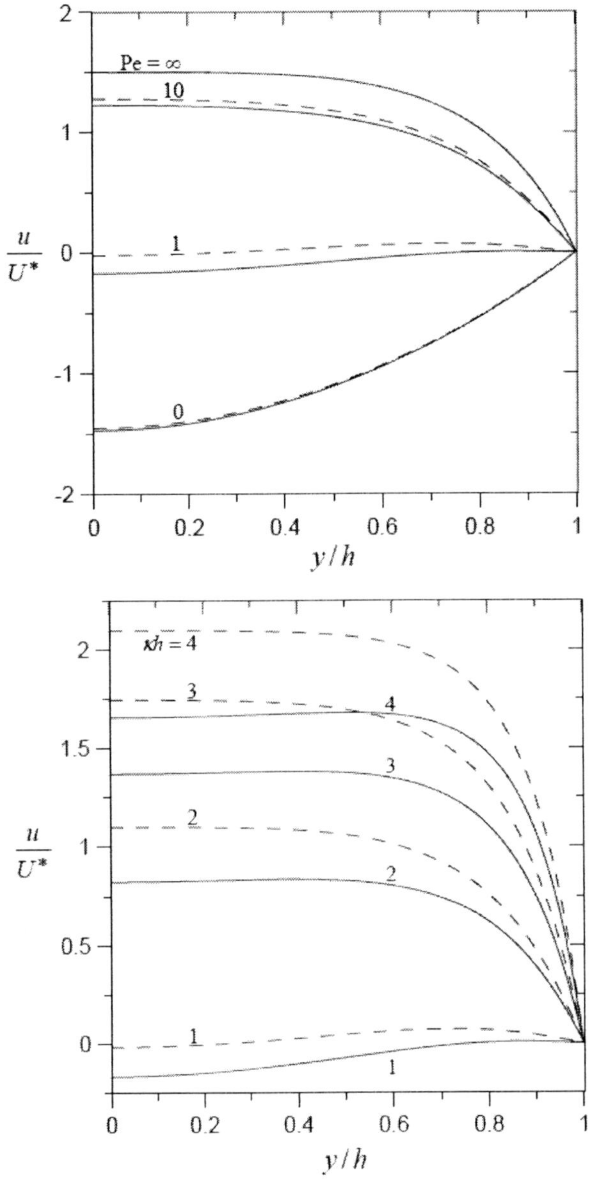

Figure 17. Plots of the normalized diffusioosmotic velocity in a slit versus the electrokinetic coordinate y/h for the case of $\bar{\zeta} = 4$: (a) $\kappa h = 1$; (b) $\mathrm{Pe} = 1$. The solid curves represent the case $\beta = -0.2$ and the dashed curves denote the case $\beta = 0$.

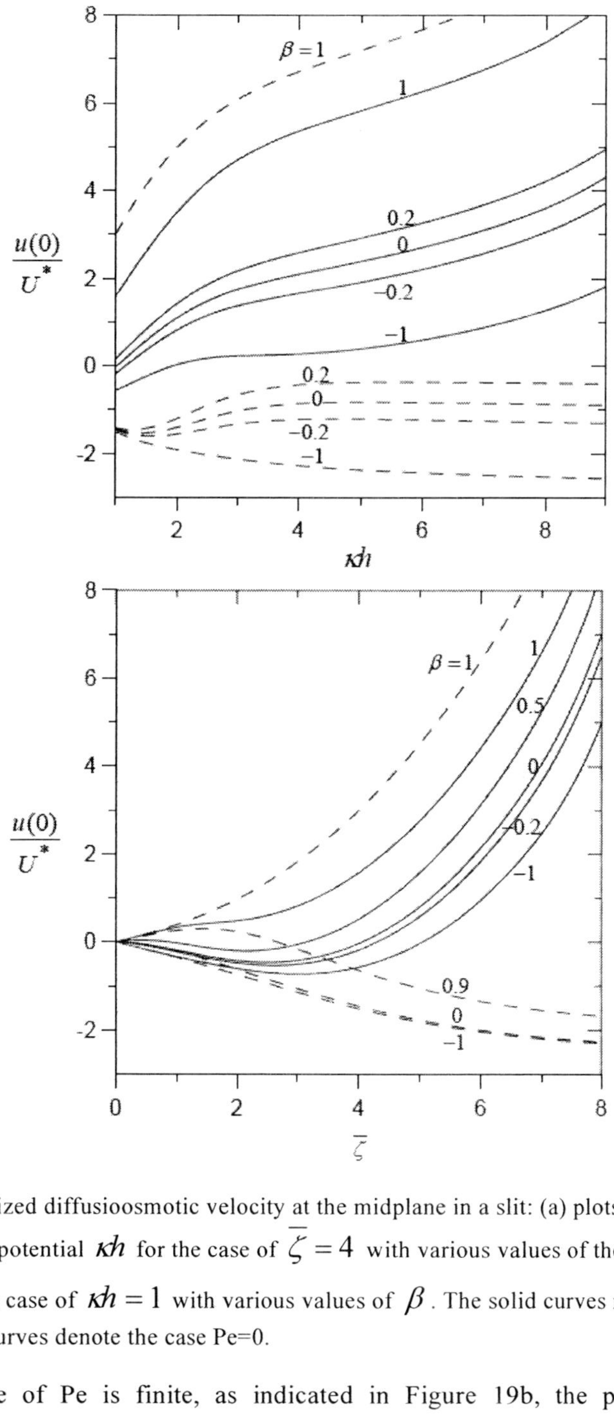

Figure 18. The normalized diffusioosmotic velocity at the midplane in a slit: (a) plots versus the dimensionless surface potential κh for the case of $\overline{\zeta} = 4$ with various values of the parameter β; (b) plots versus $\overline{\zeta}$ for the case of $\kappa h = 1$ with various values of β. The solid curves represent the case Pe=1 and the dashed curves denote the case Pe=0.

When the value of Pe is finite, as indicated in Figure 19b, the probability of the diffusioosmotic flow of the electrolyte solution in the direction toward higher concentration for a combination of β and $\overline{\zeta}$ is again greatly reduced, due to the effect of the electrolyte convection.

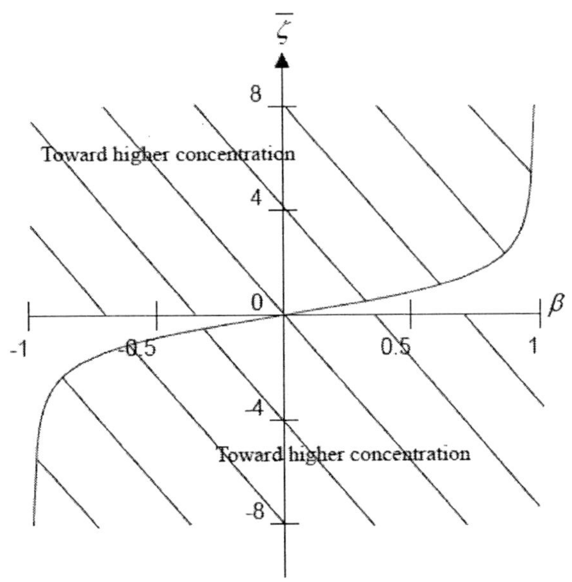

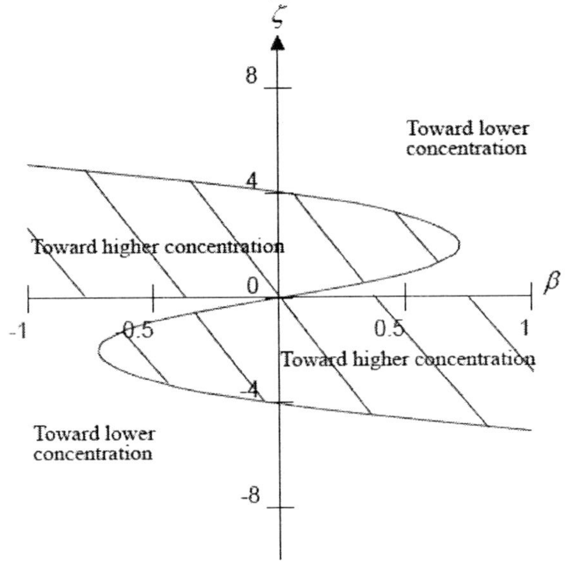

Figure 19. Maps showing the direction of the diffusioosmotic velocity at the midplane in a slit for the case of $\kappa h = 1$: (a) Pe=0; (b) Pe=1.

3.3.3. Diffusioosmosis for the Case of Low Surface Potential without Relaxation Effect

The functions Φ_1, Φ_2, and Φ_3 given by Eq. (55) determine the diffusioosmotic velocity of a symmetric electrolyte in a capillary slit correct to $O(\zeta^2)$ for the case of Pe=0. In Figure 20, these functions are plotted versus the normalized coordinate y/h for various values of κh. The values of Φ_1, Φ_2, and Φ_3 at the midplane of the capillary slit (with $y = 0$) and their averaged results over a cross section of the slit as functions of κh are presented in Table 1 and plotted in Figure 21 for comparison. All of the functions Φ_1, Φ_2

and Φ_3 decrease monotonically with an increase in y/h for a constant value of κh, from maxima at the midplane of the capillary slit to zero at the slit walls (with $y = h$).

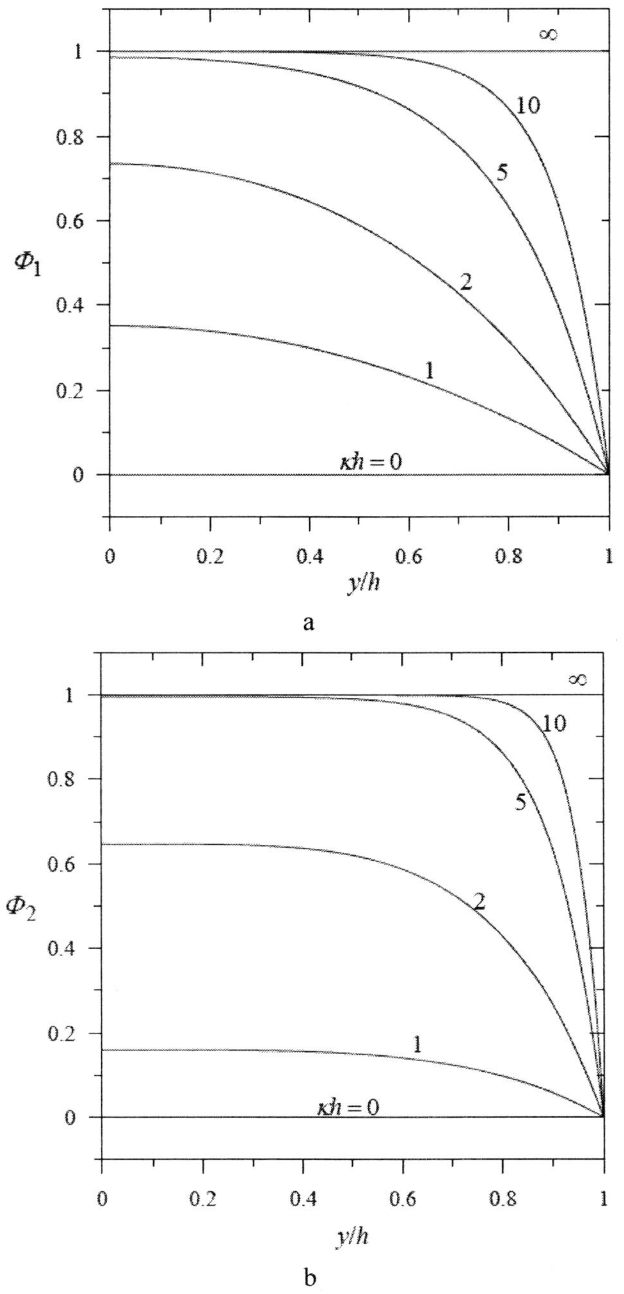

a

b

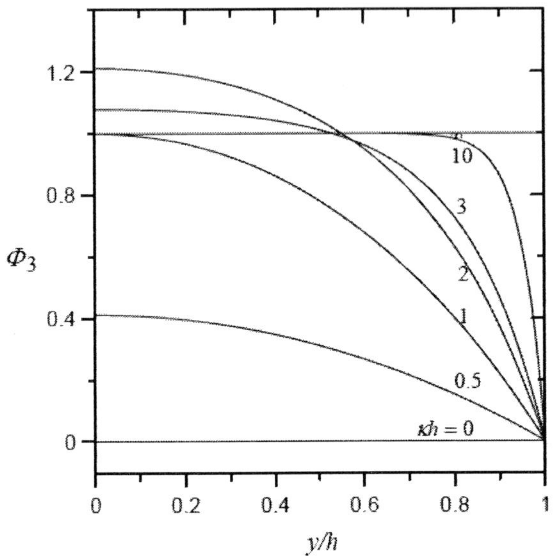

c

Figure 20. Plots of the functions Φ_1, Φ_2, and Φ_3 as calculated from Eq. (55) versus the normalized coordinate y/h for various values of κh.

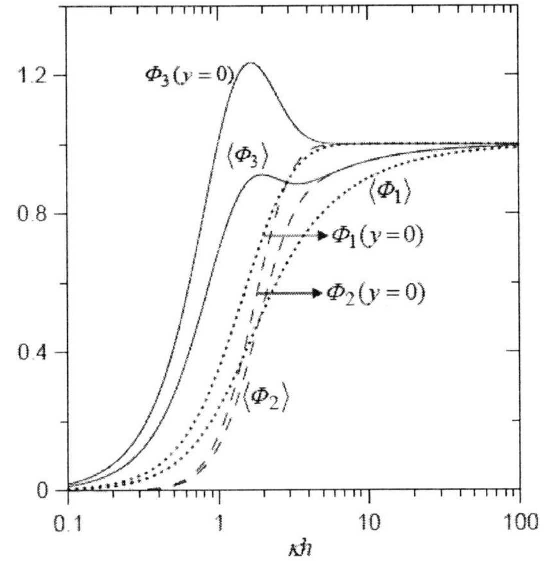

Figure 21. Plots of the functions $\Phi_1(y=0)$, $\Phi_2(y=0)$, $\Phi_3(y=0)$, $\langle\Phi_1\rangle$, $\langle\Phi_2\rangle$, and $\langle\Phi_3\rangle$ versus κh.

Table 1. Values of $\Phi_1(y=0)$, $\langle\Phi_1\rangle$, $\Phi_2(y=0)$, $\langle\Phi_2\rangle$, $\Phi_3(y=0)$, and $\langle\Phi_3\rangle$ for various values of κh

κh	$\Phi_1(y=0)$	$\langle\Phi_1\rangle$	$\Phi_2(y=0)$	$\langle\Phi_2\rangle$	$\Phi_3(y=0)$	$\langle\Phi_3\rangle$
0	0	0	0	0	0	0
0.2	0.0197	0.0131	0.0005	0.0004	0.0774	0.0517
0.5	0.1132	0.0758	0.0169	0.0136	0.4101	0.2757
1	0.3519	0.2384	0.1601	0.1292	1.0000	0.6892
2	0.7342	0.5180	0.6467	0.5353	1.2116	0.9121
3	0.9007	0.6683	0.9013	0.7700	1.0789	0.8884
4	0.9634	0.7502	0.9772	0.8601	1.0201	0.8887
5	0.9865	0.8000	0.9953	0.8969	1.0044	0.9029
10	0.9999	0.9000	1.0000	0.9500	1.0000	0.9500
20	1.0000	0.9500	1.0000	0.9750	1.0000	0.9750
50	1.0000	0.9800	1.0000	0.9900	1.0000	0.9900
∞	1	1	1	1	1	1

For a given value of y/h, different from the trend of Φ_1 or Φ_2 as an increasing function of κh, Φ_3 is not necessarily a monotonic function of κh, due to the competition between the effect of the deviation of the local induced axial electric field from its bulk-phase quantity (strong as $\kappa h \to 0$ and weak as $\kappa h \to \infty$) and the effect of the solute-wall interaction (strong as $\kappa h \to \infty$ and weak as $\kappa h \to 0$), and may have a maximal value greater than unity at some value of κh. The maximum of Φ_3 at the midplane of the slit, which is greater than unity for cases with κh greater than about 1, occurs in the vicinity of $\kappa h = 1.67$. When the slit wall is approached ($y/h \to 1$), this maximum arises at $\kappa h \to \infty$ and its value is close to unity.

In Figure 22, the diffusioosmotic velocities of electrolyte solutions at the midplane of a capillary slit normalized by their characteristic value given by Eq. (12) are plotted as functions of the dimensionless surface potential $\bar{\zeta}$ (with its magnitude less than 2) for various values of κh. For the case of a symmetric electrolyte that the cation and anion diffusivities are equal ($\beta = 0$, representing the aqueous solution of KCl if $Z = 1$) shown in Figure 22a, the magnitude of the normalized velocity $u(y=0)/U^*$ (or u/U^* at any other location, or $\langle u\rangle/U^*$, which are not illustrated here) increases monotonically with an increase in $\bar{\zeta}$ for a fixed value of κh, but is not a monotonic function of κh for a constant value of $\bar{\zeta}$. Only the results at positive surface potentials are shown because the fluid velocity, which is due to the $O(\zeta^2)$ contribution entirely as illustrated by Eq. (54), is an even function of the zeta potential. Since our analysis is based on the assumption of small surface potential, the magnitudes of $\bar{\zeta}$ considered are less than 2. For the case of a symmetric electrolyte whose cation and anion have different diffusion coefficients ($\beta = -0.2$ is chosen, representing the aqueous solution of NaCl if $Z = 1$) as illustrated in Figure 22b, the fluid velocity is not necessarily a monotonic function of $\bar{\zeta}$ for a fixed value of κh and the fluid flow can reverse its direction more than once as the surface potential of the slit walls varies from negative to positive values.

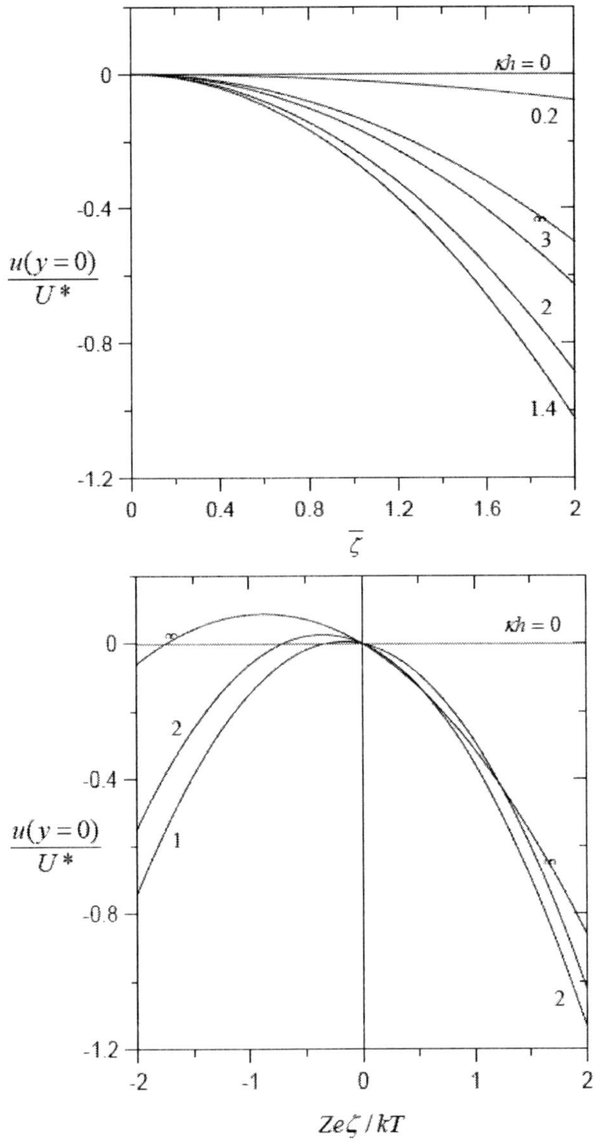

Figure 22. Plots of the normalized diffusioosmotic velocity at the midplane of a capillary slit versus the dimensionless surface potential $\bar{\zeta}$ for various values of κh : (a) $\beta = 0$; (b) $\beta = -0.2$.

In this case, the diffusioosmotic velocity is neither an even nor an odd function of the zeta potential of the capillary walls. In general, the local or average fluid velocity is not a monotonic function of κh for a constant value of $\overline{\zeta}$. Both Figures 22a and 22b demonstrate that the electroosmotic effect of $O(\zeta^2)$ (involving the function Φ_3) gives a dominant contribution to the net diffusioosmotic flow.

4. DIFFUSIOOSMOSIS IN A CAPILLARY TUBE

In this section, we consider the steady diffusioosmotic flow of a solution of a symmetric electrolyte in a straight capillary tube of radius *R*, as illustrated in Figure 23. The applied electrolyte concentration gradient ∇n^{∞} is a constant along the axial (*z*) direction in the capillary, where $n^{\infty}(z)$ is the linear concentration distribution of the electrolyte in the bulk solution phase in equilibrium with the fluid inside the tube. Again, the end effects are neglected. The analysis for this case is similar to those for the cases of diffusioosmosis along a single plane wall and in a capillary slit presented in the previous sections.

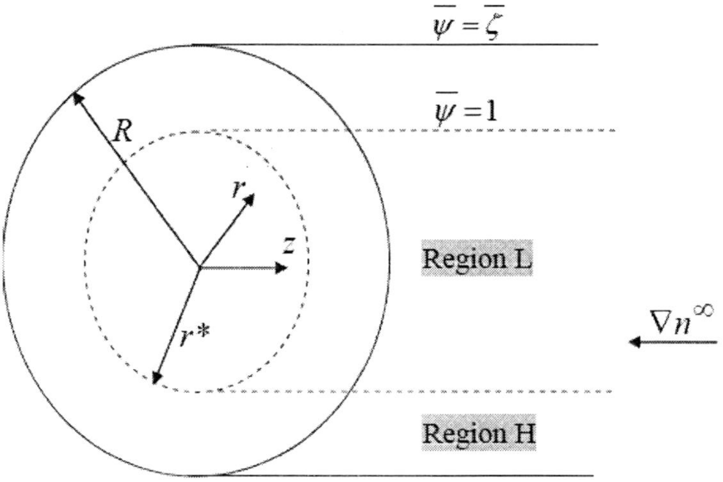

Figure 23. Geometrical sketch for the diffusioosmosis in a capillary tube due to an axially applied concentration gradient of an electrolyte.

4.1. Electrostatic Potential Distribution

If $\psi(r)$ represents the electrostatic potential at a point with distance *r* from the axis of the capillary tube relative to that in the bulk solution, and $n_+(r,z)$ and $n_-(r,z)$ denote the local concentrations of the cations and anions, respectively, then the Poisson equation gives

$$\frac{1}{r}\frac{\mathrm{d}}{\mathrm{d}r}\left(r\frac{\mathrm{d}\psi}{\mathrm{d}r}\right) = -\frac{4\pi Ze}{\varepsilon}[n_+(r,0) - n_-(r,0)]. \tag{59}$$

Substitution of Eq. (2) into Eq. (59) results in the Poisson-Boltzmann equation,

$$\frac{1}{r}\frac{\mathrm{d}}{\mathrm{d}r}\left(r\frac{\mathrm{d}\bar{\psi}}{\mathrm{d}r}\right)=\kappa^2\sinh\bar{\psi}\,. \tag{60}$$

For the case of constant surface potential, the boundary conditions for ψ are

$$r=0:\ \frac{\mathrm{d}\bar{\psi}}{\mathrm{d}r}=0\,, \tag{61a}$$

$$r=R:\ \bar{\psi}=\bar{\zeta}\,. \tag{61b}$$

Since there is no simple analytical solution of Eq. (60) available for the case of cylindrical symmetry, we follow a previous approach [27,48] and use an approximation to provide a good representation of $\sinh\bar{\psi}$ throughout the range $\bar{\psi}\geq 0$,

$$\sinh\bar{\psi}=\bar{\psi} \qquad \text{if } 0\leq\bar{\psi}<1\,, \tag{62a}$$

$$\sinh\bar{\psi}=\frac{1}{2}\exp(\bar{\psi}) \text{ if } \qquad \bar{\psi}>1\,, \tag{62b}$$

and replace Eq. (60) by a pair of equations,

$$\frac{1}{r}\frac{\mathrm{d}}{\mathrm{d}r}\left(r\frac{\mathrm{d}\bar{\psi}_{\mathrm{L}}}{\mathrm{d}r}\right)=\kappa^2\bar{\psi}_{\mathrm{L}} \text{ if } 0\leq r\leq r^*\,, \tag{63a}$$

$$\frac{1}{r}\frac{\mathrm{d}}{\mathrm{d}r}\left(r\frac{\mathrm{d}\bar{\psi}_{\mathrm{H}}}{\mathrm{d}r}\right)=\frac{1}{2}\kappa^2\exp(\bar{\psi}_{\mathrm{H}}) \text{ if } r^*\leq r\leq R\,. \tag{63b}$$

Only the positive values of $\bar{\psi}$ are considered here without the loss of generality. Evidently, we have divided the capillary into two hypothetical concentric regions such that $\bar{\psi}=1$ ($\sinh\bar{\psi}=1.175$ and $\exp(\bar{\psi})/2=1.359$) at their junction $r=r^*$, where the subscripts L and H designate the inner (or low potential) and outer (or high potential) regions, respectively, as shown in Figure 23.

If $\overline{\zeta} \leq 1$, then region L comprises the whole of the capillary tube. If $\overline{\psi}(0) \geq 1$, then region H occupies the whole tube. In other cases, Eq. (63) is subject to Eq. (61) and the additional boundary conditions

$$r = r^* : \overline{\psi}_{\mathrm{L}} = \overline{\psi}_{\mathrm{H}} = 1, \tag{64a}$$

$$\frac{\mathrm{d}\overline{\psi}_{\mathrm{L}}}{\mathrm{d}r} = \frac{\mathrm{d}\overline{\psi}_{\mathrm{H}}}{\mathrm{d}r}, \tag{64b}$$

which together ensure that the calculated $\overline{\psi}(r)$ profile, albeit approximate, will be a smooth continuous function in the neighborhood of $r = r^*$.

The relation among the dimensionless parameters κr^*, $\overline{\zeta}$, and κR is displayed in Figure 24 and the analytical solutions for $\overline{\psi}_{\mathrm{L}}(r)$ and $\overline{\psi}_{\mathrm{H}}(r)$ are outlined below. In the Subdomain I ($\overline{\zeta} \leq 1$), the low potential region fills the capillary entirely, and the electrostatic potential distribution is

$$\overline{\psi}_{\mathrm{L}}(r) = \overline{\zeta}\frac{I_0(\kappa r)}{I_0(\kappa R)} \quad \text{for } 0 \leq r \leq R, \tag{65}$$

where I_n is the modified Bessel function of the first kind of order n. As expected, Figure 24 illustrates that $r^* = R$ as $\overline{\zeta} = 1$.

In the Subdomain IIA ($r^* > r_0^*$), the solution of Eqs. (63), (64), and (61) results in

$$\overline{\psi}_{\mathrm{L}}(r) = \frac{I_0(\kappa r)}{I_0(\kappa r^*)} \quad \text{for} \quad 0 \leq r \leq r^*, \tag{66a}$$

$$\overline{\psi}_{\mathrm{H}}(r) = \ln\{\frac{-C}{(\kappa r)^2 \cos^2[\cos^{-1}\sqrt{-C/(\kappa r^*)^2 \mathrm{e}} + \sqrt{-C/4}\ln(r/r^*)]}\}$$

$$\text{for } r^* \leq r \leq R, \tag{66b}$$

where C is an integration constant dependent on the parameter κr^*,

$$C = [2 + \kappa r^* \frac{I_1(\kappa r^*)}{I_0(\kappa r^*)}]^2 - (\kappa r^*)^2 e . \tag{67}$$

The above equation gives $C < 0$ for $r^* > r_0^*$ and $C = 0$ at $r^* = r_0^*$ (the junction of the Subdomains IIA and IIB), where $\kappa r_0^* = 2.15852$; in the latter case Eq. (66b) becomes

$$\overline{\psi}_H(r) = -2\ln(\frac{r}{\sqrt{e}r_0^*} - \frac{\kappa r}{2}\ln\frac{r}{r_0^*}) \text{ for } r^* \le r \le R . \tag{68}$$

In the Subdomain IIB ($0 < r^* < r_0^*$ and $0 < C < 4$), $\overline{\psi}_L$ is also given by Eq. (66a), while

$$\overline{\psi}_H(r) = \ln\{\frac{4AC(r/r^*)^{\sqrt{C}}}{(\kappa r)^2[1 - A(r/r^*)^{\sqrt{C}}]^2}\} \text{ for } r^* \le r \le R , \tag{69}$$

where

$$A = \frac{\sqrt{C + (\kappa r^*)^2 e} - \sqrt{C}}{\sqrt{C + (\kappa r^*)^2 e} + \sqrt{C}} . \tag{70}$$

In the Subdomain III ($r^* = 0$, $C = 4$, and $\overline{\psi}_H(0) \ge 1$), the region of the potential $\overline{\psi}_H$ fills the capillary entirely, and

$$\overline{\psi}_H(r) = \overline{\psi}_H(0) - 2\ln[1 - \frac{(\kappa r)^2}{16} e^{\overline{\psi}_H(0)}] \text{ for } 0 \le r \le R . \tag{71}$$

Here, the quantity $\overline{\psi}_H(0)$ can be determined as a function of the parameters $\overline{\zeta}$ and κR from the above equation setting $r = R$ and $\overline{\psi}_H(R) = \overline{\zeta}$.

If the constant surface charge density σ , instead of the surface potential ζ , is known at the capillary wall, the boundary condition specified by Eq. 4b should be replaced by the Gauss condition,

$$r = R: \frac{d\psi}{dr} = \frac{4\pi\sigma}{\varepsilon}. \tag{72}$$

The solutions for ψ given by Eqs. (65)-(71) still hold for this condition, with the connection between ζ and σ for an arbitrary value of κR as

$$\bar{\zeta} = \frac{\bar{\sigma}_2}{\kappa R} \frac{I_0(\kappa R)}{I_1(\kappa R)} \text{ if } \bar{\zeta} \le 1, \tag{73a}$$

$$\bar{\zeta} = \ln \frac{(\bar{\sigma}_2 + 2)^2 - C}{(\kappa R)^2} \text{ otherwise,} \tag{73b}$$

where $\bar{\sigma}_2 = 4\pi RZe\sigma / \varepsilon kT$ is the dimensionless surface charge density. The relation among the parameters $\bar{\sigma}_2$, $\bar{\zeta}$, and κR given by Eq. (73) is displayed in Figure 25. Similar to the case of a capillary slit, for a given electrolyte solution in a capillary tube with a specified radius, σ increases (almost linearly) with an increase in κ or $[n^\infty(0)]^{1/2}$ for the case of constant surface potential, and ζ decreases with an increase in κ or $[n^\infty(0)]^{1/2}$ for the case of constant surface charge density.

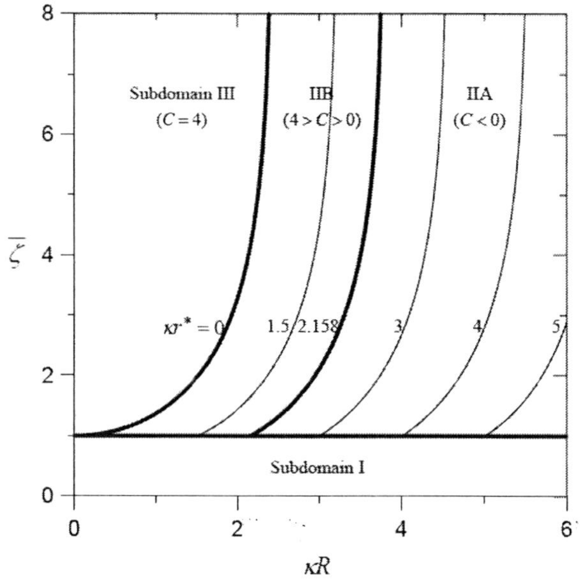

Figure 24. A plot of the relation among the dimensionless parameters κr^*, $\bar{\zeta}$, and κR, and displaying the fundamental subdomains for the solution of $\bar{\psi}(r)$.

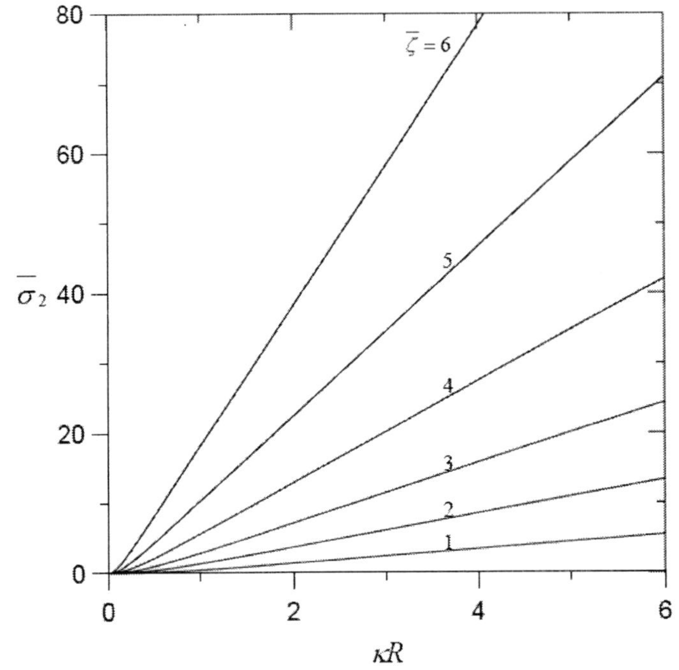

Figure 25. A plot of the relation among the dimensionless parameters $\bar{\sigma}_2$, $\bar{\zeta}$, and κR.

4.2. Fluid Velocity Distribution

4.2.1. General Analysis

We now deal with the steady diffusioosmotic flow of a symmetric electrolyte solution in a capillary tube under the influence of a constant concentration gradient of the electrolyte prescribed axially. The momentum balances on the fluid in the r and z directions give

$$\frac{\partial p}{\partial r} + Ze(n_+ - n_-)\frac{\mathrm{d}\psi}{\mathrm{d}r} = 0, \tag{74a}$$

$$\frac{\eta}{r}\frac{\mathrm{d}}{\mathrm{d}r}\left(r\frac{\mathrm{d}u}{\mathrm{d}r}\right) = \frac{\partial p}{\partial z} - Ze(n_+ - n_-)E \tag{74b}$$

where $p(r,z)$ is the dynamic pressure distribution. The boundary conditions for u at the axis and at the no-slip wall of the capillary tube are

$$r = 0: \ \frac{\mathrm{d}u}{\mathrm{d}r} = 0, \tag{75a}$$

$$r = R : u = 0 . \tag{75b}$$

After the substitution of Eq. (2) into Eq. (74a) based on the assumption that the equilibrium ionic distributions are not affected by the net electrolyte flux $\mathbf{J}$, which is warranted if $|\nabla n^\infty|/\kappa n^\infty(0) << 1$, the pressure distribution can be determined as

$$p = p_0 + 2kTn^\infty(z)[\cosh(\overline{\psi}) - \cosh(\overline{\psi}(0))]. \tag{76}$$

Here, p_0 is the pressure on the axis of the capillary tube, which is a constant in the absence of the applied pressure gradient, and the electric potential distribution $\overline{\psi}(r)$ is given by Eqs. (65)-(71).

Substituting the ionic concentration distributions of Eq. (2) and the pressure profile of Eq. (76) into Eq. (74b) and then performing the integration with respect to *r* twice subject to the boundary conditions in Eq. (75), we obtain

$$\frac{u}{U^*} = (\kappa R)^2 \int_1^{r/R} \frac{R}{r} \int_0^{r/R} \frac{r}{R}[\cosh\overline{\psi} - \cosh\overline{\psi}(0) + \frac{Zen^\infty(0)}{kT|\nabla n^\infty|} E\sinh\overline{\psi}]\mathrm{d}(\frac{r}{R})\mathrm{d}(\frac{r}{R}) \tag{77}$$

After the substitution of Eq. (77) for *u* and Eqs. (65)-(71) for $\overline{\psi}$ into Eq. (11), the induced electric field distribution *E* can be numerically solved. With the known results of $\overline{\psi}$ and *E*, the diffusioosmotic velocity distribution of the electrolyte solution as a function of the dimensionless parameters κR, $\overline{\zeta}$, β, and Pe can be determined from Eq. (77) with the numerical integrations.

4.2.2. Case of Low Surface Potential without Relaxation Effect

In this subsection, we consider the case of low zeta potential at the tube wall and Pe=0. Now, the momentum equations for the steady flow of an electrolyte solution in the capillary tube with an axially applied concentration gradient of the electrolyte can be solved by substituting the linearized form of Eq. (2) into Eq. (74). The pressure distribution can still be expressed as Eq. (53) with $\overline{\psi}_c = \overline{\psi}(0)$.

When the effect of the ionic convection on the induced axial electric field caused by the diffusioosmotic flow itself (relaxation effect) is negligible, the fluid velocity can be obtained by substituting the ionic concentration distributions of Eq. (2), the electrostatic potential distribution of Eq. (65), the pressure profile of Eq. (53), and the electric field profile of Eq. (25) into Eq. (74b), and solving for the fluid velocity subject to the boundary conditions in Eq. (75). The result is [41,43]

$$u = -\frac{2Ze}{\eta\kappa^2}|\nabla n^\infty|\{\beta\zeta\Theta_1(\kappa R, r/R) + \frac{Ze}{8kT}\zeta^2[\Theta_2(\kappa R, r/R)$$

$$+2(\beta^2-1)\Theta_3(\kappa R, r/R)] + O(\zeta^3)\}, \quad (78)$$

where Θ_1, Θ_2, and Θ_3 are functions given by

$$\Theta_1 = 1 - \frac{I_0(\kappa r)}{I_0(\kappa R)}, \quad (79a)$$

$$\Theta_2 = \frac{1}{[I_0(\kappa R)]^2}[B(\kappa R) - B(\kappa r) - \kappa^2(R^2 - r^2)], \quad (79b)$$

$$\Theta_3 = \frac{1}{[I_0(\kappa R)]^2}[B(\kappa R) - B(\kappa r)], \quad (79c)$$

and the function B is defined as

$$B(x) = x^2 I_0(x)[I_0(x) + I_2(x)] - 2[xI_1(x)]^2. \quad (80)$$

The average fluid velocity $\langle u \rangle$ over a cross section of the capillary tube can be expressed as

$$\langle u \rangle = -\frac{2Ze}{\eta\kappa^2}|\nabla n^\infty|\{\beta\zeta\langle\Theta_1\rangle + \frac{Ze}{8kT}\zeta^2[\langle\Theta_2\rangle + 2(\beta^2-1)\langle\Theta_3\rangle] + O(\zeta^3)\}. \quad (81)$$

Here the definition of the angle brackets becomes

$$\langle u \rangle = \frac{2}{R^2}\int_0^R u(r) r \, \mathrm{d}r. \quad (82)$$

It is easy to find that

$$\langle \Theta_1 \rangle = 1 - \frac{2I_1(\kappa R)}{\kappa R \, I_0(\kappa R)}, \tag{83}$$

but the explicit forms of $\langle \Theta_2 \rangle$ and $\langle \Theta_3 \rangle$ are quite complicated and they can be determined by numerical integration.

It can be found from Eq. (79) that all the functions Θ_1, Θ_2, and Θ_3 are always positive. Evidently, $\Theta_1 = \Theta_2 = \Theta_3 = 0$ and $u = 0$ at the capillary wall ($r/R = 1$). In the limiting situation that $\kappa R = 0$, Eq. (79) gives $\Theta_1 = \Theta_2 = \Theta_3 = 0$ at any position in the tube, which also leads to $\langle \Theta_1 \rangle = \langle \Theta_2 \rangle = \langle \Theta_3 \rangle = 0$. Note that $\Theta_3 /(\kappa R)^2 = 4\Theta_1 /(\kappa R)^2 = 1-(r/R)^2$ and $\langle \Theta_3 \rangle /(\kappa R)^2 = 4\langle \Theta_1 \rangle /(\kappa R)^2 = 1/2$, although $\Theta_2 /(\kappa R)^2 = \langle \Theta_2 \rangle /(\kappa R)^2 = 0$, as $\kappa R = 0$; thus, the diffusioosmotic flow caused by the induced electric field $\mathbf{E}$ still exists for this case with a finite radius R of the capillary, as shown in Eq. (78). In the other limiting situation that $\kappa R \to \infty$, Eq. (79) yields $\Theta_1 = \Theta_2 = \Theta_3 = 1$ and $\langle \Theta_1 \rangle = \langle \Theta_2 \rangle = \langle \Theta_3 \rangle = 1$ at an arbitrary position with $r/R \neq 1$. It is surprising and interesting that Θ_3, which reflects the effect of the radial distribution of the induced axial electric field in the electric double layer, does not vanish as $\kappa R \to \infty$, where $\psi = 0$ at any position other than $r/R = 1$. For all cases of finite κR, Θ_3 and $\langle \Theta_3 \rangle$ are comparable with but always greater than Θ_2 and $\langle \Theta_2 \rangle$, respectively.

The terms in Eqs. (78) and (81) involving the functions Θ_1 and Θ_3 represent the contribution from electroosmosis, while the remainder terms (containing the function Θ_2) are the chemiosmotic component. Note that the direction of the local or average fluid velocity caused by electroosmosis is determined by the combination of the parameters κR, $\bar{\zeta}$, and β, while the fluid flow generated by chemiosmosis for the case of small values of ζ and Pe=0 is always in the direction of increasing electrolyte concentration.

4.3. Results and Discussion

4.3.1. Induced Electric Field

The distribution of the macroscopic electric field $E(r)$ induced by a concentration gradient of a symmetric electrolyte prescribed axially in a capillary tube can be numerically determined after substituting $u(r)/U^*$ in the form of Eq. (77) and $\bar{\psi}(r)$ calculated from Eqs. (64)-(71) into Eq. (11). This induced electric field normalized by its quantity at the axis of the capillary, $E(0)$, as a function of the normalized coordinate r/R is plotted in Figures

26 and 27 for several values of the dimensionless parameters $\bar{\zeta}$, β, Pe, and κR. Note that each curve with specified values of $-\bar{\zeta}$ and β in the figures would be identical to that with the values $\bar{\zeta}$ and $-\beta$. As expected, the magnitude of the normalized induced electric field in general is a sensitive function of r/R and can deviate much from its bulk-phase value.

When Pe=0, $E(r)/E(0)$ is positive and its value increases with an increase in r/R from unity at the axis of the tube to a maximum at the capillary wall, increases with an increase in κR, decreases with an increase in $|\bar{\zeta}|$, equals unity in the limit of $\beta\zeta/|\zeta| = \pm 1$, and increases with an increase in $\beta\zeta/|\zeta|$ if it is not too close to unity (depending on the value of κR), for an otherwise specified condition. When the value of Pe is finite, the value of $E(r)/E(0)$ is larger than that for the case of Pe=0 if the magnitude of $\bar{\zeta}$ is small (depending on the values of κR and β), but it may not be a monotonic function of r/R and may become negative if the magnitude of $\bar{\zeta}$ is large. In general, the effect of the electrolyte convection on the local induced electric field in the electric double layer adjacent to the capillary wall can be quite significant even for the case of low zeta potential at the wall.

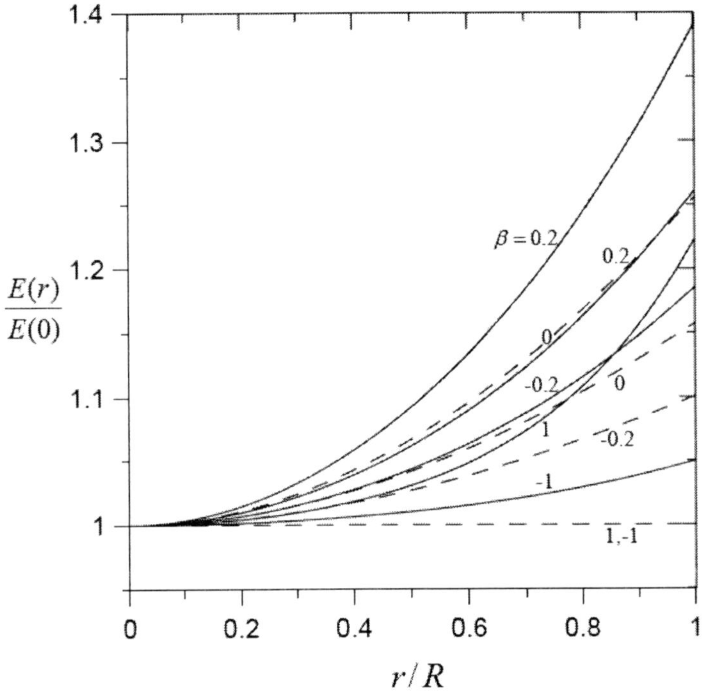

Figure 26. (Continued)

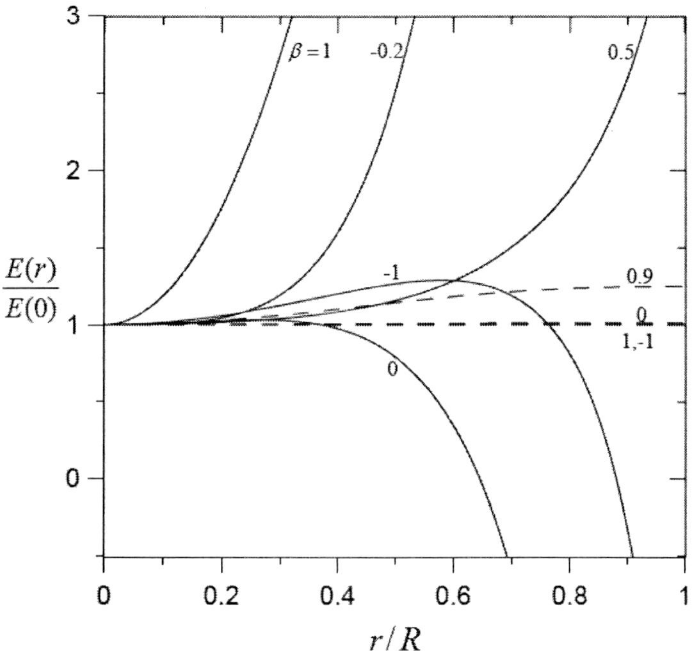

Figure 26. Plots of the normalized electric field induced by an electrolyte gradient in the axial direction of a capillary tube versus the dimensionless coordinate r/R for the case of $\kappa R = 1$ with various values of the parameter β: (a) $\overline{\zeta} = 1$; (b) $\overline{\zeta} = 6$. The solid curves represent the case Pe=1 and the dashed curves denote the case Pe=0.

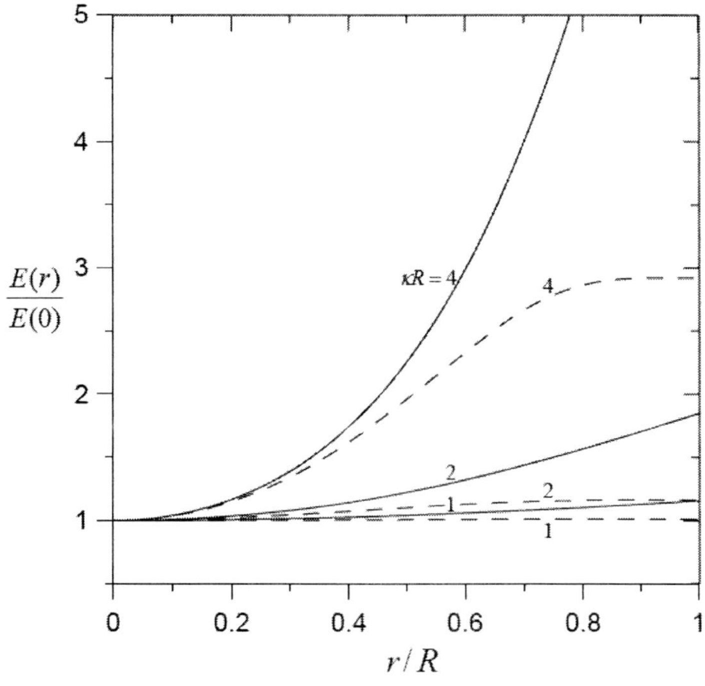

Figure 27. (Continued)

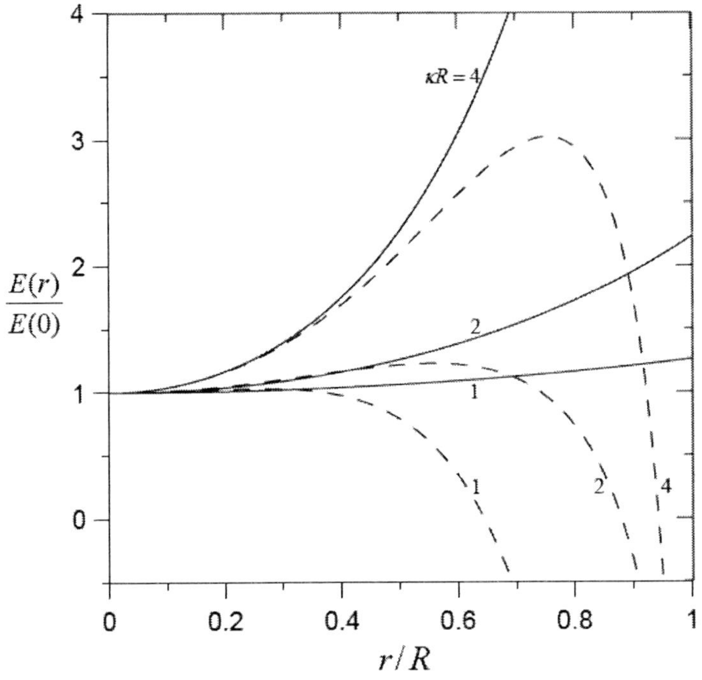

Figure 27. Plots of the normalized electric field induced by an electrolyte gradient in the axial direction of a capillary tube versus the dimensionless coordinate r/R for the case of $\beta = 0$ with various values of the parameter κR : (a) Pe=0; (b) Pe=1. The solid curves represent the case $\bar{\zeta} = 1$ and the dashed curves denote the case $\bar{\zeta} = 6$.

4.3.2. Diffusioosmotic Velocity

The dimensionless diffusioosmotic velocity distribution $u(r)/U^*$ of an electrolyte solution in a capillary tube can be numerically calculated using Eq. (77) with the known results of $\bar{\psi}$ and *E*. This distribution is plotted in Figures 28 and 29 for several values of the parameters $\bar{\zeta}$, β, Pe, and κR. The diffusioosmotic velocity u/U^* can be either positive or negative and is a monotonic increasing function of $\beta\zeta/|\zeta|$. In general, the magnitude of u/U^* decreases monotonically with an increase in the normalized coordinate r/R (there are exceptions), but it is not necessarily a monotonic function of κR for given values of β, Pe, and r/R. When Pe=0 and $\beta\zeta/|\zeta|$ is not too close to unity, u is negative, meaning that the diffusioosmotic flow is in the direction of increasing electrolyte concentration, and the magnitude of u/U^* increases with an increase in $|\bar{\zeta}|$ and with a decrease in $\beta\zeta/|\zeta|$, for an otherwise specified condition.

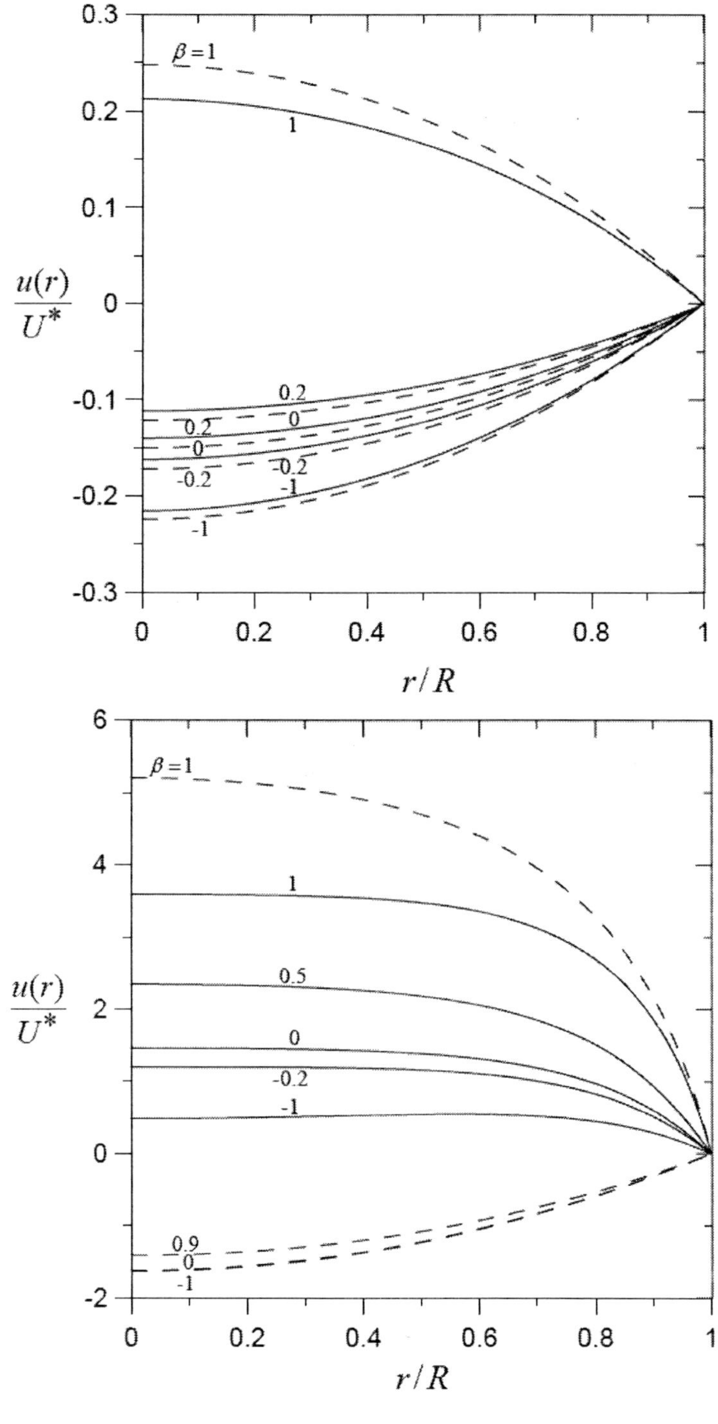

Figure 28. Plots of the normalized diffusioosmotic velocity in a capillary tube versus the dimensionless coordinate r / R for the case of $\kappa R = 1$ with various values of the parameter β: (a) $\overline{\zeta} = 1$; (b) $\overline{\zeta} = 6$. The solid curves represent the case Pe=1 and the dashed curves denote the case Pe=0.

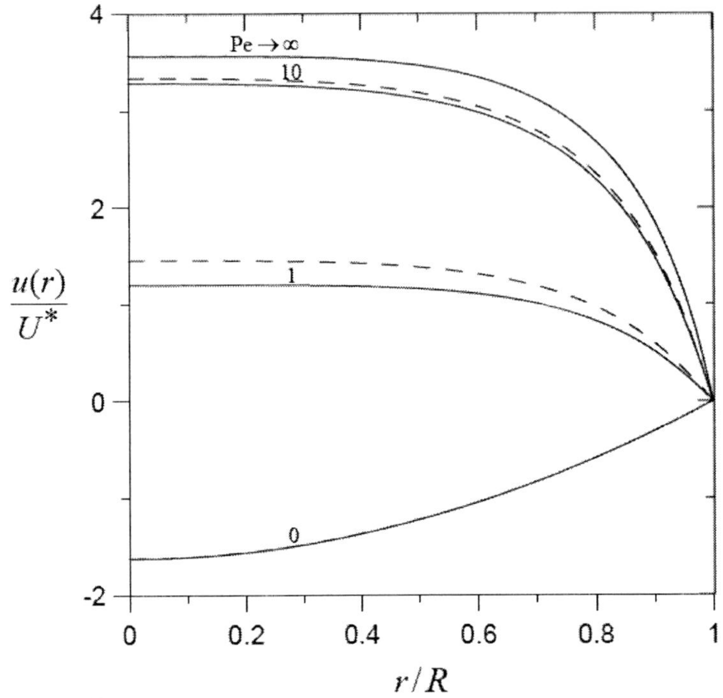

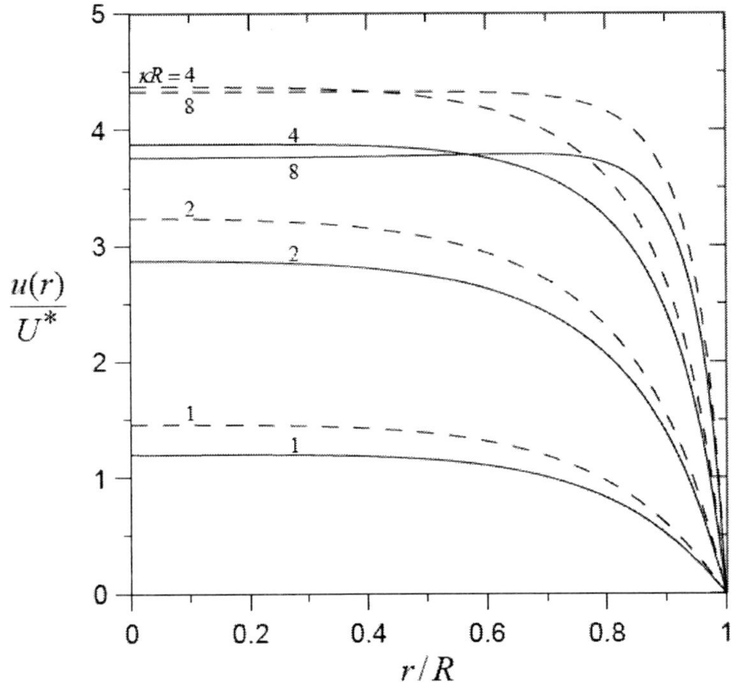

Figure 29. Plots of the normalized diffusioosmotic velocity in a capillary tube versus the dimensionless coordinate r/R for the case of $\bar{\zeta} = 6$: (a) $\kappa R = 1$; (b) Pe=1. The solid curves represent the case $\beta = -0.2$ and the dashed curves denote the case $\beta = 0$.

When Pe=0 and $\beta\zeta/|\zeta|$ approaches unity, the fluid flows against the electrolyte concentration gradient (u is positive) and u/U^* is a monotonic increasing function of $|\overline{\zeta}|$. When the value of Pe is finite, the dependence of u on r/R is similar to that for the case of Pe=0 if the value of $|\overline{\zeta}|$ is small, but u can be positive for any given value of β if the value of $|\overline{\zeta}|$ is large. In general, the value of u/U^* increases monotonically and remarkably with an increase in the value of Pe (the relaxation effect on the diffusioosmotic flow is quite significant) for specified values of κR, $\overline{\zeta}$, and β except for the situation that the value of $\beta\zeta/|\zeta|$ is close to unity. In the limit of $\mathrm{Pe} \to \infty$, u/U^* is finite.

In Figure 30, the normalized diffusioosmotic velocity $u(0)/U^*$ of the electrolyte solution at the axis of the capillary tube is plotted versus the parameters κR and $\overline{\zeta}$ at specified values of Pe and β. Maps showing the direction of this velocity for a typical value of κR are also drawn in Figure 31. The dependence of $u(0)/U^*$ on $\overline{\zeta}$, β, κR, and Pe is quite similar to that of u/U^* for a given value of r/R, and it is not necessarily a monotonic function of κR for given values of β and Pe. When Pe=0 and the product of ζ and β is negative (inside the second and fourth quadrants in Figure 31a), $u(0)$ is negative and the electrolyte solution flows toward higher concentration. When Pe=0 and the product of ζ and β is positive (inside the first and third quadrants in Figure 31a), the diffusioosmotic velocity $u(0)$ may reverse its direction from against the concentration gradient to along with it as $|\overline{\zeta}|$ increases not much from zero for all practical cases of β (in addition to a reversal occurring at $\overline{\zeta} = 0$), or as $|\beta|$ decreases from 1 to 0 for a small magnitude of $\overline{\zeta}$. When the value of Pe is finite, as indicated in Figure 31b, the probability of the diffusioosmotic flow of the electrolyte solution in the direction toward higher concentration for a combination of β and $\overline{\zeta}$ is greatly reduced, due to the effect of the electrolyte convection. This effect is quite significant, irrespective of the thickness of the electric double layer adjacent to the capillary wall.

Throughout this section we have used the mathematical approximation presented in Subsection 4.1.1 for the solution of the electrostatic potential distribution on a cross section of the capillary tube. To check the accuracy of this approximation for the resulting diffusioosmotic velocity profile, in some typical cases, we numerically solve Eqs. (60) and (61) for $\overline{\psi}(r)$, curve-fit its values in terms of a tenth-order polynomial, and then substitute it into Eqs. (11) and (77) to numerically determine the 'exact' solution for the diffusioosmotic velocity $u(r)$.

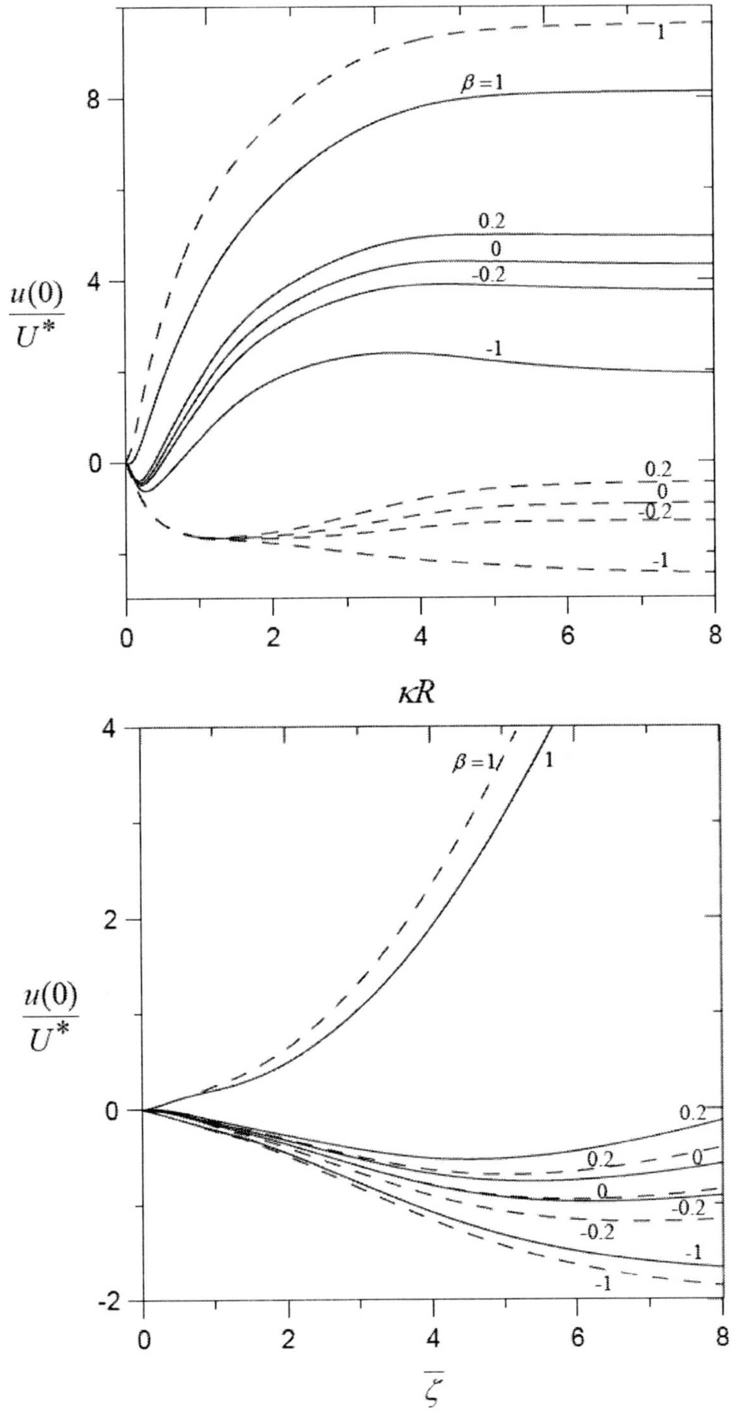

Figure 30. The normalized diffusioosmotic velocity at the axis of a capillary tube for various values of the parameter β: (a) plots versus κR for the case of $\bar{\zeta} = 6$; (b) plots versus $\bar{\zeta}$ for the case of $\kappa R = 1$. The solid curves represent the case Pe=1 and the dashed curves denote the case Pe=0.

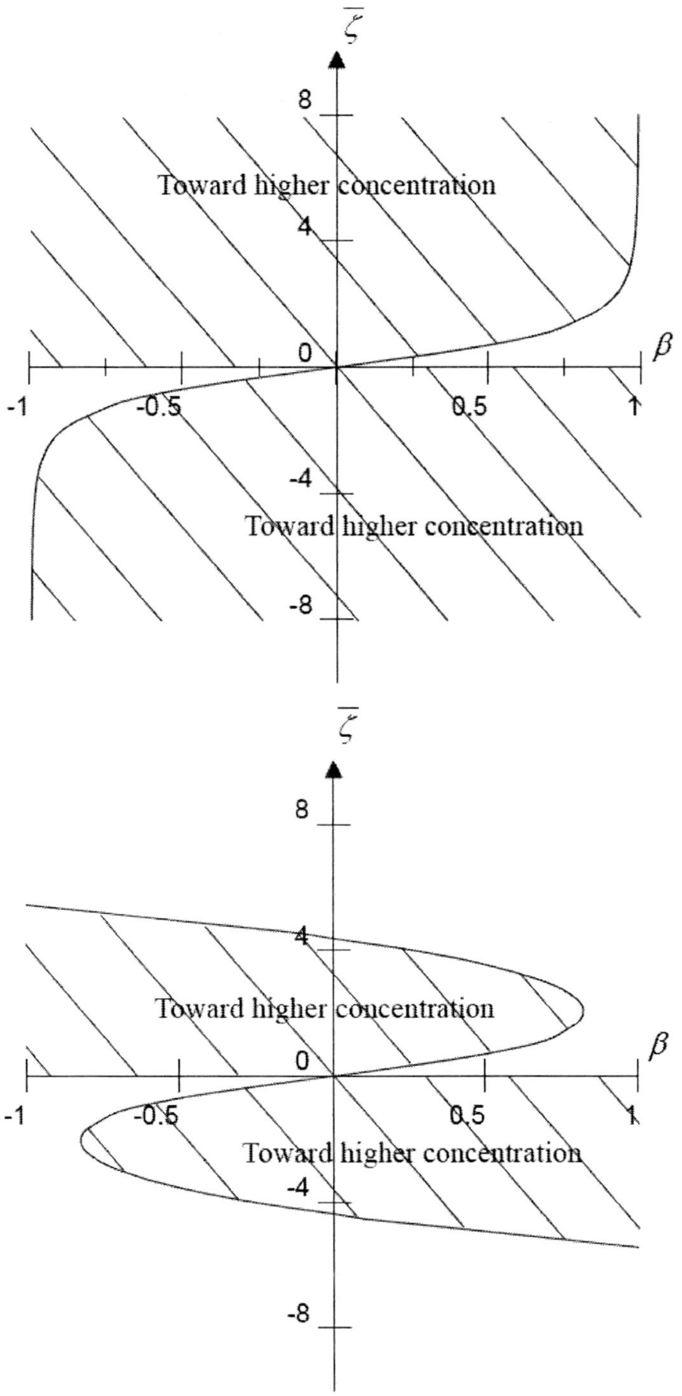

Figure 31. Maps showing the direction of the diffusioosmotic velocity at the axis of a capillary tube for the case of $\kappa R = 1$: (a) Pe=0; (b) Pe=1.

Figure 32 gives a comparison of the approximate solution with the 'exact' solution. It can be found that the difference between the two solutions for $u(r)$ increases with an increase in the normalized coordinate r/R but is less than 5% if $r/R < 0.9$. This outcome means that the mathematical approximation presented in Subsection 4.1.1 for the solution of $\overline{\psi}(r)$ is generally acceptable in the evaluation of the diffusioosmotic velocity of electrolyte solutions in a fine capillary tube when compared with the relevant experimental data.

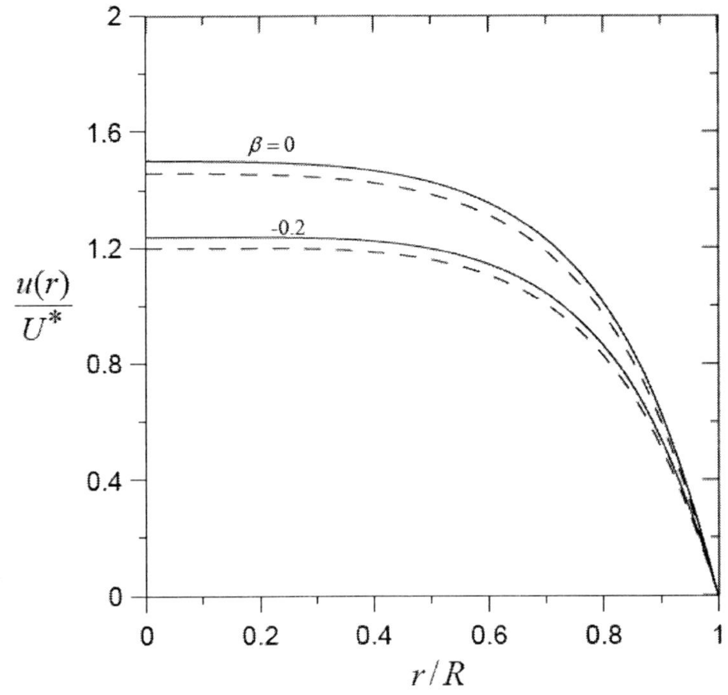

Figure 32. The normalized diffusioosmotic velocity distribution in a capillary tube for the case of $\overline{\zeta} = 6$, $\kappa R = 1$, and Pe=1. The dashed curves represent the result obtained by using the mathematical approximation for the evaluation of $\overline{\psi}(r)$ in Section 4.1.1 and the solid curves denote the numerical 'exact' solution.

4.3.3. Diffusioosmosis for the Case of Low Surface Potential without Relaxation Effect

The functions Θ_1, Θ_2, and Θ_3 given by Eq. (79) for the diffusioosmotic velocity of a symmetric electrolyte in a capillary tube correct to $O(\zeta^2)$ for the case of Pe=0 have the same physical meanings as those of the functions Φ_1, Φ_2, and Φ_3 for the corresponding flow in a capillary slit. In Figure 33, the function Θ_1, Θ_2, and Θ_3 are plotted versus the normalized coordinate r/R for various values of κR. The values of Θ_1, Θ_2, and Θ_3 at the axis of the capillary tube (with $r = 0$) and their averaged results over a cross section of the tube as functions of κR are presented in Table 2 and plotted in Figure 34 for comparison.

As expected, for a specified value of κR, Θ_1, Θ_2, and Θ_3 are monotonically decreasing functions of r/R from a maximum at the axis of the tube (with $r = 0$) to zero at the tube wall (with $r = R$). For a fixed value of r/R, Θ_2 and Θ_3 are not necessarily monotonic functions of κR, due to the competition between the effect of the deviation of the local induced axial electric field from its bulk-phase quantity and the effect of the solute-wall interaction, and have a maximal value which is greater than unity at some value of κR. The maxima for Θ_2 and Θ_3 at the axis of the tube occur in the vicinity of $\kappa R = 4.8$ and 2.48, respectively, and their locations shift to larger κR as r/R increases. When the tube wall is approached ($r/R \to 1$), these maxima take place at $\kappa R \to \infty$ and their values are close to unity.

Table 2. Values of $\Theta_1(r=0)$, $\langle\Theta_1\rangle$, $\Theta_2(r=0)$, $\langle\Theta_2\rangle$, $\Theta_3(r=0)$, and $\langle\Theta_3\rangle$ for various values of κR

κR	$\Theta_1(r=0)$	$\langle\Theta_1\rangle$	$\Theta_2(r=0)$	$\langle\Theta_2\rangle$	$\Theta_3(r=0)$	$\langle\Theta_3\rangle$
0	0	0	0	0	0	0
0.2	0.0099	0.0050	0.0002	0.0001	0.0394	0.0197
0.5	0.0596	0.0300	0.0071	0.0047	0.2281	0.1152
1	0.2101	0.1072	0.0848	0.0571	0.7087	0.3691
2	0.5613	0.3022	0.5440	0.3777	1.3138	0.7625
3	0.7951	0.4600	0.9529	0.6921	1.3307	0.8810
4	0.9115	0.5682	1.1051	0.8446	1.2303	0.9072
5	0.9633	0.6427	1.1258	0.9017	1.1595	0.9186
10	0.9997	0.8103	1.0597	0.9531	1.0597	0.9531
20	1.0000	0.9025	1.0271	0.9757	1.0271	0.9757
50	1.0000	0.9604	1.0103	0.9901	1.0103	0.9901
∞	1	1	1	1	1	1

The dependence of the diffusioosmotic velocity of an electrolyte solution at the axis of a capillary tube on the dimensionless zeta potential of the tube wall at various values of κR is displayed in Figure 35. The value of this fluid velocity calculated by Eq. (78) is normalized by a characteristic value given by Eq. (12). Figure 35a is drawn for the case of an electrolyte that the cation and anion diffusivities are equal ($\beta = 0$). It can be seen from the $O(\zeta^2)$ term of Eq. (78) that the fluid flows contributed from chemiosmosis and electroosmosis are always in the opposite directions, and for the case of $\beta = 0$, the net flow is dominated by the electroosmotic effect, even when the electric double layer is very thin or $\kappa R \to \infty$. Figure 35a shows that the magnitude of the reduced diffusioosmotic velocity $u(r=0)/U^*$ increases monotonically with the increase of $\bar{\zeta}$ for a given value of κR and, similar to the function $\Theta_3(r=0)$ illustrated in Figure 33 and 34, is not a monotonic function of κR for a fixed value of $\bar{\zeta}$. In Figure 35b, the reduced diffusioosmotic velocity $u(r=0)/U^*$ as a function of $\bar{\zeta}$ at various values of κR is plotted for the case of a symmetric electrolyte whose cation and anion have different diffusion coefficients ($\beta = -0.2$). It can be seen that the fluid velocity is not necessarily a monotonic function of $\overline{\zeta}$ given a constant value of κR. Again, for a specified value of $\overline{\zeta}$, $u(r=0)/U^*$ is not necessarily a monotonic function of κR

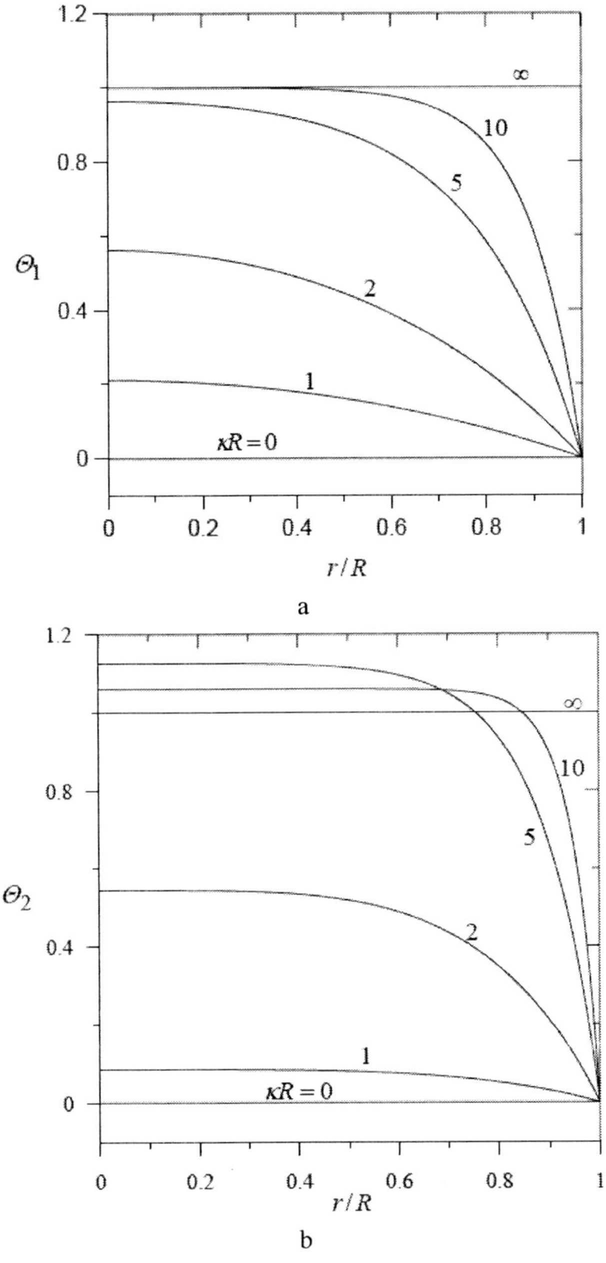

Figure 33. (Continued)

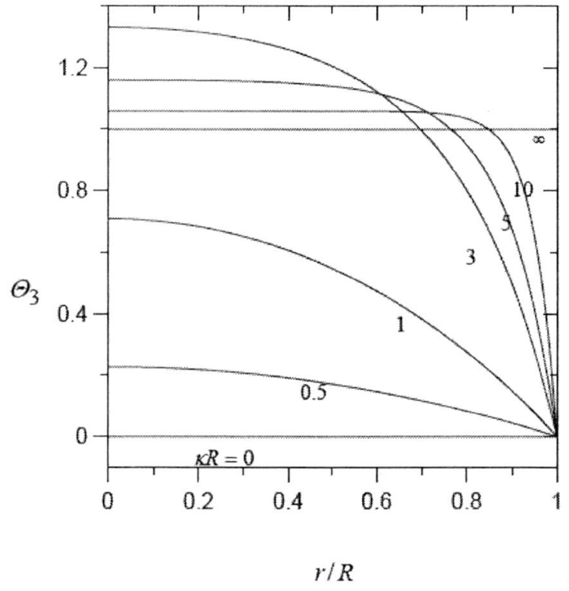

c

Figure 33. Plots of the functions Θ_1, Θ_2, and Θ_3 as calculated from Eq. (79) versus the normalized coordinate r/R for various values of the electrokinetic radius κR.

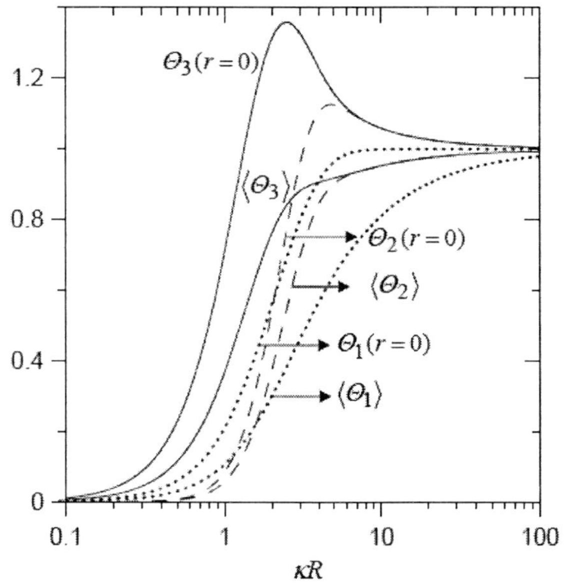

Figure 34. Plots of the functions $\Theta_1(r=0)$, $\Theta_2(r=0)$, $\Theta_3(r=0)$, $\langle\Theta_1\rangle$, $\langle\Theta_2\rangle$, and $\langle\Theta_3\rangle$ versus the electrokinetic radius κR.

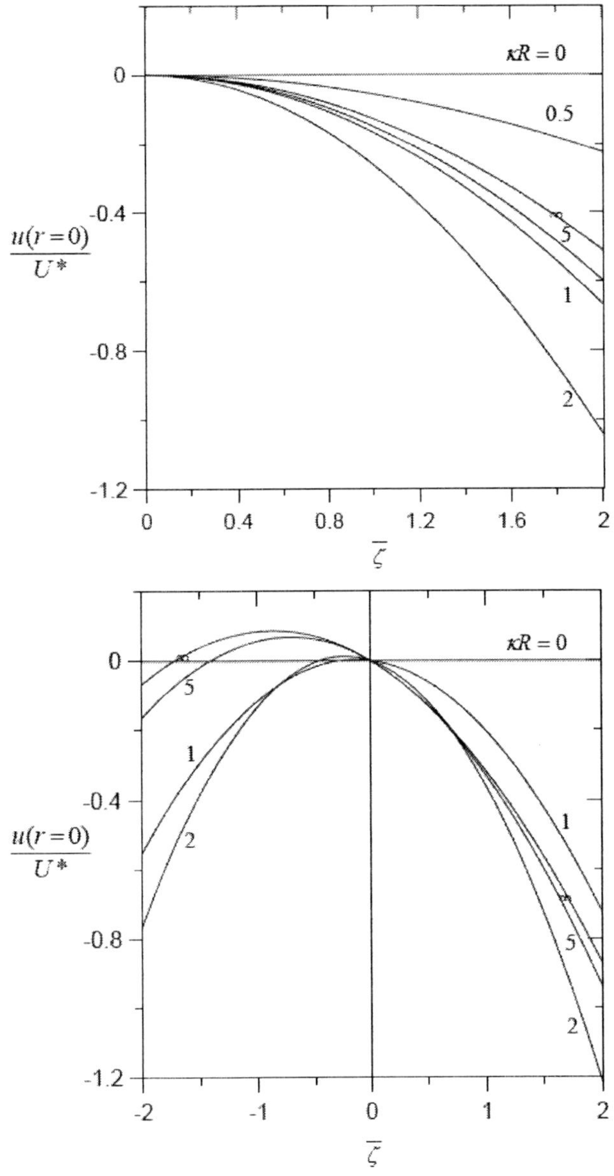

Figure 35. Plots of the normalized diffusioosmotic velocity at the axis of a capillary tube versus the dimensionless surface potential $\bar{\zeta}$ for various values of κR : (a) $\beta = 0$; (b) $\beta = -0.2$.

5. Conclusion

A theoretical study of the steady diffusioosmotic flows of solutions of symmetric electrolytes parallel to a charged plane wall and in a capillary slit or tube is presented in this work. It is assumed that the fluid is only slightly nonuniform in the electrolyte concentration along the charged walls, but no assumption is made about the thickness of the electric double layers adjacent to the walls and the magnitude of the surface potential. The effect of lateral distributions of the electrolyte ions (or of the electrostatic potential) and the effect of ionic convection caused by the diffusioosmotic flow (relaxation effect) on the tangential electric field induced by the applied concentration gradient are taken into account. The plane walls may have either a constant surface potential or a constant surface charge density of an arbitrary quantity.

By solving the Poisson-Boltzmann equation and the modified Navier-Stokes equation applicable to the system, the electrostatic potential distribution, the induced electric field distribution, and the pressure distribution under the influence of the imposed electrolyte gradient are determined either analytically or semi-analytically. The Debye-Huckel approximation is adopted only in the case of low surface potential. Close-form formulas or numerical results for the local and bulk-phase diffusioosmotic velocities as functions of relevant parameters are presented in detail. The results show that the effect of the deviation of the local induced tangential electric field inside the double layer from its bulk-phase quantity and the relaxation effect can not be neglected in the evaluation of the diffusioosmotic velocity of electrolyte solutions along the walls, even for the case of very thin double layer.

In contrast to diffusioosmosis, diffusiophoresis refers to the migration of colloidal particles in response to a macroscopic gradient of solute concentration in a solution. Without considering the variation of the induced tangential electric field with the normal position and the effect of ionic convection inside the electric double layer, u_∞ given by Eq. (20) has been used as a slip velocity at the surface of a nonconductive charged particle of arbitrary shape with a very thin double layer to obtain the steady diffusiophoretic velocity of the particle [2,36]. This terminal velocity turned out to be equal in magnitude but opposite in direction to u_∞ expressed by Eq. (20). When the effect of the lateral distribution of the induced electric field and the relaxation effect in the double layer are taken into account, the correct diffusiophoretic velocity of the dielectric particle should be equal in magnitude but opposite in direction to u_∞ obtained in this work on the basis of the same derivation.

It is worth repeating that all the results in this study are obtained on the basis of a small external gradient of the electrolyte concentration along the walls. If the imposed concentration gradient $|\nabla n^\infty|$ is relatively large, then the effect of variation of the electrostatic potential ψ in the double layer adjacent to the walls with the tangential position may not be neglected. However, it is reasonable for one to expect that this effect will lead to quantitatively rather than qualitatively different results.

ACKNOWLEDGEMENT

This work was supported by the National Science Council of the Republic of China under grants NSC94-2214-E-002-019 and NSC95-2221-E-002-281.

REFERENCES

[1] Dukhin S. S.; Derjaguin B. V. In *Surface and Colloid Science*; Matijevic, E.; Ed.; Wiley: New York, NY, 1974; Vol. 7.

[2] Prieve, D. C.; Anderson, J. L.; Ebel, J. P.; Lowell, M. E. *J. Fluid. Mech.* 1984, *148*, 247.

[3] Anderson, J. L.; Prieve, D. C. *Langmuir*. 1991, *7*, 403.

[4] Ebel, J. P.; Anderson, J. L.; Prieve, D. C. *Langmuir*. 1988, *4*, 396.

[5] Staffeld, P. O.; Quinn, J. A. *J. Colloid Interface Sci*. 1989, *130*, 88.

[6] Smith, R. E.; Prieve, D. C. *Chem. Eng. Sci*. 1982, *37*, 1213.

[7] Prieve, D. C. *Adv. Colloid Interface Sci*. 1982, *16*, 321.

[8] Prieve, D. C.; Roman, R. *J. Chem. Soc. Faraday Trans*. 1987, *283*, 1287.

[9] Pawar, Y.; Solomentsev, Y. E.; Anderson, J. L. *J. Colloid Interface Sci*. 1993, *155*, 488.

[10] Keh, H. J.; Chen, S. B. *Langmuir* .1993, *9*, 1142.

[11] Keh, H. J.; Luo, S. C. *Langmuir*. 1996, *12*, 657.

[12] Keh, H. J.; Jan, J. S. *J. Colloid Interface Sci.* 1996, *183*, 458.

[13] Tu, H. J.; Keh, H. J. *J. Colloid Interface Sci.* 2000, *231*, 265.

[14] Wei, Y. K.; Keh, H. J. *Langmuir* 2001, *17*, 1437.

[15] Chen, P. Y.; Keh, H. J. *J. Colloid Interface Sci.* 2005, *286*, 774.

[16] Keh, H. J.; Wei, Y. K. *J. Colloid Interface Sci.* 2002, *252*, 354.

[17] Keh, H. J.; Huang, T. Y. *J. Colloid Interface Sci.* 1993, *160*, 354.

[18] Keh, H. J.; Wei, Y. K. *Langmuir* 2000, *16*, 5289.

[19] Wei, Y. K.; Keh, H. J. *J. Colloid Interface Sci.* 2002, *248*, 76.

[20] Wei, Y. K.; Keh, H. J. *J. Colloid Interface Sci.* 2004, *269*, 240.

[21] Keh, H. J.; Wei, Y. K. *Langmuir* 2002, *18*, 10475.

[22] Wei, Y. K.; Keh, H. J. *Colloids Surfaces A* 2003, *222*, 301.

[23] Helmholtz, H. *Ann.* 1879, *7*, 337.

[24] Smoluchowski, M. In *Handbuch der Elektrizitat und des Magnetismus*; Graetz, I.; Ed.; Barth: Leipzig, 1921; Vol. 2, pp 336.

[25] Burgreen, D.; Nakache, F. R. *J. Phys. Chem.* 1964, *68*, 1084.

[26] Rice, C. L.; Whitehead, R. *J. Phys. Chem.* 1965, *69*, 4017.

[27] Levine, S.; Marriott, J. R.; Neale, G.; Epstein, N. *J. Colloid Interface Sci.* 1975, *52*, 136.

[28] Ohshima, H; Kondo, T. *J. Colloid Interface Sci.* 1990, *135*, 443.

[29] Masliyah, J. H. *Electrokinetic Transport Phenomena*; AOSTRA: Edmonton, Alberta, Canada, 1994.

[30] Keh, H. J.; Liu. Y. C. *J. Colloid Interface Sci*. 1995, *172*, 222.

[31] Yang, C.; Li, D. *J. Colloid Interface Sci.* 1997, *194*, 95.

[32] Szymczyk, A.; Aoubiza, B.; Fievet, P.; Pagetti, J. *J. Colloid Interface Sci.* 1999, *216*, 285.

[33] Keh, H. J.; Tseng, H. C. *J. Colloid Interface Sci*. 2001, *242*, 450.

[34] Keh, H. J.; Ding, J. M. *J. Colloid Interface Sci*. 2003, *263*, 645.

[35] Derjaguin, B. V.; Dukhin, S. S.; Korotkova, A. A. *Kolloidn. Zh*. 1961, *23*, 53.

[36] Anderson, J. L. *Annu. Rev. Fluid Mech*. 1989, *21*, 61.

[37] Keh, H. J.; Ma, H. C. *Langmuir* 2005, *21*, 5461.

[38] Fair, J. C.; Osterle, J. F. *J. Chem. Phys.* 1971, *54*, 3007.

[39] Sasidhar, V.; Ruckenstein, E. *J. Colloid Interface Sci.* 1982, *85*, 332.

[40] Westermann-Clark, G. B.; Anderson, J. L. *J. Electrochem. Soc.* 1983, *130*, 839.

[41] Keh, H. J.; Wu, J. H. *Langmuir* 2001, *17*, 4216.

[42] Wu, J. H.; Keh, H. J. *Colloids Surfaces A* 2003, *212*, 27.

[43] Keh, H. J.; Ma, H. C. *Colloids Surfaces A* 2004, *233*, 87.

[44] Ma, H. C.; Keh, H. J. *Colloids Surfaces A* 2005, *267*, 4.

[45] Dukhin, S. S. *Adv. Colloid Interface Sci.* 1993, *44*, 1.

[46] Ma, H. C.; Keh, H. J., *J. Colloid Interface Sci.* 2006, *298*, 476.

[47] Prudnikov, A. P.; Brychkov, Y. A.; Marichev, O. I. *Integrals and Series*; Gordon and Breach Science Publishers: New York, NY, 1990; Vol. 3.

[48] Philip, J. R.; Wooding, R. A. *J. Chem. Phys.* 1970, *52*, 953.

In: New Colloid and Surface Science Research
Editor: Lawrence V. Schwartz, pp. 131-163
ISBN: 978-1-60021-594-0

Chapter 3

FLUID INTERFACES FOR SELECTIVE BIOPARTICLE RECOVERY

***Pim van Hee**[*1], **Martijn A Hoeben**[2], **Rob GJM van der Lans**[3] and **Luuk AM van der Wielen**[3]*

[1]DSM Food Specialties, A. Fleminglaan 1, 2600 MA, Delft, The Netherlands
[2]Tebodin Consultants and Engineers, Mauritsstraat 76, Eindhoven, The Netherlands
[3]Delft University of Technology, Department of Biotechnology, Julianalaan 67, 2628BC, Delft, The Netherlands

ABSTRACT

Biological products are often produced as or included in particles, such as inclusion bodies, crystals, cell organelles, viruses and virus-like particles. Selective harvesting of these specific particles is crucial for efficient product recovery. It requires separation techniques that have the ability to separate the target particles from the liquid phase and in general also from other particles, such as cells, cell debris, by-products or immobilised catalysts. In many cases the target particles contain the product already at high purity, making selective particle recovery advantageous. Conventional separation methods such as centrifugation and filtration are inefficient in many particle-particle separations due to the small density differences and overlapping particle size distributions of the particles that require separation. New physical separation methods are therefore desirable.

Particles can adsorb to fluid interfaces. Under specific conditions this process is selective for one type of particle in a particle mixture, which makes the phenomenon useful for particle-particle separation. Selectivity in the adsorption process can be created in several ways.

This chapter deals with the use of fluid interfaces for particle separation both from an experimental and theoretical point of view. It shows that interfacial separation is very valuable for separation of biotechnological particle mixtures. Two principles for creating the fluid interfaces are discussed: 1) injection of the dispersed phase, and 2) emulsification. Selectivity in these processes can be created by selective particle-interface collision, selective particle attachment, selective particle detachment, or a combination of

[*]E-mail: pim.hee-van@dsm.com

these sub-processes. Experimental examples from literature are given for all of these sub-processes. For example, ampicillin and phenylglycine crystals were separated by making use of solvent-water interfaces (Jauregi et al., 2001; 2002; Hoeben et al., 2005). In this case the crystals showed competition for the interface and under certain conditions partitioning of the phenylglycine crystals occurred from one liquid phase to the other. A second example is the separation of polyhydroxyalkanoate inclusion bodies from ruptured fermentation broth making use of the air-water interface (Van Hee et al., 2006b). The selectivity was created by selective aggregation of the inclusion bodies, which enhanced their transport to the interface (more favourable hydrodynamics) relative to that of the other particles in the mixture. Finally, the flotation behaviour of synthetic micrometer-sized bioparticles was investigated by determining the relation between the properties of the surface groups of the particles and their adsorption behaviour to the air-water interface (Van Hee et al., 2006c). Particle aggregation again showed to be essential for these small particles. In addition, the floatability of the particles was dominated by the adsorption kinetics of their surface groups. Besides these examples, many other examples from literature are used to illustrate the mechanisms that can be applied for interfacial bioparticle separation.

List Of Symbols

A_{132}	Hamaker constant (J)
C_j	concentration of ionic species j (mol/m^3)
$C_{\text{surfactant}}$	surfactant concentration (mol/L)
d_d	diameter of the droplets/bubbles of the dispersed phase (m)
d_i	diameter of particle i (m)
d_p	particle diameter (m)
e	elementary charge (C)
E_B	energy barrier between particle and interface (J)
E_C	colloidal interaction energy barrier between particle and interface (J)
$E_{k\text{-}A}$	kinetic energy of attachment (J)
$E_{k\text{-}D}$	kinetic energy of detachment (J)
ΔE_{ads}	adsorption energy (J)
F_d	force required for drainage of liquid between particles (N)
F_E	electrostatic interaction force (N)
F_H	hydrophobic force (N)
F_{vdW}	Van der Waals force (N)
g	gravitational acceleration (m/s^2)
h	separation distance (m)
k	Boltzmann constant (J/K)
K_{132}	constant for hydrophobic force (J)
K_i	force constant for hydrophobic force (N/m)
k_i^{eff}	effective collection rate constant of particle type i (1/s)
L	scale of the large energy containing eddies (m)
N	particle concentration (#/m^3)
N_{av}	Avogadro number (mol^{-1})
n_i	concentration of particle type i in the suspension (#/m^3)
Re_b	bubble Reynolds number (-)

R_{pore}	pore radius (m)
T	temperature (K)
t	time (s)
V_d	bubble/droplet volume (m^3)
v_d	superficial velocity of the dispersed phase droplets/bubbles (m/s)
Z_{12}	collision rate of particles 1 and 2 ($m^{-3}\,s^{-1}$)
z_j	ion valency of ionic species j (-)
α	angle between the three-phase contact line and the particle axis (refer to figure 7)
$\widehat{\varepsilon}$	mechanical energy input per time and unit volume (J/m^3s)
ε	local rate of energy dissipation (m^2/s^3)
ε_p	dielectric permittivity of the fluid (F/m)
$<\varepsilon>$	local average energy dissipation rate (m^2/s^3)
ϕ	equilibrium contact angle (-)
γ	interfacial tension (N/m)
η	Kolmogoroff length scale (m)
η_i^c	collision efficiency (-)
η_i^{ic}	interception collision efficiency (-)
η_i^a	attachment efficiency (-)
η_i^d	detachment efficiency (-)
κ^{-1}	Debye decay length (m)
λ_i	decay length for hydrophobic force (m)
μ	dynamic viscosity (Pa.s)
ν	kinematic viscosity (m^2/s)
$\Delta\rho$	density difference between the dispersed phase and the continuous phase (kg/m^3)
ρ_c	density of the continuous phase (kg/m^3)
$\Delta\rho_{interface}$	density difference particle and displaced fluid phases (kg/m^3)
ζ	zeta-potential (V)

1. INTRODUCTION TO PARTICLE SEPARATION IN BIOTECHNOLOGY

Many products in biotechnological processes are included in particles, such as inclusion bodies, crystals, precipitates, cell organelles, viruses and virus-like particles. These particles are formed because of supersaturation of the liquid phase, improper folding of intracellular compounds, or as the result of a storage mechanism within a microorganism. Besides the product particles other particles are often present in the liquid phase. Examples are microbial cells, cell debris, by-products or immobilised catalysts (figure 1). Product purification therefore requires the separation of the product particles from the liquid phase and the other particles in the mixture.

There are two strategies for product purification when dealing with particulate products. First of all, the product particles can be separated from the other particles by physical separation, e.g. centrifugation and filtration. A second possibility is to solubilise the product after which it can be selectively recrystallised or extracted. In many cases the target particles contain the product already at high purity, making selective particle recovery through physical separation advantageous because it reduces the number of operations for product purification.

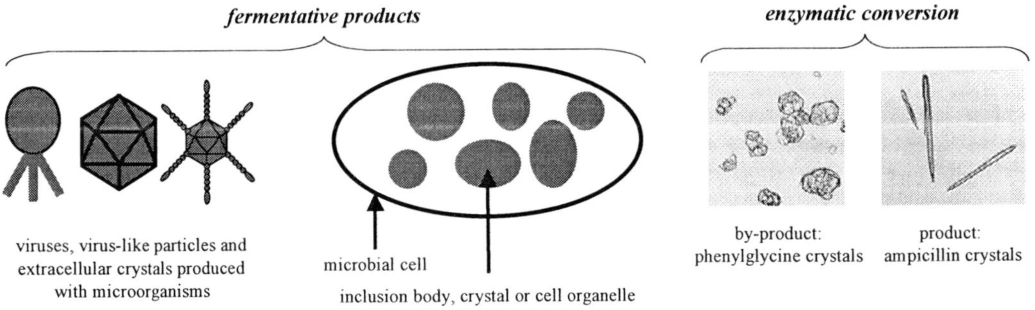

Figure 1. Particles in biotechnological processes.

Currently, centrifugation and filtration are the main physical separation techniques that are applied in biotechnology. Filtration is by no means ideal for bioparticle separation because the particle size distributions of the product particles and other particles overlap in most cases. Centrifugation on the other hand can deal with overlapping size distributions if the settling velocity differences in the centrifuge are large enough. However, most bioparticle mixtures contain small particles (0.05-100 μm) with densities (900-1540 kg/m^3) near the density of the often aqueous liquid phase. This results in low settling velocities and small settling velocity differences between the product particles and the other particles. Centrifugal separation is therefore inefficient for many separations (Van Hee et al., 2006a). An alternative for these physical separation techniques is therefore desirable.

Particles can adsorb to fluid interfaces. This phenomenon can be used for particle separation when the adsorption process is selective for one type of particle in a mixture. The adsorption process depends on many factors. First of all, the total energy of the system should get lower by particle adsorption (figure 2). The energy reduction per particle that is related to the adsorption process can be described with the following equation (Binks, 2002):

$$\Delta E_{ads} = -\pi/4 \cdot d_p^{\ 2} \cdot \gamma \cdot (1+\cos\phi)^2 \qquad \text{(equation 1)},$$

where ΔE_{ads} is the adsorption energy (J), d_p is the particle diameter (m), γ is the interfacial tension (N/m) and ϕ is the equilibrium contact angle between the particle-fluid interface and the fluid-fluid interface.

Particle adsorption does not necessarily have to take place when the interfacial energy is lowered by particle adsorption. Under certain conditions the particles may not be able to reach the fluid interface because of other interactions, e.g. colloidal interaction forces and hydrodynamics, that play a role. Additionally, when particles do adsorb at the interface they may detach when for instance the drag force on the particle exceeds the interfacial tension force. All these different aspects of the adsorption process could be employed for creating a selective particle adsorption process.

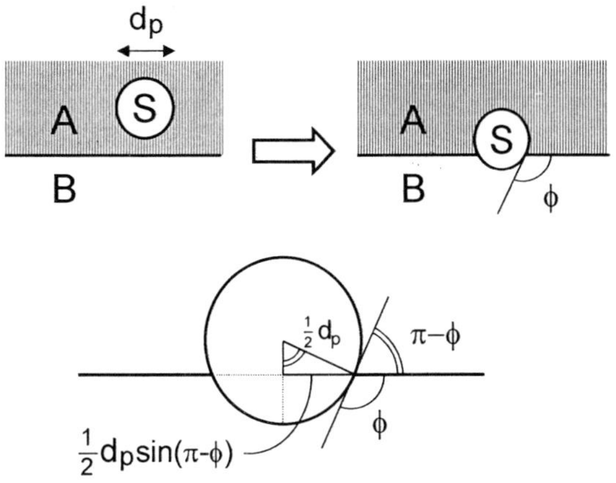

Figure 2. Schematic representation of particle adsorption at a fluid-fluid interface.

In biotechnology fluid interfaces have been applied for the recovery and/or separation of bioparticles and solutes from mixtures. Aqueous two-phase systems (Andrews, et al., 1995; Guereca, et al., 1994; Heywood-Waddington, et al., 1986; Kula, 1993; Walker and Lyddiatt, 1999) and other liquid-liquid systems (Borbas, et al., 2001; Dennison and Lovrien, 1997; Hoeben et al., 2004; Jauregi, et al., 2001; Jauregi, et al., 2002; Kiss, et al., 1998; Pike and Denisson, 1989; Tan and Lovrien, 1972) were used to capture the desired products at the interface between the two liquids. Andrews et al. (1995) showed that virus-like particles (0.06 μm) can be purified from a yeast cell homogenate (diameter not reported but in order of 1 μm) by interfacial partitioning in an aqueous two-phase system. The target particles were accumulated in an interphase, which is an emulsion layer that is located between the top and bottom phase. Walker and Lyddiatt (1999) reported on the purification of inclusion bodies (0.15 μm) from yeast cell homogenates (0.15 – 0.30 μm) with aqueous two-phase systems. In this particular case, the interphase contained mostly cell debris. More recently, Jauregi et al. (2001, 2002) and Hoeben et al. (2004, 2005) have shown that the ampicillin and phenylglycine crystal mixture (6 - 62 μm), that is formed during the semi-synthetic production of ampicillin (Bruggink et al., 1998), can be separated by selective adsorption of these crystals to various liquid-liquid interfaces, while this separation is impossible with centrifugation or filtration due to the small size and density differences of the crystals. In their procedure the particles were brought into contact with the liquid-liquid interface by mixing two immiscible liquids of which one contained the particle mixture. Hoeben et al. (2006) later showed that this separation is also feasible when the crystals and interface are contacted by injecting a dispersed liquid phase into the continuous phase that contains the ampicillin and phenylglycine crystals.

Besides liquid two-phase systems, air flotation has been applied for the recovery of whole cells (Bahr and Schügerl, 1992; De Dousa, et al., 2003; Gahr and Schügerl, 1992; Gaudin, 1957; Grieves and Wang, 1966; Grieves and Wang, 1967; Kalyuzhnyi, et al., 1965; Palmieri, et al., 1996; Sadowski and Golab, 1991; Tybussek, et al., 1994; Vlaski, et al., 1996; Wang, et al., 1994). This method makes use of the adsorption of particles and/or molecules to air bubbles that rise in the liquid phase due to buoyancy. In previous work we have shown that this method can be applied for the separation of polyhydroxyalkanoate inclusion bodies from ruptured fermentation broth (Van Hee et al., 2006b) and the separation of other bioparticle mixtures (Van Hee et al., 2006c). This chapter deals with the use of gas-liquid and liquid-liquid interfaces for bioparticle separation. Section 2 deals with bioparticles and their properties. Section 3 describes the principle of interfacial separation methods. The various mechanisms for creating selectivity in the separation process are discussed from a theoretical point of view in section 4 and on the basis of literature examples in section 5. Finally, process design is briefly discussed in section 6.

2. Bioparticles

The product particles in bioprocesses can be produced inside microbial cells or free in suspension. There is a wide variety of these bioparticles and often very little is known about the properties of these particles. We therefore only briefly discuss the origin of bioparticles and the properties that may have an influence on the selectivity in interfacial adsorption, such as contact angle and surface charge. For a more complete overview the reader should refer to one of our previous publications (Van Hee et al., 2006a).

2.1. Particle Mixtures

Intracellular Particulate Microbial Products

Non-excreted microbial products accumulate inside the microorganism and often cause the formation of crystals or inclusion bodies that have sizes in the range of 0.05-1.2 μm. Crystals are formed when the product is produced above its saturation concentration, such as in the case of β-carotene production by *Blakeslea trispora* (Jeaong, et al., 1999) and the production of xanthophylls by microorganisms or microalgae (Eonseon, et al., 2003). Inclusion bodies (IBs) are formed when the product is accumulated in a specific compartment inside the cells, such as with polyhydroxyalkanoates (Lee, et al., 1999), or when it aggregates inside the microorganism. This often happens with peptides and proteins that are folded incorrectly, for instance due to lacking or insufficient chaperones (Buchner, et al., 1992; Honda, et al., 2000; Wong, et al., 1997). Besides protein inclusion bodies, crystals and PHA granules, many other forms of inclusion bodies are found in nature (Van Hee, et al., 2006a).

Viruses and virus-like particles are also produced intracellularly. These particles typically have sizes between 20-200 nm. They often consist of an inactivated virus or a surface antigen of a virus that is produced by a genetically modified microorganism.

Physical separation of particulate intracellular microbial products from biomass requires product release through cell disintegration.

Product Particles Free in Suspension

Several cases have been reported where an excreted microbial product forms crystals in the range of 5-100 μm (Beards, et al., 1993; Buque-Taboada, et al., 2004; Michielsen, et al., 2000; Spassov, et al., 1996). For instance, both substrate and product are crystals in the conversion of Ca-maleate to Ca-D-malate by permeabilised *Pseudomonas pseudoalcaligenes* (Michielsen, et al., 2000).

Enzymatic reactions yielding solid products require particle-particle separation in subsequent process steps when there are other particles present. These other particles may be particulate biocatalysts, particulate by-products or crystals of unconverted enantiomer in case of conversion of a mixture of racemic crystals into chiral crystals (Straathof, et al., 1998). When an immobilised catalyst is the only other particle present in the mixture, separation may be very easy by careful selection of the catalyst carrier material with respect to size, density and structure, e.g. catalytic membranes (Kasche and Galunsky, 1995) and catalytic expanded bed media (Van der Wielen, et al., 1997). With other particle mixtures in enzymatic conversion processes physical particle separation may be a key step in the purification process.

2.2. Particle Properties

Surface Charge

Bioparticles have a surface charge density that is dependent on their surface chemistry and the composition of the liquid phase. The surfaces of many bioparticles contain carboxyl, amino and hydroxide groups. All of these groups can exchange H^+ with the surrounding medium, which causes a change in surface charge. The relation between pH, ionic strength and the zeta-potential of bioparticles, such as microbial cells and inclusion bodies, has been studied to a reasonable extent. The typical zeta-potential for bioparticles lies in the range of –100 mV to 50 mV (Van der Wal, et al., 1997; Yan, et al., 1992; Van Hee et al., 2006c).

Contact Angle

Contact angles of bioparticles depend on the properties of the two fluid phases at the interface. Theoretically this contact angle can be between 0° and 180° but in practice less extreme values are encountered. The air-water contact angle for bacterial cells ranges from 5° to 107° (Absolom, et al., 1986; Busscher, et al., 2000; Chattopadhyay, et al., 1995). The large variation in these contact angles is caused by variation of the surface properties of the bacterial cells, as well as the variation in composition of the aqueous phase.

3. Principle of Separation Methods

Separation of particles with the aid of fluid interfaces requires 1) the formation of interface, 2) contact between particles and interface, and 3) separation of particle-laden interface from the bulk liquids. Fluid interfaces can be created by emulsification of two immiscible fluid phases or by injection of a dispersed phase (gas or liquid) into a continuous phase. After the interface has been formed, the particles need to be brought into contact with it. In literature this process is often split up into three sub-processes. First an effective collision between a particle and the interface has to take place. This process is governed by Brownian motion and by hydrodynamics. Then, attachment of the particle to the interface can take place. Counteracting, when a particle-interface complex has been formed, the particle may detach from the interface.

This section focuses on methods for creating fluid interfaces and separation of the particle-laden interface from the bulk liquids that contain the remaining particles.

3.1. Emulsification

Figure 3 shows a schematic representation of an emulsification process that can be applied for particle separation. The first step in the process is the introduction of a particle mixture and two immiscible liquid phases to a container. In most cases the particles in biotechnological processes are suspended in water. The particle suspension will then serve as one of the liquid phases. The second step in the process is to mix the two immiscible liquid phases to form an emulsion. The phase in which the stirrer is placed will in many cases become the continuous phase unless when phase inversion occurs, as is described by Kralchevsky et al. (2005). The droplet size that is created during mixing can be approximated with the following equation for the turbulent inertial regime (Tcholakova et al., 2004):

$$d_d \approx \hat{\varepsilon}^{-2/5} \cdot \gamma^{3/5} \cdot \rho_c^{-1/5} \qquad \text{(equation 2)},$$

where $\hat{\varepsilon}$ is the mechanical energy input per time and unit volume (J/m^{3}s), d_d is the droplet diameter (m) and ρ_c is the density of the continuous phase (kg/m^3). This equation does not take into account the impact of the particles on droplet size. After mixing is stopped the emulsion will cream or sediment. If this proceeds together with coalescence a layer of top and bottom phase will appear. The particles that are present at the interface will form a steric barrier that (partly) prevents coalescence. Hence a particle-rich stable emulsion layer will remain. This emulsion layer is called the interphase, referring to its position in between the two liquid phase layers. The particles that do not end up in the interphase layer will be present in the top and/or bottom phase. Separation of different types of particles from mixtures is thus possible when the composition of the particle mixture in the interphase differs from the particle mixture in the bottom and/or top liquid.

The sub-processes of collision, attachment and detachment have been briefly described at the start of this section. In the emulsification process we assume that, when Brownian motion does not play a significant role, partitioning of the particles from the bulk liquids to the interface occurs only during the agitation step. During the stirring process the particles collide with the interface and depending on their interactions with the interface they may attach. When stirring is stopped the emulsion will cream during which coalescence and particle detachment can take place. The coalescence process is strongly dependent on the stabilising effect of the particles on the emulsion droplets, as the result of Pickering stabilisation (Pickering, 1907). Pickering stabilisation can lead to very stable emulsions with micrometer-sized drops. To apply an emulsification process for particle separation very small drop sizes and strong emulsion stabilisation are undesirable in view of long sedimentation/creaming times and difficulties in breaking the emulsion in a later stage.

The interphase will often contain particles that are not adsorbed at the interface. These particles are either entrained in the continuous phase of the emulsion or they are included in the emulsion droplets. The first effect is undesired when the entrainment process is unselective. The latter effect is strongly dependent on the affinity of the particles for the dispersed phase.

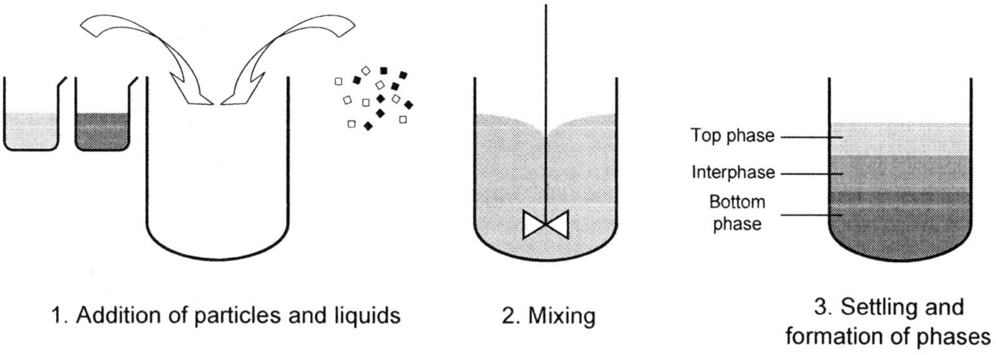

Figure 3. General method for the use of emulsification for particle separation.

The use of the interfacial tension force for complete particle-particle separation in a single stage requires one type of particle to adsorb at the interface while the other particles remain in the top or bottom phase (and vice versa). When this is accomplished the adsorbed particles can be selectively removed from the system by separating the interphase from the top and bottom phase. In many systems the process is less selective causing the interphase to contain both product particles and other particles. Single stage purification is then not sufficient to obtain the required product purity. In this case multistage interfacial partitioning can be applied. In figure 4 an example is given of a multistage interfacial partitioning process in which the interphase is enriched with product particles and the bottom phase contains the remaining particles. The interphase is moved counter-current to the bottom phase. In this fashion the purified product can be collected from the first stage (left) and the waste at the last stage (right). A reflux of particles (V_{N+1} and L_0) can be introduced to achieve higher purity, but the impact refluxes strongly depends on the partitioning mechanism of the particles.

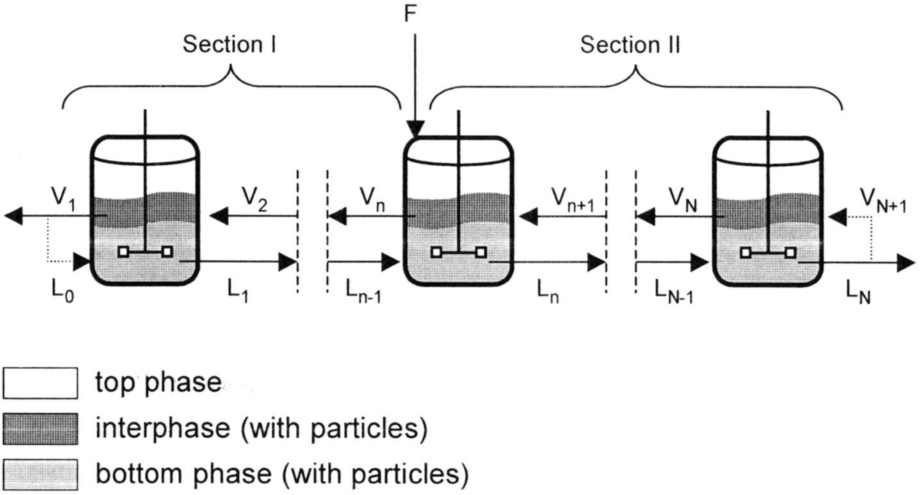

Figure 4. Multistage counter-current interfacial partitioning process.

In this section we only discussed the use of two immiscible liquid phases. In theory this approach should also work with gas-liquid interfaces. Nevertheless, injection of a gas phase, as is discussed in the next section, may then be more convenient.

3.2. Dispersed Phase Injection

Droplets and bubbles can be created by injecting a gas or an immiscible liquid into a particle suspension (figure 5). Separation of particle mixtures with this process has many similarities to the emulsification process. In both processes transport of particles to the interface is determined by the relative movement of the particles and the dispersed phase. In absence of mixing the transport process is a function of density differences, the viscosities of the fluid phases, interface coverage and mobility, droplet/bubble size, and particle size. This will be the case for dispersed phase injection when the system is not stirred.

Particles in biotechnological processes are in most cases suspended in water. Injection of a gas phase will therefore cause the dispersed phase to rise through the particle suspension. The same is true for the use of many organic solvents in an aqueous medium. Downward movement of the dispersed phase will only occur when a denser liquid phase is injected. This can be accomplished by working with aqueous two-phase systems.

The dispersed phase droplets/bubbles that can be created have sizes down to 100 nm depending on the interfacial tension and the injection system that is used. Examples of methods for creating small gas bubbles (10-150 μm) are electroflotation and dissolved-air flotation (Burns et al., 1997). Electroflotation makes use of electrodes to split water into hydrogen gas and oxygen gas. In dissolved-air flotation air-oversaturated water is introduced into the separation device for the formation of air bubbles. A pressure drop over pores in a frit can also be used to produce bubbles/droplets. Nguyen and co-workers (2006) applied this method to produce gas bubbles of approximately 150 μm with a glass frit.

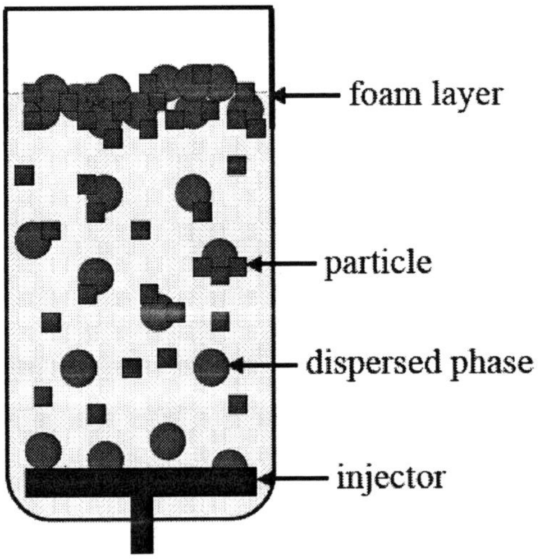

Figure 5. Interfacial separation by injection of a dispersed phase.

At low pressure drop the bubble/drop size is related to the interfacial tension, the pore diameter and the buoyancy force that acts on the dispersed phase. The following equation gives an approximation of the bubble/droplet volume (V_d) as a function of the pore radius (R_{pore}):

$$V_d = \frac{2 \cdot \pi \cdot R_{pore} \cdot \gamma}{\Delta\rho \cdot g} \qquad \text{(equation 3),}$$

where $\Delta\rho$ is the density difference between the dispersed phase and the continuous phase (kg/m^3), and g is the gravitational acceleration (m/s^2). When a significant pressure drop over pores is used to create bubbles/droplets, the bubble/drop size will be directly related to the pore size, provided that the flow of dispersed phase is high enough. In case the dispersed phase inertia are larger than the capillary forces, a jet of dispersed phase is formed that breaks up into bubbles/drops as a result of Rayleigh instability. It can be shown that in that case (Clift et al., 1978):

$$R_d = 1.89 \cdot R_{pore} \qquad \text{(equation 4),}$$

where R_d and R_{pore} are the radii of the drops/bubbles and pores, respectively. This equation applies in case v_{inj}^2 is larger than approximately $2 \cdot \gamma / (\rho_d \cdot R_{pore})$, where v_{inj} is the injection velocity of the dispersed phase.

The particle collection process for dispersed phase injection is different in comparison to the emulsification process. The dispersed phase rises through the continuous phase and at the surface of the continuous phase the bubbles/droplets coalesce and/or form a stable foam/emulsion layer. When coalescence takes place particles may detach from the interface. This leads to inefficient use of the interface. On the other hand, coalescence at the surface may have a positive effect because a "natural" particle reflux is generated. The released particles return to the bulk liquid where they again encounter dispersed phase droplets/bubbles that rise through the continuous phase. These particles thus get another chance to adsorb at the interface and to enter the foam/emulsion layer. This "natural" reflux can enhance the selectivity of certain processes depending on the particle adsorption mechanism. When the selectivity of the separator is insufficient a multistage process can be applied similar as in figure 4, but in this case with dispersed phase injection.

The continuous phase that is present in the foam/emulsion can contain particles. This liquid is transported along with the dispersed phase and may therefore cause non-selective particle transport. Thereby it reduces the overall selectivity of the process. Entrainment of particles in this fashion should thus be minimised. A liquid flow with the same composition as the continuous phase can be introduced counter-current to the flow of the dispersed phase. In this fashion the particles that have a tendency for entrainment are washed out of the foam/emulsion layer (Yoon et al., 1998; Ata et al., 2002).

In interfacial separation regeneration of the fluid phases is easier when a dispersed gas phase is used instead of a dispersed liquid phase. Dissolved biomolecules that are present in the particle suspension will not partition to the gas phase as the result of their low volatility at moderate temperatures, whereas in liquid-liquid systems partitioning will occur. Additionally, in liquid-liquid systems the particulate product will partly dissolve in the second liquid phase that is introduced. This causes product loss and/or requires an additional recovery process. The use of gas-liquid interfaces is thus advantageous.

4. Particle-Interface Interactions

The performance of interfacial separation processes is often expressed by a collection rate constant that is the first order rate constant, k_i^{eff}, for the change of particle concentration in time.

$$\frac{dn_i}{dt} = -k_i^{eff} \cdot n_i \qquad \text{(equation 5)},$$

where n_i is the concentration of particle type i in the suspension (#/m^3), t is time (s) and k_i^{eff} is the effective collection rate constant of particle type i (1/s). Separation of one type of particle from a particle mixture requires that these particles adsorb more efficiently to the interface than the other particles, such that enrichment takes place at the interface (or visa versa). Thus, particle separation by interfacial separation requires a difference in collection rate constants for the particles that need to be separated.

In literature, interfacial separation processes are generally described using three sub-processes. First an effective collision between a particle and the interface has to take place (section 4.1). This process is governed by Brownian motion and/or hydrodynamics. Then, if there is a net attractive force between the particle and the interface, attachment can take place (section 4.2). Counteracting, when a particle-interface complex has been formed it may be broken up again (section 4.3). The contributions of these sub-processes are included in k_i^{eff} according to equation 6.

$$k_i^{eff} = \frac{3 \cdot v_d \cdot \left(\eta_i^c \cdot \eta_i^a \cdot \left(1 - \eta_i^d\right)\right)}{2 \cdot d_d} \qquad \text{(equation 6)},$$

where v_d is the superficial velocity of the dispersed phase droplets/bubbles (m/s), d_d is the diameter of the droplets/bubbles of the dispersed phase (m), η_i^c is the collision efficiency (-), η_i^a is the attachment efficiency (-) and η_i^d is the detachment efficiency (-). Besides these sub-processes, particle aggregation may also play an important role because particle/aggregate size strongly influences the hydrodynamics in the sub-processes (especially collision).

In the following sections (4.1-4.3) the underlying theory for each of the sub-processes is presented.

4.1. Collision

Yoon (2000) and Dai et al. (2000) have reviewed various collision models for air flotation systems. They showed that comparison of these models is difficult because of the different assumptions (mobile/rigid bubble surface, collision mechanism, inertial forces, bubble size range) and hydrodynamic conditions (Stokes flow, potential flow, intermediate flow, turbulent flow) on which the models are based. Nevertheless, there are many similarities between the trends of the models with respect to bubble size and particle size. Generally, an increase of the bubble size will result in a lower collision efficiency and an increase of particle size for particles between approximately 1 μm and 100 μm results in an increase of the collision efficiency. In this section we will briefly discuss the collision process in turbulent and quiescent systems. The overview of these processes is incomplete, but it gives sufficient theoretical background to understand how particle separation can be accomplished on the basis of a selective particle-interface collision process.

Interface Mobility

In biological systems surfactants, such as proteins and lipids, are often present. These compounds adsorb onto the bubble surface, which becomes rigid when the pressure resulting from the surface tension gradient along the surface counterbalances or exceeds the shear stress. It is expected that this is the case in many biological systems due to the high concentration of surface-active compounds.

Brownian Motion

Brownian motion can play an important role in the collision process for particle sizes below 1 μm. Ramirez et al. (1999) performed calculations to determine what conditions, i.e. the Péclet number of dispersed phase droplets/bubbles and particle size, are required for the collision process to be dominated by 1) hydrodynamics, 2) Brownian motion, or 3) a combination of Brownian motion and hydrodynamics. They showed that the particle collection efficiency will not be greater than 10% when convective transport dominates particle-bubble collision, while efficiencies up to 100% can be observed when Brownian motion dominates.

Turbulent Conditions

In turbulent systems, energy is being transferred from large eddies to smaller ones up to the smallest-scale eddies in which energy is converted into heat by viscous dissipation. According to Kolmogoroff's theory on local isotropy, the small energy dissipating eddies are statistically independent of the main flow pattern. These statistically independent fluctuations of the velocity components force particles in a turbulent flow field to make random movements with respect to each other, which leads to particle-particle collisions. These random movements are of course influenced by hydrodynamic particle-fluid interactions, such as particle inertia (Caulet et al., 1996).

In Kolmogoroff's theory, the turbulent flow field is characterised by two parameters, namely the kinematic viscosity ν and the local rate of energy dissipation in the fluid ε. Together these define the Kolmogoroff length scale η, which is the scale of the small energy dissipating eddies:

$$\eta = (\nu^3 / \varepsilon)^{1/4} \qquad \text{(equation 7)}$$

In literature, several expressions can be found for particle-particle collision rates that are based on expressions for turbulent velocity fluctuations. For particles in the inertial sub-range of turbulence (i.e. $\eta << d_1, d_2 << L$, where L is the scale of the large energy-containing eddies), the collision rate can be shown to be (Podgorska and Baldyga, 2001):

$$Z_{12} = \sqrt{\frac{8\pi}{3}} \cdot \langle \varepsilon \rangle^{1/3} \cdot (0.5 \cdot (d_1 + d_2))^{7/3} (0.5 \cdot (d_1 + d_2) / L)^{0.027} \cdot N_1 \cdot N_2 \qquad \text{(equation 8)}$$

where Z_{12} is the collision rate of particles 1 and 2 ($m^{-3}\,s^{-1}$), $<\varepsilon>$ is the local average energy dissipation rate (m^2/s^3), d is the particle diameter (m) and N is the particle concentration ($\#/m^3$). There are many other expression for turbulent particle-particle collisions that can be found in literature, each having their own assumptions. For example, the model of Kruis and Kusters (1997) describes the collision between two particles with the same density for a large range of hydrodynamic conditions taking into account shear and inertial effects. This model will, however, not be useful for modelling particle-bubble and particle-droplet collisions because the particle density and dispersed phase density differ in most cases. A less accurate model, which can be used for modelling turbulent collisions between objects with different densities, was presented by Abrahamson (1975). It is thus important to select the proper model for describing the collisions for the system of interest.

Quiescent Conditions

In quiescent systems the collision efficiency is dictated by the hydrodynamic flow pattern around the dispersed phase droplets/bubbles. Yoon and Luttrell (1989) derived an empirical stream function for air flotation systems at intermediate Reynolds numbers assuming a rigid bubble surface. They used this relation to obtain an expression for interception particle-bubble collision in case the inertial force is negligible (equation 9).

$$\eta_i^{ic} = \left[\frac{3}{2} + \frac{4 \cdot \mathrm{Re}_b^{\,0.72}}{15} \right] \cdot \left(\frac{d_{p,i}}{d_d} \right)^2 \qquad \text{(equation 9)}$$

where Re_b is the bubble Reynolds number. Similar expression can be derived for liquid-liquid systems. Equation 9 does not take into account contributions of particle gravity and Brownian motion. Expressions for the effect of gravity on particle-bubble collision can be found in Weber and Paddock (1983) and Schulze (1989). Brownian motion will only have a significant impact on the collision efficiency for particle sizes below approximately 1 μm.

Nguyen et al. (2006) included the effect of Brownian motion in a particle trajectory analysis model that could predict the collision efficiency between particles and air bubbles accurately. They showed that Brownian motion enhances the collision efficiency for particle sizes below approximately 20 nm to 100 nm depending on the magnitude of the colloidal interaction forces (refer to section 4.2). Figure 6 gives a qualitative relation between particle size and collision efficiency at constant bubble size. The figure is drawn on the basis of experimental data of Nguyen et al. (2006). It is expected that a similar qualitative relation can be found for the collision between particles and droplets. Figure 6 shows that Brownian motion dominates the collision process in region A, while in region B hydrodynamics is the determining factor. The crossover point in between the two regimes can move to smaller or larger particle sizes when a different dispersed phase or continuous phase is chosen. This is due to the change in velocity of the dispersed phase droplets/bubbles that is caused by the change in $\Delta\rho$ and the liquid viscosities.

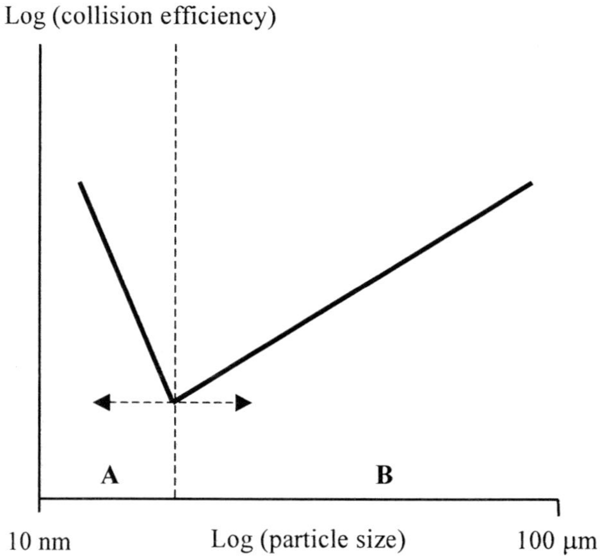

Figure 6. Collision efficiency between particles and gas bubbles under quiescent conditions.

4.2. Attachment

When a particle and bubble/droplet collide, they are at first separated by a liquid film. This liquid film needs to be drained for attachment (formation of three-phase contact line) to take place. Different film thinning models are known for different conditions with respect to geometry and interface rigidity. Most of these models are based on the assumption that the volume flux in axial direction between the particle and bubble/droplet has to be equal to the volume flux in radial direction. This way the approach velocity of the particle and the bubble/droplet is related to the radial liquid velocity in the film. Liquid film thinning requires a net force that causes particle motion in the direction of the bubble/droplet. The attachment process therefore depends on hydrodynamics and interaction forces, such as Van der Waals interaction, electrostatic interaction and hydrophobic interaction. The attachment efficiency can be expressed on the basis of these forces according the following equation (Yoon and Mao, 1996):

$$\eta_i^a = \exp\left(-\frac{E_B}{E_{k-A}}\right) \qquad \text{(equation 10)}$$

$E_{k\text{-}A}$ is the kinetic energy of attachment (J), which is dependent on the hydrodynamic conditions of the system. There are many expressions for this energy because of the variety in hydrodynamic conditions, e.g. quiescent conditions and turbulent conditions. For turbulent conditions $E_{k\text{-}A}$ can be determined from the turbulent kinetic energy of the eddies with a length scale in the order of the particle and bubble/droplet size (Gore and Crowe, 1991). For quiescent systems $E_{k\text{-}A}$ is dependent on the relative velocity of the bubbles/droplets and particles. These velocities are dependent on the densities of all phases, inertial forces and viscous forces.

E_B is the energy barrier between the particle and the interface that is caused by colloidal interaction forces and the resistance to liquid film drainage. Expressions for these forces are given below. On the basis of these forces (or energies) E_B can be obtained by determining the maximum repulsive energy from the energy-distance curve.

The colloidal interaction forces, and thus E_B, can be controlled by addition of collectors and depressants. Both agents modify the colloidal interaction forces by adsorbing onto the particle surface. Collectors enhance the attachment efficiency, while depressants reduce the attachment efficiency (Schulze, 1989). Collectors can be classified into ionising and non-ionising compounds. The non-ionising collectors, such as hydrocarbons, simply render the particles hydrophobic by covering them with a thin film. The ionising collectors consist of molecules with a polar group and a hydrophobic (carbon) tail, the latter of which causes the increase in hydrophobicity.

The attachment efficiency as defined by equation 10 only takes into account the energy barrier between the particle and the interface. The available information on the colloidal interaction forces and the liquid film drainage process in relation to the particle-interface distance is thus not completely used in this approach. Particle trajectory analysis calculations do make full use of the available information. Such calculations therefore give a more accurate approximation of the attachment efficiency, but are also more laborious (Okada et al., 1990b, Nguyen et al., 2006).

Liquid Film Drainage

Attachment of particles to bubbles/droplets is preceded by drainage of the liquid film between them. The drainage process is to sustain up to a point where a critical film thickness is reached at which spontaneous film rupture occurs. The critical film thickness depends on the colloidal interaction forces (see subsequent sections) and is typically in the order of 10 nm.

Liquid film drainage forms a barrier against particle attachment and gains importance at small separation distances. The drainage process is one of viscous flow and the contribution to the energy barrier E_B is therefore expressed as total energy transferred into heat by viscous dissipation up to the critical film thickness.

The force associated with drainage of the liquid film between particles and bubbles/droplets depends on the mobility and deformation of the fluid-fluid interface. As mentioned earlier, in biological systems the presence of surface-active material will generally lead to rigid fluid-fluid interfaces. This results in an increased resistance to film drainage. In case the bubble/drop does not deform, the force will then be given by the Taylor equation, valid for (rigid) spherical particles (Tobin and Ramkrishna, 1999):

$$F_d = \frac{3\pi\mu}{2h}\frac{1}{(\frac{1}{d_1}+\frac{1}{d_2})^2}\frac{dh}{dt} \qquad \text{(equation 11)},$$

where μ is the viscosity of the continuous phase (Pa·s), h is the separation distance (m) and dh/dt is the rate of approach of bubble/drop and particle (m/s).

When the pressure in the film starts to exceed that inside the drop/bubble, deformation will occur. The drop/bubble will then flatten or dent which slows down the drainage process. This effect will only occur at distances below approximately $F_d/2\pi\gamma$ and is generally of minor importance for small particles.

Van Der Waals Interaction

The magnitude of the Van der Waals force is related to the properties of the interacting phases as is shown by the following equation (Schulze, 1984):

$$F_{vdW} = -\frac{A_{132}}{12\cdot h^2}\cdot\left(\frac{d_1\cdot d_2}{d_1+d_2}\right)$$

$$\text{if} \quad \left(\frac{d_1\cdot d_2}{d_1+d_2}\right) >> h \qquad \text{(equation 12)},$$

where A_{132} is the Hamaker constant for the interaction between phase 1 and 2 across phase 3 (J), d is the diameter of object 1 or 2 (m) and h is the separation distance (m). This equation describes the non-retarded interaction that is active at short range. At large distances the force will be related to h^{-3} instead of h^{-2} due to retardation. The Van der Waals force assumes a negative value when two objects attract one another.

Electrostatic Interaction

The electrical double layers of charged objects overlap when the distance between their surfaces is in the order of the Debye decay length (Schulze, 1984):

$$\kappa = \sqrt{\frac{e^2\cdot N_{av}}{\varepsilon_p\cdot k\cdot T}\cdot\sum_j(C_j\cdot z_j{}^2)} \qquad \text{(equation 13)},$$

where κ^{-1} is the Debye decay length (m), e the elementary charge (C), N_{av} the Avogadro number (mol^{-1}), ε_p the dielectric permittivity of the fluid (F/m), k the Boltzmann constant (J/K), T the temperature (K), C_j concentration of ionic species j (mol/m^3) and z_j the ion valency of ionic species j (-). This double layer overlap results in an attractive (negative) or repulsive (positive) force that is called the electrostatic interaction force. For two spherical objects with zeta-potentials between -60 mV and 60 mV the electrostatic interaction force can be approximated with the following equation (Hogg et al., 1965):

$$F_E = 2 \cdot \pi \cdot \varepsilon \cdot \frac{d_1 \cdot d_2}{d_1 + d_2} \cdot \kappa \cdot \zeta_1 \cdot \zeta_2 \cdot \left[\frac{e^{-\kappa \cdot h}}{1 + e^{-\kappa \cdot h}} - \frac{(\zeta_1 - \zeta_2)^2 \cdot e^{-2 \cdot \kappa \cdot h}}{2 \cdot \zeta_1 \cdot \zeta_2 \cdot (1 - e^{-2 \cdot \kappa \cdot h})} \right]$$

$$\text{if} \left(\frac{d_1 \cdot d_2}{d_1 + d_2} \right) >> h \text{ and } \kappa \cdot d_1 >> 1 \qquad \text{(equation 14),}$$

where ζ is the zeta-potential of object 1 and 2 (V), respectively and object 1 is the smallest object. At more extreme zeta-potentials the Debye-Hückel approximation and Derjaguin's method for deriving the energy of interaction between spherical particles from that between flat plates are not valid (Hogg et al., 1965). Under these conditions other relations should be applied.

Hydrophobic Interaction

Many researchers have shown that the DLVO forces are not sufficient to describe the interaction force between two hydrophobic objects, i.e. surface-surface, particle-surface, particle-bubble, etc. The discrepancy between force measurements and the DLVO theory was ascribed to a non-DLVO force, which is often referred to as the hydrophobic force. The exact origin of this non-DLVO force is still unexplained but many mechanisms have been proposed (Pazhianur and Yoon, 2003). Examples of these mechanisms are 1) the correlation of dipoles due to domain formation of an adsorbed surfactant film (Rabinovich et al., 1991), 2) non-uniform charge distribution on net neutral charged surfaces (Miklavic et al., 1994), 3) nucleation of air bubbles on the surface (Carambassis et al., 1998), and 4) structural effects due to ordering of water molecules (Eriksson et al., 1989). In this work we will refer to this additional force as the hydrophobic force, as proposed by Pazhianur and Yoon (2003) and others, even though the effects that are included in this force may suggest a different name.

Two expressions have been proposed for the hydrophobic force (equations 15 and 16). Both expressions can be used to correct the DLVO theory for the additional force that has been observed in force measurements.

$$F_H = -\frac{K_{132}}{12 \cdot h^2} \cdot \left(\frac{d_1 \cdot d_2}{d_1 + d_2} \right) \qquad \text{(equation 15),}$$

$$F_H = \left(\frac{d_1 \cdot d_2}{d_1 + d_2} \right) \cdot \left[K_1 \cdot \exp(-h/\lambda_1) + K_2 \cdot \exp(-h/\lambda_2) \right] \qquad \text{(equation 16),}$$

where K_{132} (J) is a similar type of constant as the Hamaker constant A_{132} (J) describing the magnitude of the hydrophobic interaction between the interacting phases, K_1 and K_2 are force constants (N/m), and λ_1 and λ_2 are decay lengths (m).

Yoon and Mao (1996) have shown that K_{132} can be calculated in a similar way as A_{132} and that it is related to the interactions between the different phases according to:

$$K_{132} = \sqrt{K_{131} \cdot K_{232}} \qquad \text{(equation 17).}$$

Yoon and Aksoy (1999) measured K_{232} for the interaction between two air bubbles across water that contained dodecylammonium hydrochloride, a cationic surfactant. They found the following empirical relation between the surfactant concentration and K_{232}:

$$K_{232} = \exp\left(-39.67 - 117.7 \cdot \sqrt{C_{\text{surfactant}}}\right) \quad \text{(equation 18)},$$

where $C_{\text{surfactant}}$ is the surfactant concentration (mol/L). Yoon (2000) presented a relation between K_{131} and the contact angle of particles at the air-water interface. He made use of force measurements from many different researchers. The empirical relation that he found was:

$$\begin{aligned} &\text{for } \phi < 87^\circ && K_{131} = 2.73 \cdot 10^{-21} \cdot \exp(0.0414 \cdot \phi) \\ &\text{for } 87^\circ < \phi < 92^\circ && K_{131} = 4.89 \cdot 10^{-44} \cdot \exp(0.644 \cdot \phi) \\ &\text{for } \phi > 92^\circ && K_{131} = 6.33 \cdot 10^{-27} \cdot \exp(0.217 \cdot \phi) \end{aligned}$$

This empirical relation is based on measurements that were performed at low ionic strengths ($<10^{-2}$ mol/m^3) in absence of surfactants. It can be used to approximate the hydrophobic force between particles and the air-water interface when the contact angle is known. According to our knowledge such an empirical relation has not yet been found for other fluid-fluid systems. Nevertheless, we expect that a similar approach can be taken to obtain these relations.

4.3. Detachment

There are two mechanisms that can cause particle detachment. First of all, detachment can occur when the sum of the detachment forces, such as gravity and shear, exceed the adhesive force between the particle and the interface (mechanism 1). The second detachment process is the result of bubble/droplet coalescence, which causes the available interfacial area to decrease (mechanism 2). In this case detachment will occur when the available interfacial area is too small for all particles to remain attached. Competition between the particles for the interface will then force some particles out of the interface.

The adhesive force of the particles at the interface is dominated by the capillary force (Nguyen, 2003):

$$F_c = \pi \cdot \gamma \cdot d_p \cdot \sin\alpha \cdot \sin(\phi - \alpha) \cong \pi \cdot \gamma \cdot d_p \cdot (1 - \cos\phi) \quad \text{(equation 19)},$$

where γ is the interfacial tension (N/m), α is the angle between the three-phase contact line and the particle axis, and ϕ is the contact angle (figure 7). The approximation of the capillary force that is given in equation 19 is obtained from the derivative of equation 1 to the distance a particle needs to travel for its release from the interface. It assumes a flat bubble surface and does account for deformation of the interface during particle detachment.

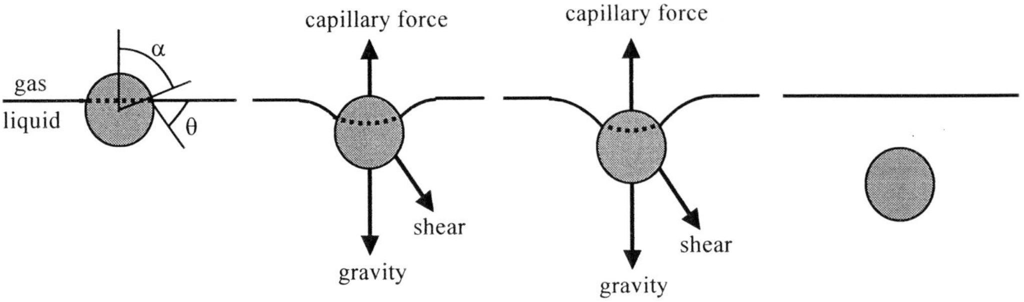

Figure 7. Schematic representation of the detachment process of a particle from a fluid interface under influence of gravity and shear.

The detachment efficiency according to mechanism 1 can be expressed with the following equation (Yoon and Mao, 1996):

$$\eta_i^d = \exp\left(-\frac{E_C + \Delta E_{ads}}{E_{k-D}}\right) \quad \text{(equation 20),}$$

where E_{k-D} is the kinetic energy of detachment (J), E_C is the colloidal force barrier (J) and ΔE_{ads} is the work of adhesion (J). E_{k-D} is dependent on the hydrodynamic conditions of the system (can be calculated in a similar fashion as E_{k-A}) and other forces that may cause detachment, such as the gravitational force and the shear force. An expression for ΔE_{ads} was given in equation 1 and E_C was discussed in section 4.2. There are two major limitations to the approach that is used in equation 20. First of all, the impact of the colloidal interaction forces on the detachment process is described by the energy barrier. This is a similar limitation as for the approach that was taken for the attachment efficiency in equation 10. Secondly, the interfacial tension force is approximated with equation 19.

Particles will remain adsorbed at the interface when the interfacial tension force exceeds the detachment forces. When we assume that gravity is the only force that counteracts the interfacial tension force and that colloidal interaction forces are negligible, a relation between particle size and particle detachment in quiescent systems can be obtained. In the equation below an expression for the gravitational force on a particle at the interface is given.

$$F_g = \frac{\pi}{6} \cdot d_p^{\,3} \cdot \Delta\rho_{\text{interface}} \cdot g \quad \text{(equation 21),}$$

where $\Delta\rho_{\text{interface}}$ is the difference between the particle density and the average density of the fluid phases that are displaced by the particle. Figure 8 presents the relation between contact angle, interfacial tension and the maximum size of particles that remain adsorbed at the interface under gravity in quiescent systems. For the construction of this graph the particle density and fluid phase densities were assumed to be 1540 and 1000 kg/m^3 respectively, and the contact angles were assumed to be in the range that was reported in section 2.2. The graph shows that particles with sizes up to approximately 0.01 meter can remain attached at gas-liquid interfaces when the contact angle is 107°. This implies that interfacial particle separation can be applied to a wide range of particle sizes. It is important to note that the contact angle depends on two-phase system that is used.

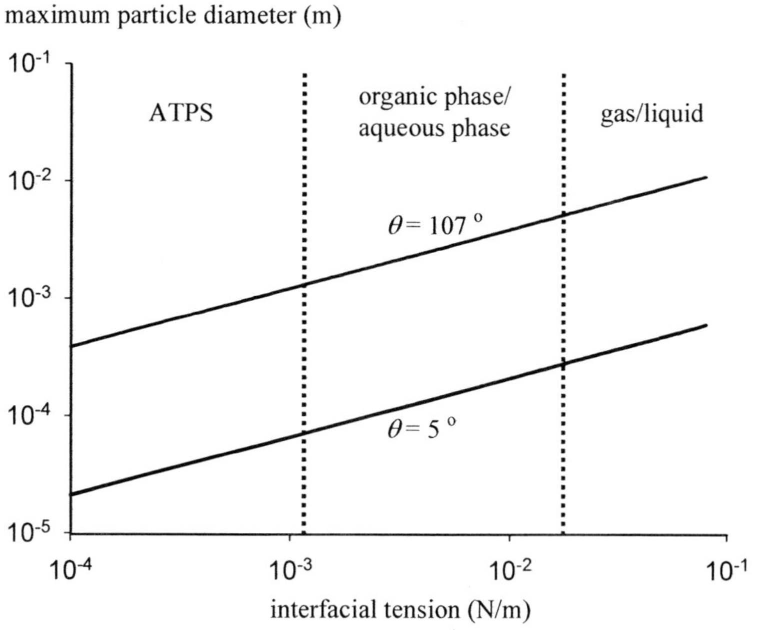

Figure 8. Relation between contact angle, interfacial tension and the maximum particle size that remains adsorbed at the interface under gravity in quiescent systems. The particle density and fluid phase densities are assumed to be 1540 and 1000 kg/m^3, respectively. In the graph approximate ranges of the interfacial tension of different types of fluid-fluid systems are given.

5. Mechanisms for Interfacial Separation of Particles

Selective enrichment of a specific class of particles requires selectivity in the sub-processes of collision, attachment and/or detachment. Complete separation can be accomplished:

1) by creating conditions that allow one type of particle to attach to the interface while the other particles in the mixture cannot (and vice versa), or
2) by allowing all particles to attach to the interface and creating conditions such that only one type of particle remains attached to (or detach from) the interface while all other particles detach from (or remain attached to) the interface.

In many cases the behaviour of the particle mixture will lie in between these two extremes. Separation of a specific class of particles from the particle mixture will then require multiple interfacial separation steps in order to obtain complete separation, as was indicated in section 3. This section deals with the selectivity of interfacial separation processes as the result of selective collision, selective attachment, and selective detachment. Experimental cases from literature are used to illustrate that selectivity can be created on the basis of each of these mechanisms. In addition a special case is discussed in section 5.5 where particles partition from one fluid phase to the other.

5.1. Selective Collision

In section 4.1 the theoretical background on the collision process was presented. Clearly, there are many factors that influence the collision process, which therefore strongly depends on the design of the separator. Two general process designs have been presented in this chapter: 1) emulsification and 2) dispersed phase injection. In this section examples from literature are presented to show that selective enrichment of a specific class of particles can be accomplished by selective collision. No clear examples of selective collision have been reported for the emulsification process. This may be due to the lack of experimental work on particle separation by emulsification. We will therefore focus on examples of dispersed phase injection.

Numerous cases of selective collision can be found in literature. Most of the reports deal with the impact of particle size and bubble/droplet size on the collision process for dispersed phase injection. Yoon (2000) has shown that the collision efficiency between air bubbles and spherical particles with a diameter of 11.4 μm decreases as the bubble size increases from 0.05 mm to 0.5 mm. Air flotation experiments with hydrophobic quartz particles of 5 to 60 μm have shown that the collision efficiency increases with particle size for bubble sizes of 0.77 and 1.52 mm. The relation between bubble size and collision efficiency was in this case qualitatively similar as in the previous case (Dai et al., 2000). Collins and Jameson (1976) showed similar trends for flotation of polystyrene particles with sizes between 4 and 20 μm and bubble sizes of 53 μm. Our research on dissolved-air flotation of protein-coated polystyrene particles (Van Hee et al., 2006c) has shown that an increase of the aggregate size of 0.9 μm protein-coated particles results in an increase of the collision efficiency. Separation of these protein-coated particles on the basis of differences in their aggregation behaviour, and therefore in effective particle size, proved to be feasible for particles with similar interactions with the air-water interface.

All cases that were presented thus far deal with particle sizes above 0.9 μm. For smaller particles Brownian motion will have an influence on the collision efficiency (section 4.1). Nguyen et al. (2006) have recently investigated the impact of particle size on the collision efficiency for particle sizes between 35 nm and 4 μm and bubble sizes between 100 and 250 μm. Their experiments showed that for particle and bubble zeta-potentials of –30 mV and -35 mV respectively, the collection efficiency increases with an increase in particle size for particles above approximately 80 nm. For particles smaller than 80 nm the collection efficiency increased with a decrease in particle size as the result of Brownian motion. The particle size at which Brownian motion starts to dominate the collision process is not easily determined from the measurement by Nguyen et al. (2006) because it is dependent on the balance between the colloidal interaction forces and Brownian motion.

The experimental examples that have been presented in this section show that selective collision can be used for separation of particles on the basis of size differences alone. Differences in aggregation behaviour of particles with similar sizes and similar interactions with the interface can therefore be applied for their separation.

5.2. Selective Attachment

Another method to obtain interfacial particle separation is by creating conditions that make attachment of a specific class of particles to the interface more (or less) favourable than for all other particles in the mixture. When this is accomplished particle separation is possible even when the collision efficiency is the same for all particles. Parameters that influence the attachment process are the colloidal interaction forces between the particles and the dispersed phase, the contact angle of the particles, and particle shape because it influences liquid film drainage.

Collins and Jameson (1976) have shown that differences in the electrostatic interaction between polystyrene particles and the air-water interface have a major impact on the flotation rate constant and thus on the attachment efficiency of these particles. In the system they investigated the air-water interface and the particle surface were both positively charged. A decrease in the electrophoretic mobility of the particles from 4.8 to 2.9 (μm/s)/(V/cm) therefore resulted in an increase of the attachment efficiency with at least a factor 3. Okada et al. (1990a, 1990b) have shown that a decrease in electrostatic repulsion between polystyrene particles and air bubbles in dissolved-air flotation causes an increase in attachment efficiency. For air flotation of oil droplets they showed that the attachment efficiency became the largest when the bubbles and the oil droplets were oppositely charged (Okada et al., 1988).

The attachment efficiency is, besides its dependency on the electrostatic interaction and Van der Waals interaction, also dependent on the contact angle of the particles. Many studies have been performed to relate the contact angle of particles to their air flotation behaviour. Hewitt et al. (1995) studied air flotation of methylated quartz particles with diameters ranging from 5 to 80 μm using bubbles of 0.75 and 2.0 mm diameter. The quartz particles had advancing contact angles with the air-water interface in the range of 20° to 88°. Hewitt and co-workers showed that the attachment efficiency of the particles increased with an increase of their contact angle, and thus with their hydrophobicity. Rabinovich and Yoon (1994) found a correlation between the advancing contact angle of particles and the so-called hydrophobic force the particles experience with fluid interfaces (refer to section 4.2). The attractive force increased with the contact angle of the particles even though the Van der Waals interaction remained constant. The researchers suggested that this correlation is the result of hydrophobic attraction between the particles and the interface. This explains the findings by Hewitt et al. (1995) that the attachment efficiency increases with an increase in contact angle. Hewitt and co-workers also found that a decrease of the electrostatic repulsion between the particles and air bubbles resulted in an increase of the attachment efficiency. This effect showed to be stronger for less hydrophobic particles, which is probably due to the balance between the colloidal interaction forces, i.e. the Van der Waals, electrostatic, and hydrophobic interactions.

The examples of attachment selectivity that have been given thus far deal with selectivity on the basis of the Van der Waals, electrostatic and hydrophobic interactions (the extended DLVO forces). However, there is another important factor that can play a role in the attachment of particles to interfaces when working with particles that have extended surface groups. The extended surface groups can cause steric hindrance between the particles and the interface when the interface contains adsorbed molecules. Additionally, the adsorption behaviour of the surface groups at the interface will influence the attachment efficiency of the particles. We have recently shown that protein-coated polystyrene particles qualitatively have the same air flotation behaviour as the proteins in solution (Van Hee et al., 2006c). The flotation behaviour of the particles could not be explained on the basis of the extended DLVO forces. The protein adsorption kinetics completely dominated the attachment of the protein-coated particles at the air-water interface. This is a very important finding because such influences have not been reported for air flotation of mineral particles. Separation of bioparticles using fluid interfaces will therefore be more complex than particle separations in the mining industry.

The experimental examples in this section show that selective attachment can be realised by controlling the surface charge of the particles, their contact angle, and the properties of the extended surface groups. Differences in one of these particle properties may be sufficient for their separation.

5.3. Combination of Selective Collision and Selective Attachment

In many cases it is unclear whether selectivity is created on the basis of selective collision, selective attachment or a combination of both. This is often due to the lack of information on the particle properties, which makes discrimination between both processes impossible.

In previous work we have shown that separation of polyhydroxyalkanoate (PHA) inclusion bodies (IBs) from disrupted fermentation broth by dissolved-air flotation was feasible near the iso-electric point of the PHA inclusion bodies and cell debris (Van Hee et al., 2006b). Both particles had iso-electric points near pH 3. At this pH the particles experienced no electrostatic interaction with the air bubbles. The selectivity was thus caused by 1) differences in the Van der Waals and hydrophobic interactions between the particles and the interface and/or 2) differences in effective particle size as the IBs were selectively aggregated near their iso-electric point. This example proves that selective aggregation of a specific type of particle can be used to create selectivity in the interfacial separation process.

Our work on protein-coated particles has shown that particles with different proteins on their surface have different aggregation behaviours and different particle-bubble interactions (Van Hee et al., 2006c). These particles can therefore be separated on the basis of differences in 1) collision efficiency because of differences in aggregate size (section 5.1), 2) attachment efficiency due to their differences in surface properties (section 5.2), or 3) a combination of both sub-processes.

To our knowledge, obvious examples of emulsification processes where particle-particle separation is achieved solely as a result of selective attachment have not been reported in literature. Most studies concern separation problems of particles that are different in surface properties, size (distribution) and shape at the same time and therefore differ in both collision and attachment efficiency. Examples of these are the separation of protein inclusion bodies from cell debris (Walker and Lyddiatt, 1999) and the isolation of proteins from crude mixtures containing e.g. lipids, hemoglobin or cell debris (Dennison and Lovrien, 1997).

5.4. Selective Detachment

Mechanism 1: Detachment Forces > Interfacial Tension Force

Finch and Smith (1979) investigated the relation between particle size and particle detachment. They calculated the size of the maximum floatable particles in air flotation on the basis of the balance between the interfacial tension force and gravity (equation 19 and 21). The expression that was applied for the interfacial tension force is different from equation 19, but the estimated values are similar. There was reasonable agreement between particle detachment in the calculations and the air flotation measurements under quiescent conditions. Deviations were ascribed to the accuracy of the data that were used in the calculations, i.e. contact angle, and the assumption that the particles were spherical.

The use of selective detachment for interfacial particle separation was clearly demonstrated by Jauregi et al. (2001). They determined the impact of the centrifugal force on particle detachment in a water-butanol system that contained a 1:1 mixture of ampicillin crystals and phenylglycine crystals. Under gravity the interphase contained a crystal mixture that consisted for 93 % of phenylglycine. The selectivity of the interphase was reversed when a centrifugal force was applied. In this case the crystal mixture at the interphase consisted for only 40-41 % of phenylglycine crystals. These results show that the balance between the centrifugal force and the interfacial tension force can be used to control the selectivity in interfacial separation.

Mechanism 2: Reduction of Available Interfacial Area by Coalescence

Moys (1978) reported that particle detachment in the froth of an air flotation process is selective for the weakest hydrophobic particles. These particles are released from the interface first when bubble coalescence causes the available interfacial area to become too small for all particles to remain adsorbed. The detachment process showed to be dependent both on particle hydrophobicity and particle size. Yianatos et al. (1988) later showed that under certain conditions the detachment rate for pyrite in flotation froth was approximately 37% higher than for the more hydrophobic chalcopyrite. This detachment selectivity can be used to improve the flotation selectivity significantly. Honaker and Ozsever (2003) suggested that external refluxing of part of the flotation product into the froth (foam) phase could provide significant improvement in product purity because of the competition of the particles for the interface.

In emulsification processes a determining factor for selectivity is the reduction of interfacial area by coalescence. The continuous break-up and coalescence of bubbles/droplets brings about an ongoing increase and decrease of interfacial area. Selective detachment from the interface upon coalescence will affect the overall degree of separation.

Jauregi et al. (2001) demonstrated selective particle detachment as a result of droplet coalescence for ampicillin and phenylglycine crystals. Experiments showed that rather unstable n-butanol/water emulsions containing both crystals could be (partly) broken up by manual shaking, upon the release of predominantly ampicillin crystals. Beside a difference in the hydrophobicity of the crystals (see also section 5.5), the deviant shape of the ampicillin crystals (needles) in comparison with the phenylglycine crystals (discs) is a possible explanation for the selective detachment. Upon collision with neighbouring particles, the needles are more likely to slide over, and hence loose contact with the interface, eventually leading to detachment.

5.5. Particle Extraction

In liquid two-phase systems particles can adsorb to the interface, but also partition to the bulk auxiliary phase. In case the particles that are to be separated distribute differently over the bulk auxiliary phase and bulk raffinate phase, separation can be achieved by selective extraction, similar to selective extraction of solutes.

Selective particle extraction has been shown to be very effective for the purification of sub-micron bioparticles. For instance, Walker and Lyddiatt (1999) demonstrated that partitioning in aqueous two-phase systems can be used for the purification of protein inclusion bodies, where impurities (mainly *E. Coli* cell debris) are accumulated in the other liquid layer. Similar selective extraction methods were presented by Andrews et al. (1995) for the purification of virus-like particles from yeast cell homogenate in aqueous two-phase systems. For even smaller-sized particles (e.g. single proteins) the differences with single molecules become less obvious and liquid-liquid extraction is common use (e.g. Liu et al, 1995).

The liquid phase to which the particles partition will normally be the one by which they are preferently wetted. Recently, we demonstrated this for phenylglycine and ampicillin crystals in water/organic systems. For instance, figures 9a and 9b show ampicillin and phenylglycine crystals at the water/n-butanol interface. The ampicillin crystals are clearly preferently wetted by the aqueous phase, whereas the phenylglycine crystals are positioned for the larger part in the n-butanol phase. This observation was reflected in their partition behaviour, since ampicillin and phenylglycine were shown to mainly end up in the aqueous phase and the organic phase, respectively. This deviation in partition behaviour led to a good separation when working with mixtures of these crystals.

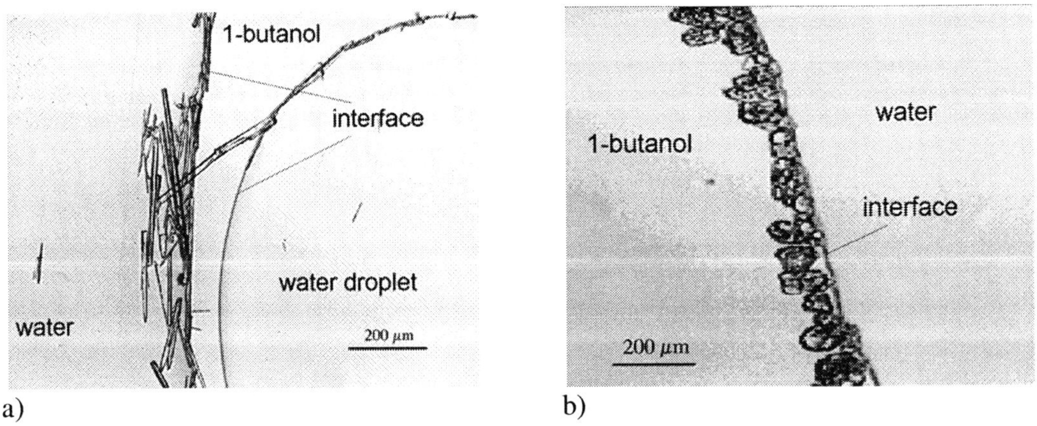

Figure 9. a) Microscopic image of ampicillin crystals at the water/1-butanol interface. The sample was prepared by adding a suspension of crystals in 1-butanol (saturated with water, ampicillin and phenylglycine) to water (saturated with 1-butanol, ampicillin and phenylglycine). b) As image a, where phenylglycine crystals are shown instead of ampicillin and all crystals were initially suspended in the aqueous phase.

Unless the contact angles of the particles are (virtually) below 0° or above 180°, selective extraction of particles in liquid-liquid systems will be accompanied by interfacial adsorption. While interfacial adsorption is not of much importance in extraction of dissolved compounds, it is likely to play a significant role with particles. Levine et al. (1989a,b) showed that for micron-sized particles with contact angles between 0 and 180 degrees, the gain of energy upon adsorption to the interface is typically so large that it dominates the favourable entropic effect of distribution over bulk phases and interface. Therefore, partitioning of particles in liquid two-phase systems is likely to be accompanied with high concentrations of particles at the interface. Considering that the thickness of a monolayer of particles is much higher than that of a monolayer of molecules, the total mass of the interfacially adsorbed fraction will always be much higher with particles than with dissolved compounds. This aspect has to be reckoned with in the design and modelling of selective two-phase particle extraction.

6. Process Design

Section 3 discussed the general methodology for interfacial particle separation by 1) dispersed phase injection and 2) emulsification. In this section more details on process design are given for both strategies. There is a difficulty in doing this, however. First of all, dispersed phase injection is commonly applied for particle separation in the mining industry and for removal of ink in paper recycling. The processes that are used on large scale in these industries are most likely applicable for bioparticle separation as well. Therefore we do not elaborate on the design of such processes and refer to published examples (Rousseau, 1987; Perry and Green, 1997; Finch and Dobby, 1990; Kawamura, 2000; Kiuru and Vahala, 2001). Secondly, emulsification has not yet been applied for large-scale particle separation according to our knowledge. It is thus not feasible to present a thorough discussion on large-scale application of this process. The important aspects of this process will be highlighted without going into too much detail.

6.1. Emulsification

Figure 3 gives an example of an emulsification process that can be applied for interfacial particle separation. In general any type of emulsification process may be used for this approach, e.g. mixer-settler or membrane emulsification. Important aspects of the process design are emulsion stability and recycling of the auxiliary phase.

Recycling of the auxiliary phase is mainly important for economical and environmental reasons. Using a gas as the auxiliary phase makes recycling unnecessary in most cases. The low volatility of dissolved biological compounds limits the quantity of compounds that enter the gas phase. Hence, when recycling of the gas phase is required the gas can simply be compressed without the need for extensive purification steps. An auxiliary liquid makes the recycling process more complex. Recycling may then involve removal of dissolved compounds from the liquid phase.

Emulsion stability is important for technical feasibility of the process and process economics. A very stable particle-laden emulsion does not allow for particles to detach under the influence of coalescence and detachment forces, such as shear and gravity. This is beneficial for efficient use of the interface that is created in the separator. In addition, it may allow for the use of centrifugal forces for the separation of the three phases, i.e. top phase, interphase, and bottom phase, without loss of selectivity. On the other hand, high emulsion stability is unfavourable for recovery of the particles from the interphase, which requires coalescence and particle detachment. A number of methods are available to break up emulsions. First of all, centrifugation may be used to enforce phase separation. Secondly, antifoams and other tools from physical chemistry may be used to destabilise the emulsion. Finally, if these methods fail, evaporation may solve the particle recovery issue. This latter method may, however, be too costly for many products.

Unstable emulsions make the separation process less efficient because coalescence reduces the interfacial area that can be used for particle separation. The interfacial area that needs to be created for separation of a certain particle mass thus becomes larger when coalescence causes particle detachment. It is important to note that coalescence can be required for the selectivity of the process. Emulsion stability can be enhanced with baffles or with additives, such as particles and molecules that stabilize the emulsion. The effect of these additives is strongly dependent on the specific separation issue.

6.2. Dispersed Phase Injection

The process design for dispersed phase injection in the mining industry, in paper recycling and in wastewater treatment is well described in literature, especially for gas-liquid systems (Rousseau, 1987; Perry and Green, 1997; Finch and Dobby, 1990; Kawamura, 2000; Kiuru and Vahala, 2001). Two main process designs are commonly applied: 1) flotation columns and 2) flotation tanks.

In column flotation the feed suspension is fed through an inlet that is positioned in the liquid phase (figure 10). The dispersed phase is injected at the bottom of the column. Dispersed gas phases can be injected as such or by injection of gas oversaturated liquid. Many different types of injection systems are described in literature. Entrainment of particles (and molecules) can be prevented by injection of a wash flow of the continuous phase at a specific height in the foam/emulsion layer. This flow is countercurrent to the flow of the dispersed phase droplets/bubbles, thus removing particles that are entrained in the continuous phase in the foam/emulsion layer. The particle-laden emulsion/foam can be collected above the wash flow and the other particles can be collected via the bottom outlet.

Flotation tanks can be operated in a similar fashion as flotation columns, but the much larger surface area of these separators gives some complications in collecting the foam/emulsion layer. Additionally, it is more difficult to include a wash flow in these systems to prevent particle entrainment. When desired the feed suspension can be contacted with the dispersed phase droplets/bubbles outside of the flotation tank. The tank is then used as a settler in which the droplets/bubbles rise to the top of the separator. The product and waste are collected from the tank at the opposite end from the feed injection point. The dispersed phase is collected at the top of the separator and the continuous phase at the desired position in the bulk suspension.

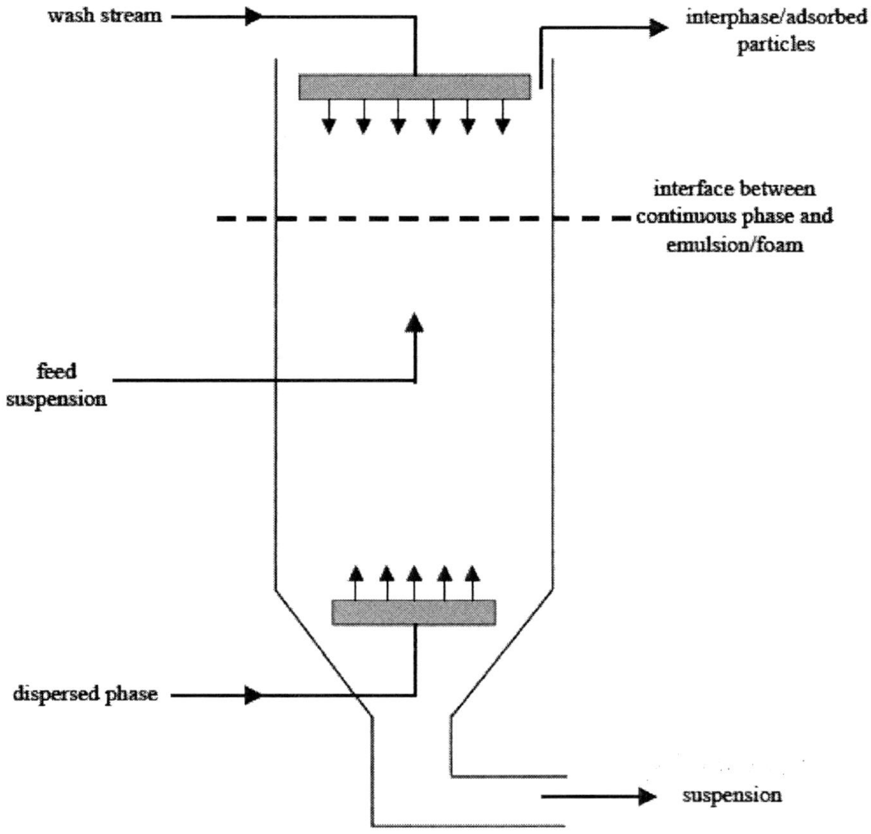

Figure 10. Design of a flotation column.

In both process designs foam/emulsion stability plays an important role. Additionally, hydrodynamics are very important for the selectivity and the collection efficiency of the process.

7. Conclusion

Separation of particles with fluid interfaces is possible by selective collision, selective attachment, and selective detachment, or a combination of these sub-processes. Selective collision can be realised by particle size differences, by differences in gravitational and inertial effects in the collision process, or by selective particle aggregation. Selective attachment is controlled by the colloidal interaction forces and liquid film thinning. Selective detachment can be accomplished by choosing the proper fluid interface and hydrodynamic conditions, and by competition of particles for the fluid interface.

This chapter shows that interfacial separation can be very valuable for separation of bioparticles from mixtures. The method can perform particle separations that cannot be accomplished with filtration and centrifugation. Two general approaches have been described: 1) dispersed phase injection, and 2) emulsification. Both methods require careful selection of the fluid phases, the hydrodynamic conditions in the separation device, and process design. There are many possible combinations of fluid phases with a wide variety of properties. This indicates that the range of possibilities for particle-particle separation using fluid interfaces is comprehensive. Nevertheless, the technique is not commonly applied in biotechnology. Air flotation is currently used in biotechnology for recovery and concentration of biomass, and the removal of particles from wastewater. Its use as selective recovery of bioparticles from particle mixtures is, however, scarcely reported in literature. The use of emulsification for bioparticle separations has been described in literature for a number of case studies, but according to our knowledge this technique has not yet been applied on large scale.

More research is required to facilitate the implementation of interfacial particle separation on large scale in biotechnological processes. This chapter shows that extensive fundamental knowledge is available from other industries, i.e. paper and mining industry. We expect that the availability of knowledge on large-scale processes from these industries will facilitate the implementation of interfacial particle separation in the biotech industry. Additionally, many models have been developed that can be implemented for interfacial bioparticle separation. An example is the work by Nguyen and co-workers (2006) that describes a particle trajectory analysis model for determination of the collection efficiency of particles to fluid interfaces.

LITERATURE

Abrahamson, J. *Chem. Eng. Sci.* 1975, *30,* 1371-1379

Absolom, D.R.; Zingg, W.A.; Neumann, A.W. *J. Colloid Interface Sci.* 1986, *112,* 599-601.

Andrews, B.A.; Huang, R.B.; Asenjo, J.A. *Bioseparation* 1995, *5,* 105-112.

Ata, S.; Ahmed, N.; Jameson, G.J. *Int. J. Miner. Process.* 2002, *64,* 101-122.

Bahr, K.H.; Schügerl, K. *Chem. Eng. Sci.* 1992, *47,* 11-20.

Beards, T.; Cohen, M.A.; Parratt, J.S.; Turner, N.J. *Tetrahedron: Assymetry* 1993, *4,* 1085-1104.

Binks, B. P. *Curr. Opin. Colloid Interface Sci.* 2002, *7,* 21-41.

Borbas, R.; Turza, S.; Szamos, J.; Kiss, E. *Colloid Polymer Sci.* 2001, *279,* 705-713.

Bruggink, A.; Roos, E.C.; Vroom, E. *Org. Process Res. Dev.* 1998, *2,* 128-133.

Buchner, J.; Brinkmann, U.; Pastan, I. *Bio/technology* 1992, *10,* 682-685.

Buque-Taboada, E.M.; Straathof, A.J.J.; Heijnen, J.J.; Van der Wielen, L.A.M. *Biotechnol. Bioeng.* 2004, *86,* 795-800.

Busscher, H.J.; Bos, R.; Van der Mei, H.C. In *Physical chemistry of biological interfaces*; Baszkin, A.; Norde, W.; Ed.; Dekker: New York, 2000; pp 431.

Carambassis, A.; Jonker, L.C.; Attard, P.; Ruthland, M. *Phys. Rev. Lett.* 1998, *80,* 5357.

Caulet, P.; Van der Lans, R.G.J.M.; Luyben, K.C.A.M. *Chem. Eng. J.* 1996. *62,* 193-206.

Chattopadhyay, D.; Rathman, J.F.; Chalmers, J.J. *Biotechnol. Bioeng.* 1995, *48,* 649-658.

Clift, R.; Grace, J.R.; Weber, M.E. *Bubbles, drops and particles;* Academic press: London, 1978.

Collins, G.L.; Jameson, G.J. *Chem. Eng. Sci.* 1976, *31,* 985-991.

Dai, Z.; Fornasiero, D.; Ralston, J. *Adv. Colloid Interface Sci.* 2000, *85,* 231-256.

Dennison, C.; Lovrien, R. *Protein Expression Purif.* 1997, *11,* 149-161.

De Dousa, S.R.; Oliveira, K.F; Souza, C.S.; Kilikian, B.V.; Laluce, C. *Colloids surf. B* 2003, i 309-319.

Eonseon, J.; Polle, J.E.W.; Lee, H.K.; Hyun, S.M.; Chang, M. *J. Microbiol. Biotechnol.* 2003, *13,* 165-174.

Eriksson, J.C.; Ljungrenn, S.; Cleasson, P.M. *J. Chem. Soc. Faraday Trans.* 1989, *2,* 163.

Finch, J.A.; Dobby, G.S. *Column flotation;* Perchamon press: Oxford, 1990.

Finch, J.A.; Smith, G.W. *Min. Sci. Eng.* 1979, *11,* 36-63.

Gahr, K.H.; Schügerl. K. *Chem. Eng. Sci.* 1992, *47,* 11-20.

Gaudin, A.M. *Flotation;*. New York: McGraw-Hill, 1957.

Gore, R.A.; Crowe, C.T. *Trans. ASME* 1991, *113,* 304-307.

Grieves, R.B.; Wang, S.L. *Biotechnol. Bioeng.* 1966, *8,* 323-336.

Grieves, R.B.; Wang, S.L. *Biotechnol. Bioeng.* 1967, *9,* 187-194.

Guereca, L.; Bravo, A.; Quintero, R. *Process. Biochem.*1994, *29,* 181-185.

Hewitt, D.; Fornasiero, D.; Ralston, J. *J. Chem. Soc. Faraday Trans.* 1995, *91,* 1997-2001.

Heywood-Waddington, D.; Peters, T.J.; Sutherland, I.A. *Biochem. J.* 1986, *235,* 245-249.

Hoeben, M.A.; Van der Lans, R.G.J.M.; Van der Wielen, L.A.M. *AIChE J.* 2004, *50,* 1156-1168.

Hoeben, M.A.; Van Hee, P.; Van der Lans, R.G.J.M.; Kwant, G.; Van der Wielen, L.A.M. *Biotechnol. Bioeng.* 2005, *93,* 607-617.

Hoeben, M.A.; Van Hee, P.; Van der Lans, R.G.J.M.; Kwant, G.; Van der Wielen, L.A.M. 2006, to be submitted.

Honaker, R.Q.; Ozsever, A.V. *Min. Eng.* 2003, *16,* 975-982.

Honda, J.; Andou, H.; Mannen, T.; Sugimoto, S. *J. Biosci. Bioeng.* 2000, *89,* 582-589.

Hogg, R.; Healy, T.W.; Fuerstenau, D.W. *Trans. Faraday Soc.* 1965, *62,* 1638-1651.

Jauregi, P.; Hoeben, M.A.; Van der Lans, R.G.J.M.; Kwant, G.; Van der Wielen, L.A.M. *Ind. Eng. Chem. Res.* 2001, *40,* 5815-5821.

Jauregi, P.; Hoeben, M.A.; Van der Lans, R.G.J.M.; Kwant, G.; Van der Wielen, L.A.M. *Biotechnol. Bioeng.* 2002, *78,* 355-364.

Jeaong, J-C.; Lee, I-Y.; Kin, S-W.; Park, Y-H. *Biotechnol. Lett.* 1999, *21,* 683-686.

Kalyuzhnyi, M.Y.; Petrushko, G.M.; Novikova, G.P. *Microbiol.* 1965, *34,* 918-924.

Kasche, V.; Galunsky, B. *Biotechnol. Bioeng.* 1995, *45,* 261-267.

Kawamura, S. *Integrated Design and Operation of Water Treatment Facilities;* John Wiley: New York, 2000.

Kiss, E.; Szamos, J.; Tamas, B.; Borbas, R. *Colloids Surf. A* 1998, *142,* 295-302.

Kiuru, H.; Vahala, R. *Water Sci. Technol.,* 2001, *43,* 1-208.

Kralchevsky, P.A.; Ivanov, I.B.; Ananthapadmanabhan, K.P.; Lips, A. *Langmuir* 2005, *21,* 50-63.

Kruis, F.E.; Kusters, K.A. *Chem. Eng. Comm.* 1997, *158,* 201-230.

Kula, M.R. *J. Cell Biochem. Suppl.* 1993, *17A,* 40.

Lee, SY.; Choi, J-I.; Wong, H.H. *Int. J. Biol. Macromol.*1999, *25,* 31-36.

Levine, S.; Bowen, B.D.; Partridge, S.J. *Colloids Surf.* 1989a, *38,* 325-343.

Levine, S.; Bowen, B.D.; Partridge, S.J. *Colloids Surf.* 1989a, *38,* 345-364.

Liu, C.; Niklas, Y.; Blankstein, D. *AIChE J.* 1995, *41,* 991-995.

Michielsen, M.J.F.; Frielink, C.; Wijffels, R.H.; Tramper, J.; Beeftink, H.H. *Biotechnol. Bioeng.* 2000. *69,* 548-558.

Milavic, S.J.; Chan, D.Y.C.; White, L.R.; Healy, T.W. *J. Phys. Chem.* 1994, *98,* 9022.

Moys, M.H. *Int. J. Miner. Process.* 1978, *5,* 21-28.

Nguyen, A.V. *Int. J. Miner. Process.* 2003, *68* 167-182.

Nguyen, A.V.; George, P.; Jameson, G.J. *Chem. Eng. Sci.* 2006, *61,* 2494-2509.

Okada, K.; Akagi, Y.; Yoshioka, N. *Can. J. Chem. Eng.* 1988 *66,* 276-281.

Okada, K.; Akagi, Y.; Kogure, M.; Yoshioka, N. *Can. J. Chem. Eng.* 1990a *68,* 393-399.

Okada, K.; Akagi, Y.; Kogure, M.; Yoshioka, N. *Can. J. Chem. Eng.* 1990b *68,* 614-621.

Palmieri, M.C.; Greenhalf, W.; Laluce, C. *Biotechnol. Bioeng.* 1996, *50,* 248-256.

Pazhianur, R.; Yoon, R.H. *Miner. Metall. Process.* 2003, *20,* 178-184.

Perry, R.H.; Green, D.W. *Perry's Chemical Engineers' Handbook;* McGraw-Hill: New York, 1997; Vol. 7.

Pickering, S. U. *J. Chem. Soc. Trans.* 1907, *91,* 2001-2021.

Pike, R.N.; Denisson, C. *Biotechnol. Bioeng.* 1989, *33,* 221-228.

Podgórska, W.; Bałdyga, J. *Chem. Eng. Sci.* 2001, *56,* 741-746.

Rabinovich, Y.; Yoon, R.H. *Langmuir* 1994, *10,* 1903-1909.

Rabinovich, Y.; Guzonas, D.; Yoon, R.H. *Langmuir* 1991, *9,* 1168.

Ramirez, J.A.; Zinchenko, A.; Loewenberg, M.; Davis, R.H. *Chem. Eng. Sci.* 1999, *54,*149-157.

Rousseau, R.W. *Handbook of Separation Process Technology;* Wiley: New York, 1987.

Sadowski, Z.; Golab, Z. *Biotechnol. Bioeng.* 1991, *37,* 955-959.

Schulze, H.J. *Physico-chemical elementary processes in flotation. Developments in mineral processing;* Elsevier: Amsterdam, 1984; Vol. 4, 85, 152.

Schulze, H.J. *Min. Process. Extractive Metall. Rev.* 1989, *5,* 43-76.

Spassov, G.; Pramatarova, V.; Wandrey, C. *Appl. Microbiol. Biotechnol.* 1996, *46,* 481-488.

Straathof, A.J.J.; Wolff, A.; Heijnen, J.J. *J. Mol. Catal. B: Enzym* 1998, *5,* 55-61.

Tan, K.H.; Lovrien, R. *J. Biol. Chem.*1972, *247,* 3278-3285.

Tcholakova, S.; Denkov, N.D., Danner, T. *Langmuir* 2004, *20,* 7444-7458.

Tobin, T.; Ramkrishna, D. *Can. J. Chem. Eng.* 1999, *77,* 1090-1104.

Tybussek, R.; Linz, F.; Schügerl, K.; Moses, N.; Léonard, A.J.; Rouxhet, P.G. *Appl. Microbiol. Biotechnol.* 1994, *41,* 13-22.

Van der Wal, A.; Minor, M.; Norde, W.; Zehnder, A.J.B.; Lyklema, J. *Langmuir* 1997, *13,* 165-171.

Van der Wielen, L.A.M.; Van Dam, M.H.H.; Luyben, K.C.A.M. Chem. Eng. Sci. 1997, *52,* 553-565.

Van Hee, P.; Hoeben, M. A.; Van der Lans, R. G. J. M.; Van der Wielen, L. A. M. *Biotechnol. Bioeng.* 2006a, *94,* 689-709.

Van Hee, P.; Elumbaring, A. C. M. R.; Van der Lans, R. G. J. M.; Van der Wielen, L. A. M. *J Colloid Interface Sci.* 2006b, *297,* 595-606.

Van Hee, P.; Lin, W. K.; Benac-Vegas, L.; Koper, G. J. M.; Van der Lans, R. G. J. M.; Van der Wielen, L. A. M. *Colloids Surf. A.* 2006c, *280,* 216-231.

Vlaski, A.; Van Breemen, A.N.; Alaerts, G.J. *J. Water SRT* 1996, *45,* 253-261.

Walker, S.G.; Lyddiatt, A. *J. Chem. Technol. Biotechnol.* 1999, *74,* 250-255.

Wang, S.; Kretzemer, G.; Schügerl, K. *Appl. Microbiol. Biotechnol.* 1994, *41,* 537-543.

Weber, M.E.; Paddock, D. *J. Colloid Interface Sci.* 1983, *94,* 328-335.

Wong, H.H.; O'Neill, B.K.; Middelberg, A.P.J. *Bioseparation* 1997, *6,* 361-372.

Yan, N.; Zhang, M.; Ni, P. *J. Membr. Sci.* 1992, *72,* 163-169.

Yianatos, J.B.; Finch, J.A., Laplante, A.R. *Int. J. Miner. Process.* 1988, *23,* 279-292.
Yoon, R.H.; Aksoy, B.S. *J. Colloid Interface Sci.* 1999, *211,* 1-10.
Yoon, R.H.; Luttrell, G.H. *Miner. Process. Extr. Metall. Rev.* 1989, *5,* 101-122.
Yoon, R.H.; Mao, L. *J. Colloid Interface Sci.* 1996, *181,* 613-626.
Yoon, R.H. *Int. J. Miner. Process.* 2000, *58,* 129-143.

In: New Colloid and Surface Science Research
Editor: Lawrence V. Schwartz, pp. 165-196
ISBN: 978-1-60021-594-0

Chapter 4

RECENT ADVANCES IN THE STABILITY AND RHEOLOGY OF MAGNETIC COLLOIDS

***M. T. López-López*[1,2]*, J. D. G. Durán*[1]*, F. González-Caballero*[1] *and G. Bossis*[2]**

[1]Departamento de Física Aplicada. Facultad de Ciencias,
Universidad de Granada. 18071 Granada. Spain.
[2]LPMC, UMR 6622, University of Nice, Parc Valrose,
06108 Nice Cedex 2, France.

ABSTRACT

Magnetic colloids are complex fluids composed by magnetic and non-magnetic particles dispersed in liquid carriers. These fluids can develop intense rheological changes upon application of an external magnetic field. This remarkable property makes possible the use of these fluids in a wide range of technological applications and, therefore, they center the attention of both scientists and engineers interested in field-responsive materials. Usually, magnetic colloids are classified into: (a) ferrofluids (FF) and (b) magnetorheological (MR) fluids. FF are dispersions of magnetic nanoparticles in a carrier liquid while MR fluids contain micron-sized particles of ferro- or ferrimagnetic materials. In the case of FF, although thermal energy prevents particle sedimentation and aggregation by magnetic attraction, it does not avoid coagulation due to van der Waals forces. Therefore, the use of adequate surfactants and/or thixotropic agents is needed to overcome this problem. In MR fluids, high density magnetic particles are dispersed in low density oil carriers and, consequently, the suspensions suffer from excessive gravitational settling, which can hinder the MR response of the fluid.

In this chapter, the preparation of fluids belonging to both categories (FF and MR fluids) is described. Furthermore, the effect of different stabilizing additives (thixotropic agents, surfactants, magnetic nanoparticles) and carrier liquids in both the sedimentation and the redispersibility of these fluids is analyzed. These studies are conducted using different experimental techniques: (i) optical and electromagnetic induction techniques to monitor the stability of dilute and concentrated suspensions; (ii) contact angle and pendant drop measurements to explain the observed stability in different carriers; and (iii) steady-state rheological tests to monitor the redispersibility of the suspensions. The results of these experiments are analyzed in detail and magnetic fluids with improved stability properties are prepared accordingly. Finally, an extensive study on the rheological properties of these fluids, in absence and presence of magnetic field, is described. Particularly, the effect of additive concentration and magnetic field intensity on the magnetorheological behavior is studied in detail. For this purpose, both controlled-stress and controlled-rate rheometers are employed.

1. INTRODUCTION

Generally speaking, magnetic colloids are a group of materials that exhibit the remarkable property of changing their flow properties under the application of an external magnetic field [Phulé and Ginder (1998a)]. They can be classified into: (*i*) ferrofluids (FF), which are stable colloidal dispersions of ferro- or ferrimagnetic nanoparticles in a carrier liquid [Charles (2002)]; and (*ii*) magnetorheological (MR) fluids, which are dispersions of micron-sized magnetic particles [Phulé and Ginder (1998a)]. Nano-sized particles (diameter $\approx$ 10 nm) used in FF are magnetically single-domain and, therefore, they possess a permanent magnetic dipole moment. Because of the possible existence of magnetic interactions between these particles in the presence of an external magnetic field, structures can form in these fluids that can largely affect their flow behavior. In particular, their rheological behavior can change from that of a Newtonian fluid to a plastic one, with appearance of measurable yield stress and a plastic viscosity depending on the applied field strength. These modifications are known as magnetorheological or magnetoviscous effects. However, even under the application of high magnetic fields, FF manifest only a relatively modest magnetoviscous response and do not develop a yield stress [Odenbach (2003); Phulé and Ginder (1998b); Rosensweig (1982)]. On the contrary, micron-sized ferro- or ferrimagnetic particles are magnetically multidomain and attain large magnetic moments under the application of rather weak magnetic fields. As a consequence, MR fluids manifest a high magneto-viscous response characterized by a high yield stress [Bossis *et al* (2002); Ginder (1998)]. Due to these properties, MR fluids are field-responsive materials with a broad range of technological applications.

In order to prevent particle aggregation in FF based on non-polar liquids due to van der Waals attraction, and the subsequent gravitational settling, it is necessary to coat the particles with long-chain molecular species (e.g. fatty acids) [Odenbach (2003)]. In this chapter, a series of organic carriers, with increasing dielectric constants, are tested to determine the adequate coupling between the hydrocarbon tail and the carrier molecules that produce stable ferrofluids.

MR fluids are typically formulated using high density materials such as iron, iron alloys or metal oxides (ceramic ferrites) dispersed in low-density liquids. Therefore, their stabilization against sedimentation arises as an important challenge, facing the technological applications of these field-responsive materials [Ginder (1996)]. Approaches to improve the stability include: (*i*) addition of thixotropic agents (e.g. carbon fibers and silica nanoparticles) [Bossis *et al* (2002); Volkova *et al* (2000)]; (*ii*) addition of surfactants (e.g. oleic acid or stearate salts) [Dang *et al* (2000); van Ewijk *et al* (1999)]; (*iii*) mixing magnetic nanoparticles [Chen *et al* (1998); Chin *et al* (2001)]; and (*iv*) use of viscoplastic media or water-in-oil emulsions as continuous phases [Park *et al* (2001); Rankin *et al* (1999)]. In this chapter, we report on the possibility of stabilizing iron-MR fluids using different surfactants (oleic acid, aluminum stearate) as well as using thixotropic agents (silica nanoparticles). In addition, we describe the preparation of magnetic fluid composites (MFC) by dispersing micron-sized iron particles in ferrofluids and we analyze their stability properties. Due to the opaqueness of the MR fluids and MFC studied, the use of classical optical methods is not always suitable to characterize the sedimentation behavior of these systems. Therefore, we have also used an electromagnetic method to measure the sedimentation rate in these suspensions. Using this method, the effect of the surfactant or silica concentration (in MR fluids) and of the nano-magnetite volume fraction (in MFC) on the stability of the suspensions is analyzed and discussed.

Finally, the magnetorheological properties of MR fluids stabilized by the addition of silica nanoparticles and of MFC are studied, both in the presence and in the absence of an applied magnetic field. For this purpose, a controlled stress rheometer and a controlled rate magnetorheometer, both with a parallel-plate measuring cell, have been used to analyze the rheological properties of MR fluids and MFC, respectively.

2. FERROFLUIDS

2.1. Ferrofluid Preparation and Particle Morphology

The ferrofluids described in this chapter are composed of magnetite nanoparticles covered by oleate molecules dispersed in different organic media. The magnetite particles are synthesized by coprecipitation of Fe(II) and Fe(III) salts in aqueous solutions, and the surfactant used to stabilize the suspensions was oleic acid. This fatty acid imposes a steric barrier between the oleate-covered magnetite particles that overcomes the van der Waals and magnetic attractions, avoiding particle aggregation in non-polar carriers.

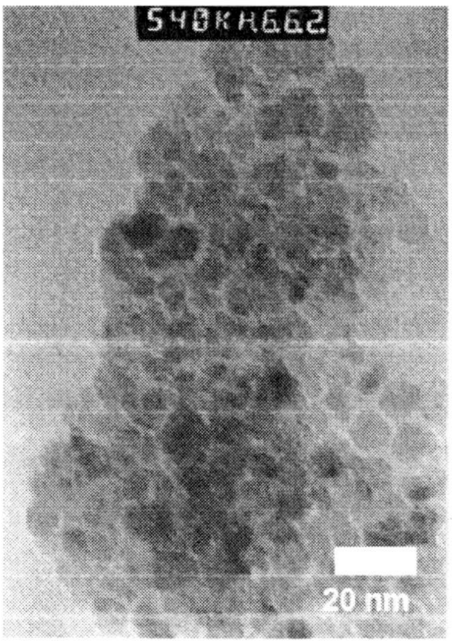

Figure 1. TEM picture of the magnetite nanoparticles remaining in suspension after centrifugation at 12000g. Liquid carrier: dodecane. Bar length 20 nm. (Reprinted from [López-López *et al* (2005a)], with permission from Elsevier).

Figure 1 shows a TEM picture of the particles. The average particle diameter after centrifugation at 12000g is shown in Table 1. The particle diameter in the dichloromethane ferrofluid is significantly smaller than in the other carriers. In this ferrofluid a clear gravitational settling is observed as a consequence of particle coagulation, leading to rather diluted suspensions in which only tiny particles remained. The dielectric constants of the liquids used in this work, measured with a Dekameter DK 300 apparatus (WTW Germany), are: 1.8 (mineral oil), 1.8 (kerosene), 2.0 (dodecane), 2.2 (carbon tetrachloride), 4.8 (chloroform), and 9.1 (dichloromethane).

The bulk crystal structure of the synthesized nanoparticles was analyzed by X-ray diffraction and the data showed excellent coincidence with the reference lines for magnetite.

2.2. Interfacial Free Energy of Interaction and Stability

The interfacial free energy of interaction between oleate-covered magnetite nanoparticles immersed in the different carriers was calculated from previous estimation of the surface free energy of the solid material and the surface tension of the liquid media, following the van Oss' approach [van Oss (1994)]. With this aim contact angle and pendant drop measurement were performed. The results are summarized in Tables 1 and 2.

This thermodynamic analysis demonstrates that the observed stability of the suspensions in liquids with $\varepsilon_r < 5$ is well correlated with the very low lyophobic (or acid-base) attraction between the particles, which can be easily surmounted by thermal agitation, since the van der Waals attraction is negligible. On the contrary, for liquids with $\varepsilon_r > 9$, the suspensions become unstable because of the combined action of the van der Waals and lyophobic attractions, the latter being dominant for very polar solvents.

Table 1. Surface free energy components (mJ/m^2) of magnetite nanoparticles, and surface tension components (mJ/m^2) of the carrier liquids at 20.0 ± 0.1 °C. γ^{LW}: van der Waals component, γ^+: electron-acceptor parameter; γ^-: electron-donor parameter. Last column: average diameter (nm) (D_{TEM}), obtained from TEM pictures, of particles dispersed in the indicated carrier liquids after centrifugation at 12000×g. (Reprinted from [López-López *et al* (2005a)], with permission from Elsevier)

	Material	γ^{LW}	γ^+	γ^-	D_{TEM}
SOLID	Oleate-covered magnetite	25.3 ± 0.6	0.15 ± 0.22	5.2 ± 0.3	- - -
	Magnetite	49.3 ± 0.2	0.17 ± 0.01	55.4 ± 0.3	- - -
CARRIER	Kerosene	25	0	0	7.8 ± 0.3
	Mineral oil	24 ± 3	0	0	7.15 ± 0.25
	Dodecane	25.35	0	0	7.9 ± 0.3
	CCl_4	27.0	0	0	6.8 ± 0.3
	Chloroform	27.15	0	0	7.8 ± 0.3
	CH_2Cl_2	53 ± 8	0	0	5.14 ± 0.14
	Water	21.8	25.5	25.5	- - -

Table 2. Total free energy of interaction, ΔG_{SLS}^{TOT}, between oleate-covered magnetite nanoparticles in the indicated carriers, and the corresponding Lifshitz-van der Waals and acid-base contributions (ΔG_{SLS}^{LW}; ΔG_{SLS}^{AB}). All quantities in mJ/m^2. (Reprinted from [López-López *et al* (2005a)], with permission from Elsevier)

Carrier	ΔG_{SLS}^{LW}	ΔG_{SLS}^{AB}	ΔG_{SLS}^{TOT}
Kerosene	- 0.002 ± 0.007	- 4 ± 3	- 4 ± 3
Mineral oil	- 0.05 ± 0.04	- 4 ± 3	- 4 ± 3
Dodecane	- 0.0001 ± 0.001	- 4 ± 3	- 4 ± 3
CCl_4	- 0.06 ± 0.04	- 1 ± 5	- 1 ± 5
Chloroform	- 0.05 ± 0.04	- 4 ± 3	- 4 ± 3
Dichloromethane	- 9.3 ± 0.5	- 4 ± 3	- 13 ± 3
Water	- 0.26 ± 0.09	- 52 ± 10	- 52 ± 10

2.3. Magnetic Properties of the Particles and the Ferrofluids

The magnetization, *M*, of the solid powder and the ferrofluids was measured at 20 °C as a function of the magnetic field strength, *H*. Figure 2 shows these curves.

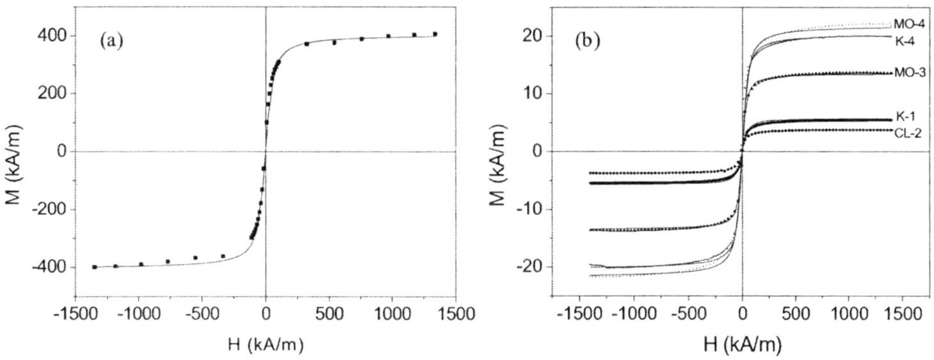

Figure 2. Magnetization data of: (a) the synthesized oleate-magnetite nanoparticles (powder); (b) the indicated ferrofluids (see sample identification in Table 3). The lines correspond to the best fit to the Langevin function (Equation 2). (Reprinted from [López-López et al (2005a)], with permission from Elsevier).

The saturation magnetization of the dry powder is M_s = 4.05 × 10^5 A/m and its initial magnetic susceptibility χ_i = 5.7. In addition, the synthesized magnetite has negligible coercivity and remanence, as expected for a superparamagnetic material above the blocking temperature. From these values, the magnetic particle diameter, D_m, can be determined by means of [Massart (1981)]:

$$D_m = \left(\frac{18kT}{\pi\mu_0} \frac{\chi_i}{M_s^2} \right)^{1/3}, \tag{1}$$

where k is the Boltzmann constant, T the absolute temperature, and μ_0 the magnetic permeability of vacuum. The calculated magnetic diameter (D_m = 7.60 ± 0.03 nm) is slightly smaller than that estimated from TEM pictures (D_{TEM} = 7.8 nm for ferrofluid in kerosene). The difference between D_m and D_{TEM} can be attributed to the presence of a magnetically "dead" layer on the particle surface, as frequently assumed in the literature [Kaiser and Miskolczy (1970)]. The magnetic moment of each particle, m, can be obtained using the Langevin function [Cullity (1974)]:

$$M = M_s \left(\coth\xi - \frac{1}{\xi} \right) \quad ; \quad \xi = \frac{\mu_0 mH}{kT}. \tag{2}$$

Fitting this function to the magnetization data in Figure 2a, we obtain a magnetic moment $m = (1.38 \pm 0.03)\times10^{-19}$ A·m^2 = (14900 ± 300)μ_B per particle, where μ_B is the Bohr magneton.

The saturation magnetization and the initial magnetic susceptibility of the ferrofluids are included in Table 3. The magnetic diameter and the magnetic moment of the particles in the different ferrofluids were determined in a similar way to that followed for dry particles using Equations 1 and 2. The data in Table 3 show that D_m and m for particles immersed in the ferrofluids are slightly larger than those obtained in the dry powder. The significant difference in the magnetic properties (D_m, m) in solid state and in suspension can be attributed to a weak structuration favored by the combined action of both interfacial and magnetic attractions between the particles immersed in the carrier liquid.

Table 3. Saturation magnetization (M_s) and initial magnetic susceptibility (χ_i) of the indicated ferrofluid samples (carrier / magnetite volume fraction ϕ). The magnetic diameter (D_m) and magnetic moment (m) of the particles in the ferrofluids were obtained by fitting the magnetization data in Figure 2b to Equations 1 and 2. (Reprinted from [López-López *et al* (2005a)] with permission from Elsevier)

Sample Carrier / ϕ (%)	M_s (A/m)	$\chi_i \times 10^3$	D_m (nm)	m (10^{-19} A·m^2)
K-1 Kerosene / 1.1 ± 0.1	5520 ± 140	337.43 ± 0.12	13.5 ± 0.4	2.02 ± 0.05
K-4 Kerosene / 4.2 ± 0.1	20400 ± 500	401.07 ± 0.25	9.12 ± 0.07	1.697 ± 0.024
MO-3 Mineral oil/ 2.8 ± 0.1	13800 ± 300	306.33 ± 0.19	9.55 ± 0.11	2.19 ± 0.06
MO-4 Mineral oil/ 4.5 ± 0.1	22000 ± 600	360.53 ± 0.09	8.60 ± 0.06	1.48 ± 0.05
CL-2 Chloroform / 1.2 ± 0.1	5740 ± 140	260.7 ± 0.3	12.0 ± 0.3	1.68 ± 0.03

3. Stability of Magnetorheological Fluids

3.1. Sedimentation Behavior in Diluted Suspensions

In this paragraph the effect of the addition of oleic acid and silica nanoparticles on the sedimentation behavior of diluted iron-based MR fluids in absence of external magnetic field is studied. For this purpose, the evolution of the optical absorbance of the suspension is monitored and the effect of additive concentration is analyzed. A similar study in the presence of external magnetic field can also be found in a previous work [López-López *et al* (2005b)].

A. Experimental. Iron powder of HQ quality (BASF, Germany) was used without further treatment. The iron particles are spherical and polydisperse (average diameter 930 ± 330 nm). Silicone oil of viscosity 35.1 ± 0.3 mPa·s and density 954 kg·m^{-3} (Fluka, Germany) and mineral oil of viscosity 39.58 ± 0.16 mPa·s and density 854 kg·m^{-3}, both supplied by Fluka (Germany), were used as liquid carriers. Oleic acid (purity 90%) (Sigma-Aldrich, Germany), and silica nanoparticles 7 nm in diameter (Aerosil-300®, Degussa-Hüls, Germany) were used as stabilizing additives.

Iron-silica suspensions were prepared as follows: (1) iron, silica and silicone oil were mixed in a polyethylene container; (2) the mixture was shaken and then introduced in a sonifier (Branson model 450, USA). The preparation of iron-OA suspensions involved the following steps: (1) different oleic acid-mineral oil solutions were prepared; (2) iron was added to the selected oil solution; (3) the suspensions were shaken and sonified (Branson model 450, USA); and (4) the samples were stirred (50 rpm; 24 h; 25 °C) to allow the adsorption of the additive. No significant changes in the viscosity of the carrier were observed as silica nanoparticles or oleic acid content increased in the concentration range employed.

The sedimentation behavior of the suspensions was inferred from the evolution of the optical absorbance of the suspensions. With this aim, a Spectronic 601 spectrophotometer (Milton Roy, USA) at 590 nm of wavelength was used. All studied suspensions contained 1.25 g/L of iron (volume fraction = 0.017%). Samples were placed in square cuvettes with 1 cm of light path. Those were placed in such a way that the center of the light beam struck 1.5 cm above their bottom.

B. Effect of oleic acid. In Figure 3 the normalized optical absorbance, $A_n = A/A_0$ (A_0 is the initial absorbance) is plotted as a function of time for suspensions containing 0.017 vol% of iron and different concentrations of oleic acid (OA). Due to the high density and size of the particles, A_n shows a clear tendency to decrease with time, as a consequence of the progressive disappearance of the iron particles from the lighted region as sedimentation proceeds. The experiments for OA concentration below 16 mM (concentration high enough to complete the first OA adsorbed monolayer, see [López-López *et al* (2005b)]) show that there is no significant change in the stability of the suspensions for this concentration range. Therefore, it seems that steric repulsion does not play any important role on the sedimentation behavior of iron suspensions at low OA concentration (< 16 mM). However, at high OA concentration (≥ 31 mM) there is a clear decrease of the settling rate, since the slope of the absorbance vs time curve decreases with the concentration of OA. Finally, at OA concentrations higher than 1.57 M, there is not any progressive improvement of the stability.

The decrease of the sedimentation rate as OA concentration increases must be originated by the progressive OA covering of iron particles. As a consequence, the stabilization mechanism must be originated by the progressive structuration of the OA molecules in adsorbed multilayers. This behavior drives steric repulsion among the OA-covered iron particles and, therefore, a decrease of settling rate.

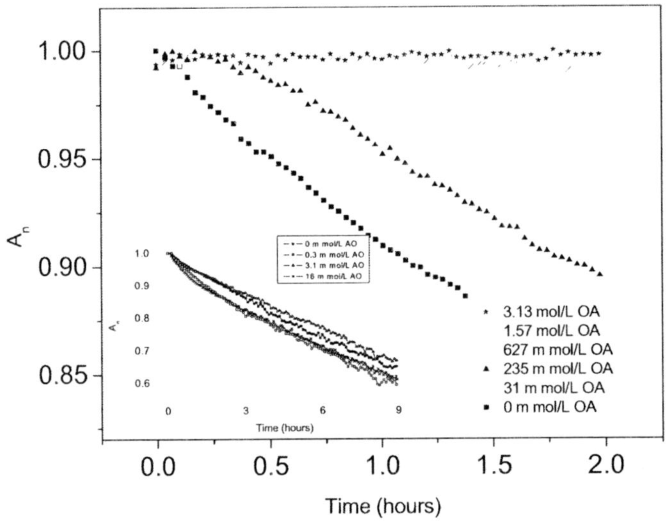

Figure 3. Normalized absorbance, $A_n = A/A_0$, as a function of time for suspensions containing 0.017 vol% of iron and the indicated OA concentration. (Reprinted from [López-López *et al* (2005b)], with permission from Elsevier).

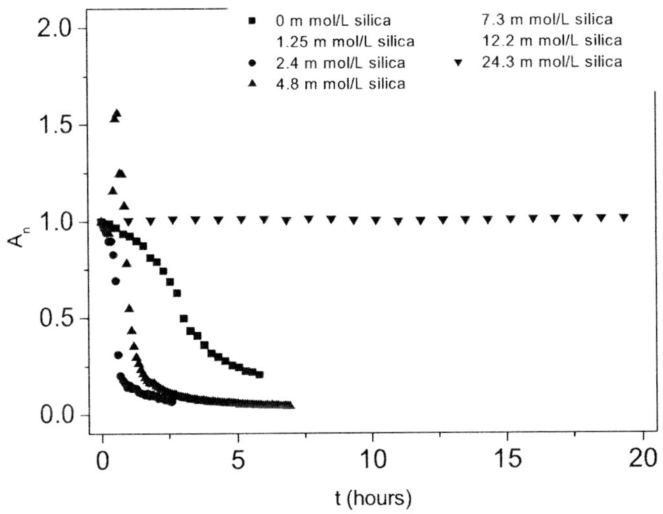

Figure 4. Normalized absorbance, $A_n = A/A_0$, as a function of time. All suspensions contained 0.017 vol% of iron and the silica concentration indicated. (Reprinted from [López-López *et al* (2005b)], with permission from Elsevier).

C. Effect of silica nanoparticles. These experiments were carried out for suspensions containing 0.017 vol% of iron and silica nanoparticles concentration up to 24.3 mM. The results of these tests are shown in Figure 4. As can be observed, suspensions with silica concentration below 2.4 mM settle rather quickly. This behavior is a direct consequence of the existence of iron-silica adhesion, which provokes a growing of the particle size and, therefore, a faster settling rate than in the absence of silica. Notice that the existence of heterogeneous aggregation among silica and magnetic particles in similar systems was proved earlier [de Vicente *et al* (2003); van Ewijk and Philipse (2001)]. On the other hand, for silica concentration higher than 2.4 mM the formation of a silica network, by inter-particle hydrogen bonding, that imparts a gel-like structure to the suspension [Bossis *et al* (2002); Volkova *et al* (2000)], is the dominant phenomenon. For 4.8, 7.3 and 12.2 silica concentration the absorbance increases at the beginning of the experiment and then it falls down. This can be explained considering that, at the beginning, the iron-silica gel scatters the light that otherwise would strike the detector of the spectrophotometer. However, after a time these iron-silica structures break and fall down, since silica concentration is still too low. Finally, as can be seen in Figure 4, for a silica concentration of 24.3 mM the iron-silica network is stiff enough to hold in suspension all the iron load (absorbance remains constant with time).

3.2. Sedimentation Behavior in Concentrated Suspensions

In the previous paragraph we analyzed the settling behavior of diluted MR fluids using a classical optical method. In this paragraph we will study the settling behavior of concentrated MR fluids. Unfortunately, due to the high particle content, the use of optical methods is not suitable to analyze sedimentation processes in concentrated suspensions. Therefore, we have used an electromagnetic method to measure the sedimentation rate in these suspensions. Using this method, the effect of the surfactant or silica concentration will be analyzed and discussed.

A. Experimental. Iron powder was used as magnetic particles. Kerosene (Sigma-Aldrich, Germany) was used as liquid carrier. Oleic acid (OA), aluminum stearate (AlSt) (technical quality) (Sigma-Aldrich, Germany) and silica nanoparticles were used as stabilizing additives. Iron-OA and iron-AlSt suspensions were prepared as described above for diluted iron-OA suspensions. Iron-silica suspensions were also prepared in the same way described above for diluted suspensions.

The method we have used for measuring the particle sedimentation rate in magnetorheological fluids is based on the time evolution of the electromotive force induced in a coil that surrounds the sample. A schematic drawing of the experimental setup is shown in Figure 5. It consisted of: (*i*) A pair of Helmholtz coils (Phywe, Germany) with an alternating current of 1.05 ± 0.01 A and frequency υ = 50 Hz. This current generates an alternating magnetic field in the vertical direction. (*ii*) A small sensing coil (1.1 cm height; 225 turns) was placed in the center of the Helmholtz coils, in such a manner that the axial directions of both coils coincided. (*iii*) A glass test tube (12 cm height; 9 mL volume) was placed in the center of the Helmholtz coils surrounded by the sensing coil. The top end of the sample was 2.3 cm above the sensing coil. During the settling process, the values of the induced potential in the sensing coil, V, were measured as a function of time. In the absence of the sample, V = 22.70 ± 0.01 mV (ν = 50 Hz).

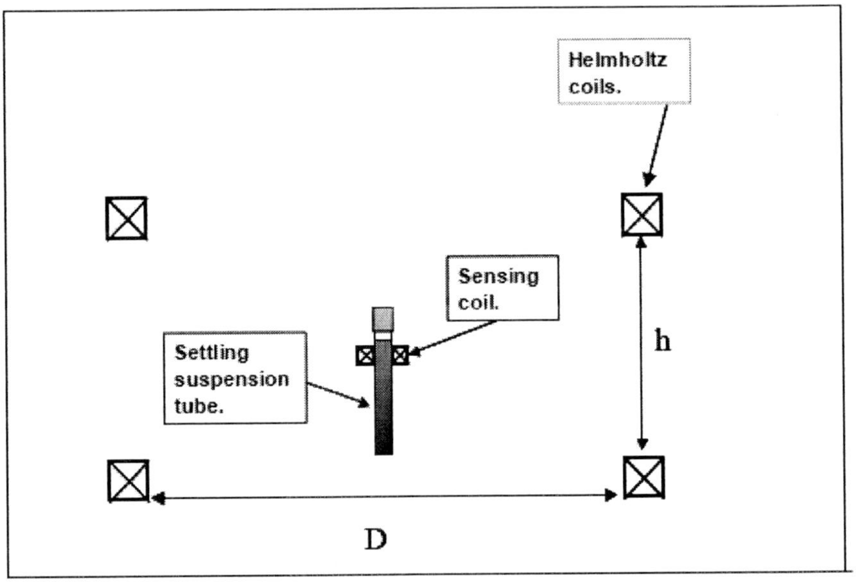

Figure 5. Scheme of the experimental setup used for sedimentation experiments. D = 40 cm; h = 20 cm. (Reprinted from [López-López *et al* (2005c)]).

In order to compare the results for the different suspensions studied in this chapter, it is useful to define the dimensionless increment of the induced potential as:

$$\Delta \upsilon = \frac{V(t) - V(t \to \infty)}{V(t=0) - V(t \to \infty)}, \tag{3}$$

where $V(t)$ is the instantaneous induced potential, $V(t \to \infty)$ is the potential in the stationary state, and $V(t = 0)$ is the initial potential. Using this expression for $V(t)$ (in S.I. units), and considering $M = \chi H$ (valid for isotropic and linear materials):

$$\Delta \upsilon = \frac{\mu_r(t) - \mu_r(t \to \infty)}{\mu_r(t=0) - \mu_r(t \to \infty)}, \tag{4}$$

where μ_r is the relative magnetic permeability of the suspension surrounded by the sensing coil. Using Maxwell-Garnett theory [Garnett (1904)], and considering that the magnetic contrast factor β $\left(\beta = \dfrac{\mu_{r,iron} - 1}{\mu_{r,iron} + 2}\right)$ is approximately equal to 1 in these experiments, we finally obtain:

$$\Delta \upsilon \propto \frac{\Phi(t)}{1 - \Phi(t)}, \tag{5}$$

where $\Phi(t)$ is the instantaneous iron volume fraction in the region surrounded by the sensing coil. The sedimentation behavior of the suspensions was hence estimated from the time evolution of the dimensionless induced potential $\Delta\upsilon$.

B. Comparison between the stabilization effects of OA and AlSt. In order to compare the stabilization efficiency of OA and AlSt, in Figure 6 are represented $\Delta\upsilon$ vs time for iron and iron/OA or AlSt suspensions. As observed, $\Delta\upsilon$ decreases faster with time the larger the OA or AlSt concentration. This can be explained by considering that, in the absence of OA (or AlSt), aggregation must be present because of van der Waals interaction and magnetic attraction due to the weak remnant magnetization of the iron particles. Presumably, large aggregates are formed that span the walls of the test tube (1 cm in diameter). It is precisely the friction with the walls that must hinder their gravitational settling. When surfactants are added, $\Delta\upsilon$ decreases faster than in their absence, this is in fact an indication of diminished aggregation: individual particles or small aggregates sediment more easily than large flocculi spanning the tube. Similar trends are obtained for OA and AlSt suspensions, although an AlSt concentration approximately seven times higher than that added in iron/OA suspensions is required to produce the same change in the stability properties, indicating that OA is a more efficient surfactant than AlSt concerning the stabilization against irreversible iron aggregation in oil-based MR fluids.

C. Effect of silica nanoparticles. Whereas OA and AlSt are surfactants that avoid iron particle aggregation by means of steric repulsion, silica nanoparticles create a gel-like structure that hinders particles settling. Therefore, it is worth to analyze the effect of silica nanoparticles on the settling behavior of concentrated suspensions and to compare it with that of OA and AlSt. Figure 7, where the dimensionless induced potential $\Delta\upsilon$ is plotted as a function of time, shows clearly that, as expected, the sedimentation rate decreases because the carrier is progressively thickened as the silica concentration increases.

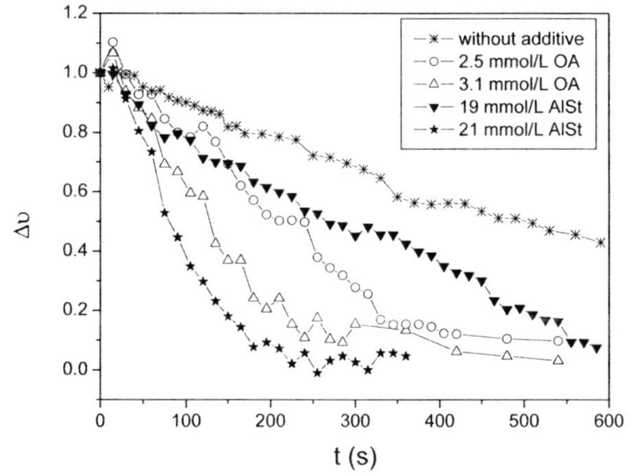

Figure 6. $\Delta\upsilon$ vs time for suspensions containing 10 % iron volume fraction and the indicated initial concentrations of oleic acid (OA) and aluminum stearate (AlSt).

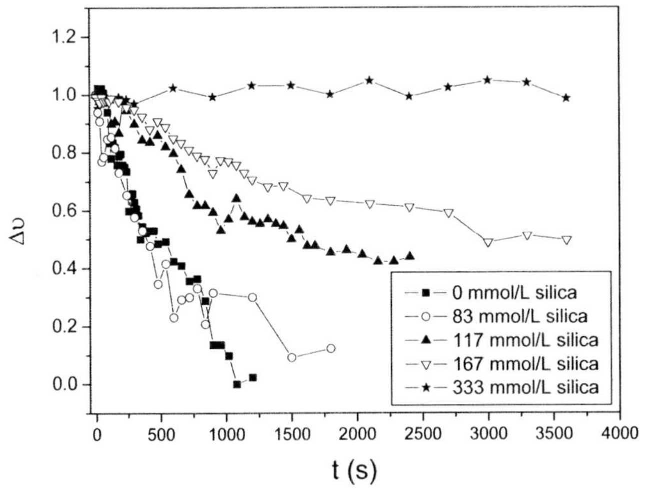

Figure 7. Dimensionless increment of the induced potential ($\Delta\upsilon$) as a function of time for suspensions containing 10 % iron volume fraction, and the indicated silica concentrations. (Reprinted with permission from [López-López *et al* (2006a)]).

For 83 mM silica concentration the thickening effect is not enough to produce any noticeable improvement on the stability properties. For higher concentrations (up to 167 mM silica) the gel formed significantly reduces the sedimentation rate, but it is not yet sufficiently thick to maintain the iron particles in suspension. Only for a concentration as high as 333 mM, the gel completely avoids particle settling, and $\Delta\upsilon$ is practically constant in the time interval studied.

3.3. Redispersibility of MR fluids

Once the particles of a MR fluid settle down, compact sediments are created, which hinder its MR response. Therefore, it is important to facilitate the redispersion of these sediments by means of appropriate additives. It is also important to have standard methods for analyzing the redispersibility of colloidal suspensions. In this paragraph we will first present a method to quantitatively study the redispersibility of colloidal suspensions and then we will use it for analyzing the effect of oleic acid and silica nanoparticles on the redispersibility of iron-based MR fluids in absence of applied magnetic field. A similar study for MR fluids in presence of applied magnetic field can be found in the work by López-López *et al* (2006a).

A. Experimental. A controlled stress rheometer (Bohlin CS10, UK) with a vane-in-cup geometry (Bohlin V-25) was used for analyzing the redispersibility of iron-based colloidal suspensions. Experiments were carried out for a gap of 1 mm between the vane end and the cup bottom, so that the sediment had to be partially redispersed when the vane rotated. Specifically, experiments were performed in the following way: *(i)* suspensions were prepared as described above and placed immediately in the measuring system; *(ii)* after 30 s of waiting time we applied a constant shear stress, high enough to induce a shear rate larger than 1 s^{-1}, and the corresponding shear rate was recorded as a function of time; *(iii)* after some minutes, the shear stress was brought down to zero and a waiting time of 15 minutes was allowed; *(iv)* afterwards, the same shear stress than in step (ii) was applied and the shear rate was monitored as a function of time. If this shear stress was not high enough to rotate the vane, it was increased in 1 Pa and so forth; and *(v)* step (iv) was repeated at 30, 45, 60, 90 and 120 minutes of waiting time.

B. Redispersibility without additives. Prior to analyzing the effects of additives on the redispersibility of iron-based colloidal suspensions, let us discuss about the redispersibility without additives. In Figure 8, the shear rate is plotted as a function of the shearing time at different time steps of the experiment (0.5, 15, and 30 minutes) and for different values of the shear stress. As can be seen, at the beginning of the experiment (0.5 minutes) a shear stress of 7 Pa is enough to move the vane at a quite high shear rate (~ 100 s^{-1}). However, after 15 minutes a higher shear stress (9 Pa) is required to produce a similar rate of rotation. This compacting tendency is confirmed as time increases: after 30 minutes a shear stress of 11 Pa only provokes a tiny movement of the vane, as observed in Figure 8. This tendency is due to the irreversible aggregation among iron particles, which together with gravitational settling provokes the formation of compact sediments. As time increases after placing the sample in the measuring system, the sediments become more compact and a higher shear stress is required to move the vane. Notice, finally, that a sharp drop of the shear rate is observed at high values of the shearing time (see for example the curve for 9 Pa; 15 min in Figure 8). This could be attributed to the gravitational settling of iron aggregates in the presence of shear, which increases the stiffness of the sediment and hinders the rotation of the vane.

C. Effect of oleic acid on the redispersibility. The effect of 2.8 mmol/L of oleic acid on the redispersibility of MR fluids containing 10 % of iron was studied in the way described above. Figure 9 shows the results of this experiment. Two evidences of the improvement of the redispersibility of iron suspensions when oleic acid is used as surfactant can be pointed out. The first one is the fact that at the beginning of the experiment a shear stress of 1 Pa is enough to move the vane in iron-OA suspensions, whereas in iron suspensions 7 Pa has to be applied (see Figure 8). Secondly, although a thickening process is induced from the data in Figure 9, 1 Pa is still enough to move the rotor at a quite large velocity even 90 minutes after the beginning of the experiment. Then, we can conclude that the addition of oleic acid reduces considerably the thickening rate in iron-based MR fluids. Finally, notice that the effect of aluminum stearate on the redispersibility was also studied [López-López *et al* (2006a)] and similar results to those for oleic acid were found.

D. Effect of silica nanoparticles on the redispersibility. As mentioned above, silica nanoparticles affect to the stability of MR fluids by means of two phenomena. One is the iron-silica adhesion, which is a very long-time phenomenon [de Vicente *et al* (2003)] and, consequently, can be rejected in the redispersibility study because suspensions were used immediately after their preparation. The other is the formation of gel-like silica networks, which can be considered a short-time phenomenon [López-López *et al* (2006a)]. Therefore, it will be assumed that the silica network is practically built after 0.5 minutes.

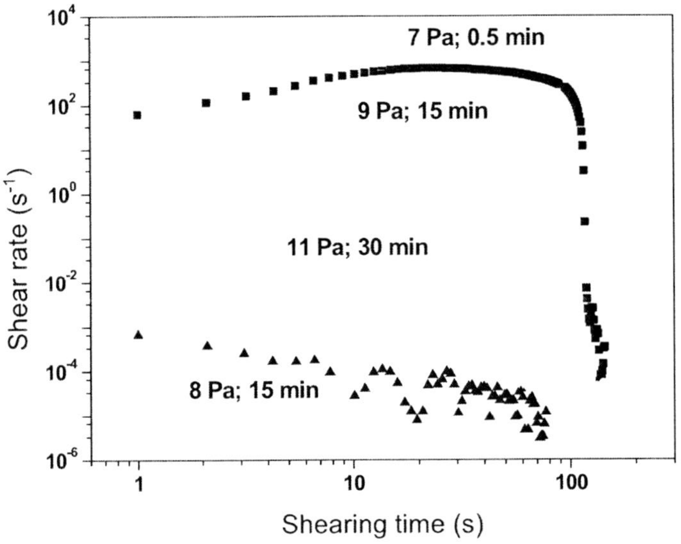

Figure 8. Shear rate as a function of shearing time for a suspension containing 10 % of iron volume fraction. Data were obtained for different values of the shear stress (σ, Pa), and at different times after placing the sample in the measuring system (T, minutes). Data labeled as: σ, T. (Reprinted with permission from [López-López *et al* (2006a)]).

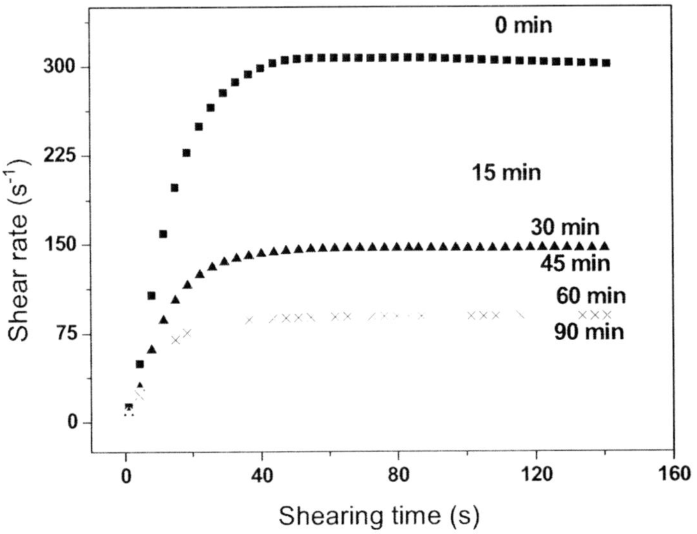

Figure 9. Similar to Figure 8, but for a suspension containing 10 % of iron and 2.8 mmol/L OA. Data were obtained for a constant shear stress of 1 Pa and at the indicated times after placing the sample in the measuring system. (Reprinted with permission from [López-López *et al* (2006a)]).

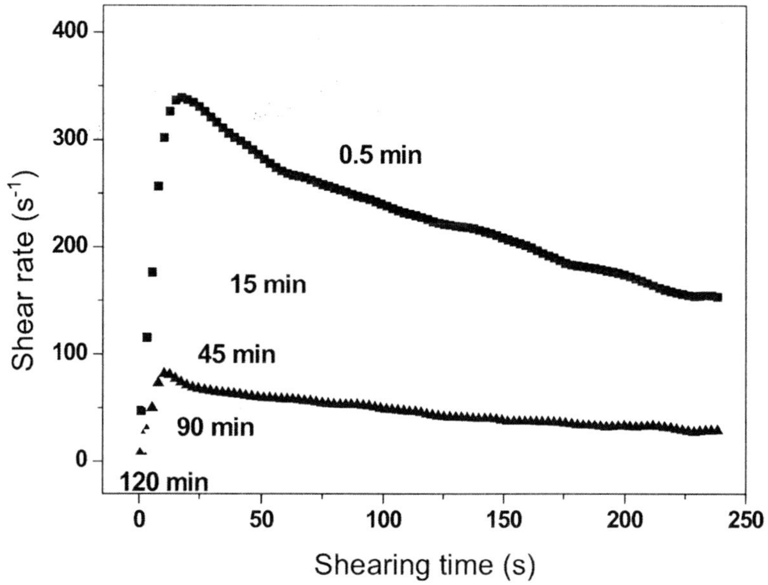

Figure 10. Shear rate as a function of shearing time for a suspension with 10 % of iron and 333 mmol/L silica. Data were obtained for σ = 10 Pa and at the indicated times after placing the sample in the measuring system. (Reprinted with permission from [López-López *et al* (2006a)]).

The results of the redispersibility study for a suspension containing 10 % of iron and 333 mmol/L silica are shown in Figure 10. Let us first explain the shape of the curves in Figure 10. As can be observed, there is an initial increment of shear rate with shearing time. Then, a maximum is reached and, finally, shear rate decreases slowly with shearing time. The initial increment corresponds to the breakage of the silica network, which culminates when the maximum in shear rate is reached. Then, iron particles sediment progressively, which provokes a thickening of the suspension (decrease of shear rate with shearing time). When the shearing is stopped the iron particles that remain in suspension are entrapped again by the silica network, but those that have settled down remain in the sediment. This is the reason why the stiffness of the suspension increases with the steps of the experiment (for the same value of the shear stress the shear rate decreases with time step –0.5, 15, …120 minutes). Therefore, we can conclude that silica nanoparticles avoid particle settling in MR fluids under rest conditions but, unfortunately, sedimentation proceeds under shearing and compact sediments are created.

4. Stability of Magnetic Fluid Composites

4.1. Preparation of Magnetic Fluid Composites

The magnetic fluid composites (MFC) studied in this chapter are composed of micron-sized iron particles (BASF, Germany) dispersed in ferrofluids, prepared as described in paragraph 2, which contain oleate-covered magnetite dispersed in kerosene. To prepare these MFC, proper amounts of iron and ferrofluids were mixed, and the mixtures were shaken and finally immersed in a Branson sonifier. All the suspensions contained the same iron volume fraction Φ= 10 %, while the magnetite volume fraction in the ferrofluids (ϕ) ranged from 0 to 24 %. The sedimentation behavior of these suspensions was studied by the electromagnetic method described in paragraph 3.

4.2. Sedimentation Study

Figure 11 shows the sedimentation behavior of iron/magnetite suspensions. Our reference is the curve in the absence of magnetite (ϕ = 0). First of all, we observe that the addition of a small amount of Fe_3O_4 implies an increase in sedimentation rate (see curves for ϕ = 0.03 and 0.06), as compared with the suspension of iron in pure kerosene (ϕ = 0). This fact can be attributed to the irreversible aggregation between iron particles in the absence of Fe_3O_4 nanoparticles, which provokes the formation of big aggregates that spread all over the tube slowing the gravitational settling. When ferrofluid carriers are used, the irreversible aggregation is prevented.

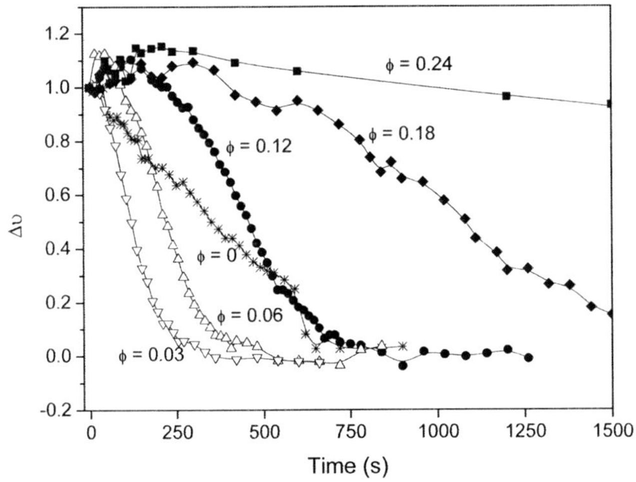

Figure 11. Dimensionless increment of the induced potential ($\Delta \upsilon$) vs time for iron suspensions (iron volume fraction Φ= 0.1) with different ferrofluid carriers. The magnetite volume fraction of the ferrofluids, ϕ, is indicated. (Reprinted from [López-López *et al* (2005c)]).

Now, we will consider only the curves corresponding to suspensions containing magnetite. As observed, an increase in magnetite volume fraction implies a decrease in the sedimentation rate that, in principle, could be attributed to the increase in the viscosity of the ferrofluid carriers. In order to check this hypothesis, a new sedimentation experiment was carried out using a Φ= 10 % iron suspension in silicone oil. The result of this experiment is shown in Figure 12. As can be seen, the sedimentation rate in silicone oil (η = 62.3 mPa·s) is even slightly higher than in the ferrofluid with ϕ= 24% (η = 40.4 mPa·s).

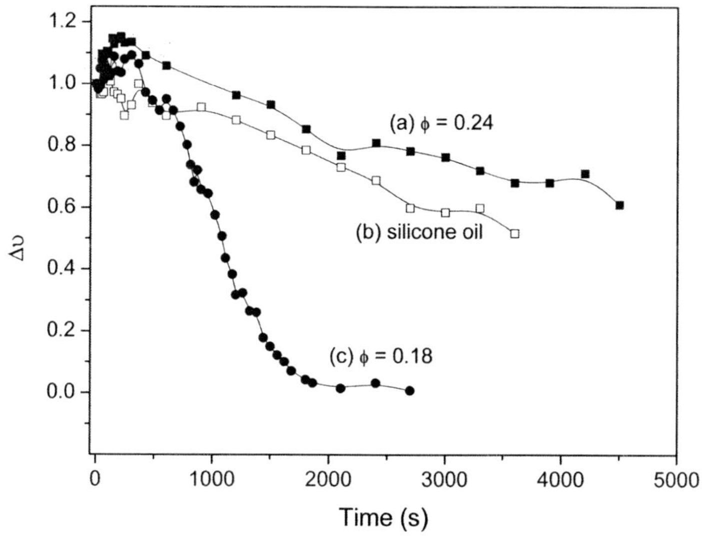

Figure 12. Similar to Figure 11, using the following carriers: (a) magnetite/kerosene ferrofluid (ϕ = 0.24; η = 40.4 mPa·s), (b) silicone oil (η = 62.3 mPa·s), and (c) magnetite/kerosene ferrofluid (ϕ = 0.18; η = 20.4 mPa·s). (Reprinted from [López-López *et al* (2005c)]).

Therefore, it seems that the progressive stabilization of the suspensions as the magnetite content increases cannot be exclusively ascribed to an increase in the drag force on the settling iron particles. As a consequence, the high stabilization achieved using a ferrofluid as continuous medium, could be associated to some kind of internal structuration of the particles in the suspension. This structure would avoid the aggregation between iron particles, favored by van der Waals and magnetic attractions.

4.3. Electron Microscopy

In order to investigate any possible internal structuration of the suspensions, TEM pictures were also taken from particles extracted from diluted (1:1000) samples. Figure 13a corresponds to particles extracted from a suspension that initially contained 10 vol% of iron. As can be seen, some iron aggregates persist even after diluting the suspension. Figure 13b corresponds to an initial iron (Φ = 10%) - magnetite (ϕ = 24%) suspension. In this case, iron aggregates have disappeared, and a "halo" of magnetite nanoparticles surrounds each particle of iron. This is presumably because of the magnetic attraction between the permanent magnetic moment of the single-domain magnetite particles and the induced magnetic moment of the iron particles.

Summarizing, the use of extremely bimodal iron-magnetite suspensions has been shown to be an efficient way to slow down settling and to prevent irreversible iron aggregation in MR fluids.

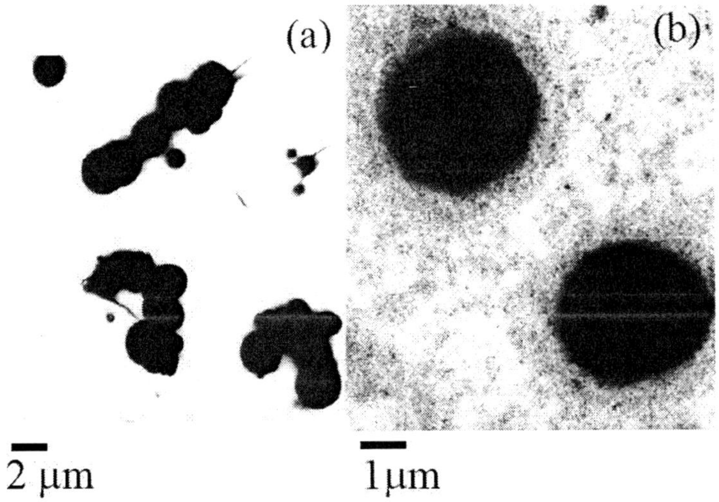

Figure 13. TEM pictures of diluted suspensions (1:1000 as compared with those used in sedimentation experiments): (a) $\Phi = 0.1$, $\phi = 0$; and (b) $\Phi = 0.1$, $\phi = 0.24$. (Reprinted from [López-López *et al* (2005c)]).

5. RHEOLOGY OF MAGNETORHEOLOGICAL FLUIDS

5.1. Experimental

Iron-silica suspensions were prepared as described in paragraph 3 using silicone oil (polidimethylsiloxane) with viscosity 31 mPa·s and density 954 kg·m^{-3} (Fluka, Germany) as liquid carrier. The magnetorheological properties of the suspensions were measured at a temperature of 25.0 ± 0.1 °C in a Bohlin CS10 controlled stress rheometer using a parallel-plate measuring cell (4 cm in diameter) (Bohlin PP-40) for gap width of 0.15 mm. The magnetic field was generated using a pair of Helmholtz coils (Phywe, Germany). The coils were placed so that their axes coincided with that of the measuring plates (see a scheme in Figure 14), thus producing a magnetic field in the vertical direction. A Hall-effect teslameter (Phywe, Germany) was used for measuring the magnetic flux density B in the air gap between the plates.

5.2. Steady-State Regime. Viscometry

To ensure reproducible results, samples were presheared for 30 s at a shear rate in the post-yield regime and zero magnetic flux density. Then, a magnetic field was applied during a 30 s waiting time (WT) with no shear applied. Finally, samples were subjected to up/down shear stress ramps, in the presence of the same magnetic field that was applied during the WT, and the corresponding shear rate, and dynamic viscosity, were measured.

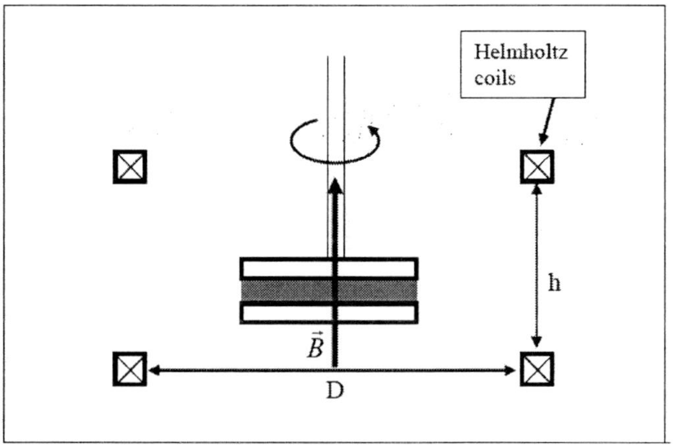

Figure 14. Schematical drawing of the rheometer used in magnetorheological experiments. Coil diameter $D = 39.5$ cm; distance between the two coils $h = 19.5$ cm. (Reprinted with permission from [de Vicente *et al* (2003)]).

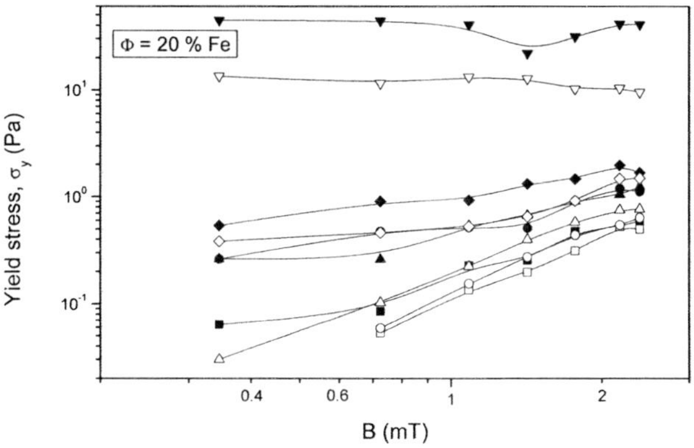

Figure 15. Yield stress as a function of magnetic flux density B for suspensions containing 20 % iron and the following silica concentrations (mmol/L): ■, □: 0; ●, ○: 122; ▲, △: 243; ◆, ◇: 486; ▼, ▽: 729. (Full symbols: increasing shear ramp; open symbols: decreasing shear ramp). (Reprinted with permission from [de Vicente *et al* (2003)]).

The values of the yield stress σ_y were determined from the central plateau in a log-log shear stress-shear rate plot. Figure 15 shows the results for $\Phi = 20$ vol%. As observed, σ_y increases with silica concentration whatever the intensity of the magnetic field. The existence of MR effect is clear for low silica content. However, the yield stress becomes independent of the external field for high silica concentration (> 486 mmol/L). As mentioned above, a silica network, which imparts a gel-like structure to the suspension, is formed. At high silica content, the network is so stiff that it hinders field-induced iron chaining and, therefore, the suspension does not show MR effect. Notice as well that in all cases the yield stress measured from the "down" curve (decreasing shear stress) is smaller than the one measured increasing the stress. This is a consequence of the gel-sol transition in these silica thixotropic suspensions.

The above mentioned iron-silica adhesion is also an additional mechanism that contributes to hindering the MR response of these fluids, since silica particles adhere on the iron ones, which limits iron approximation (under field) and, therefore, the energy of interaction between particles. A well contrasted property of MR fluids is that at low external magnetic field the yield stress can be scaled by the square of the magnetic flux density [Rankin *et al* (1998)]. The normalized yield stress (σ_y/B^2) is plotted (in a log-log scale) as a function of silica concentration for different iron-silica suspensions in Figure 16. For the lowest iron particle concentration (Figure 16a, Φ = 10 %) and at low silica content, the yield stress is effectively scaled by the square of the magnetic flux density, as observed from the initial plateau and the collapse of data corresponding to different fields. However, this collapse fails at high silica concentration and, therefore, the scaling with B^2 also fails.

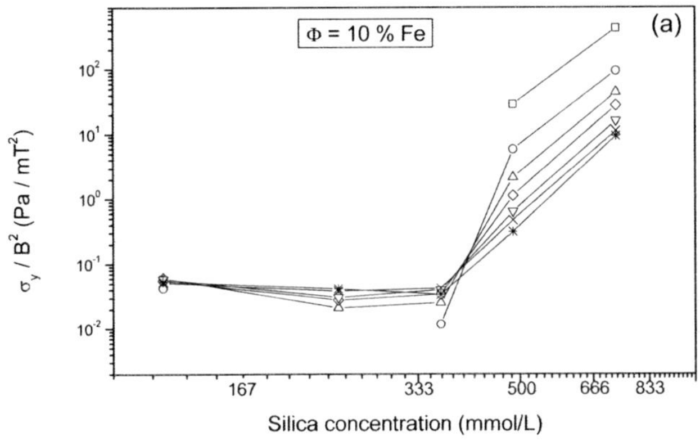

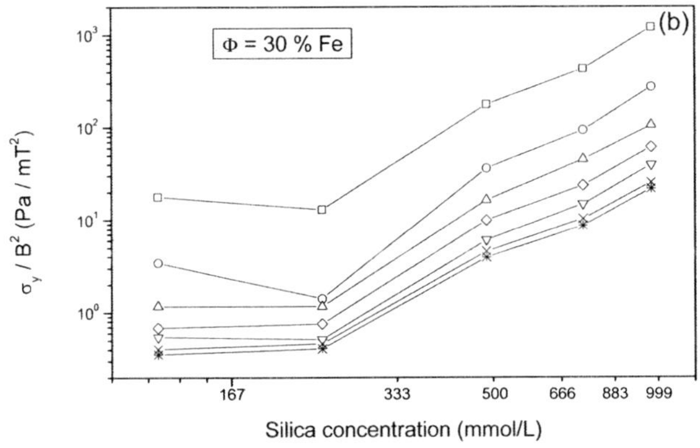

Figure 16. Normalized yield stress (σ_y/B^2) as a function of silica particle concentration for different magnetic fields applied. The iron volume fractions are: (a) 10 %; (b) 30 %. The symbols correspond to the following magnetic flux densities (mT): □: 0.345; O: 0.725; Δ: 1.085; ◇: 1.42; ∇: 1.77; ×: 2.17; *: 2.38. (Reprinted with permission from [de Vicente *et al* (2003)]).

Consequently, it seems that there is a threshold silica concentration above which the MR effect is hindered. For the highest iron particle concentration (see Figure 16b, $\Phi = 30$ %) the scaling with B^2 fails even for the smallest silica content.

The free chain model of magnetorheology [Bossis *et al* (2002); Martin and Anderson (1996)] is based on the formation of a cubic network of infinite chains of particles aligned in the direction of the field. These chains are supposed to deform affinely with the strain. The model predicts the following relation between the yield stress, σ_y, the magnetic field strength, H, and the volume fraction of multidomain particles in suspension, Φ [Bossis *et al* (2002)]:

$$\sigma_y = 2.31\Phi\mu_0 M_s^{1/2} H^{3/2}. \tag{6}$$

The linear relationship between the yield stress and the volume fraction of particles has been corroborated experimentally for not very high particle content ($\Phi < 30$) [de Gans *et al* (1999); Jolly *et al* (1996)]. Figure 17 shows the dependence of the yield stress on the particle concentration for different iron-silica suspensions.

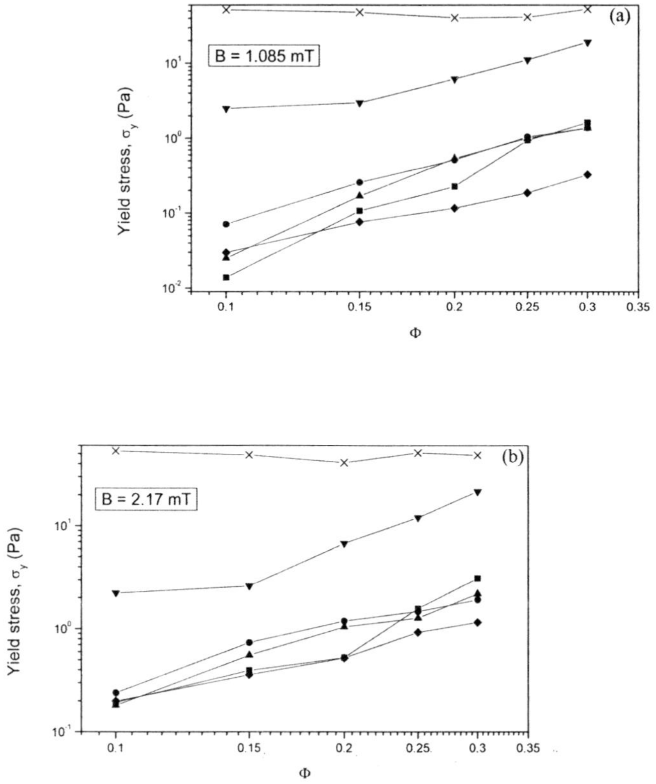

Figure 17. Yield stress as a function of the iron volume fraction for different silica contents. The magnetic flux density is fixed at: (a) 1.085 mT, and (b) 2.17 mT. The concentrations of silica (mmol/L) are: ■: 0; ●: 122; ▲: 243; ◆: 365; ▼: 486; ×: 729. (Reprinted with permission from [de Vicente *et al* (2003)]).

As can be seen, the yield stress scales linearly with iron concentration for low silica content (≤ 486 mmol/L). However, at high silica concentration (729 mmol/L) the yield stress is independent on iron concentration. This is another manifestation of the fact that for high enough silica concentration the magnetic interactions between silica-covered iron particles are strongly hindered.

5.3. Dynamic Analysis. Oscillometry

In dynamic rheological experiments, to ensure reproducibility, samples were presheared for 30 s at a large shear rate in the post-yield regime and zero magnetic flux density. Then, a magnetic field was applied during a 30 s waiting time with no shear applied. Finally, the rheological measurements were started in the presence of the same magnetic field: the low shear absolute complex viscosity plateau was determined at a frequency of 1 Hz.

The viscoelastic linear region (VLR) can be found by measuring the storage and loss moduli as a function of the stress amplitude at constant frequency. From representations of the complex viscosity, $|\eta^*|$, as a function of the amplitude of the shear stress, σ_0, we obtained the values of the absolute complex viscosity (corresponding to the low shear plateau) as a function of the external magnetic field, for different silica concentrations in the suspension. In Figure 18 we show the results obtained for suspensions containing 10 vol% iron and different silica concentrations. As can be observed, for low silica concentration the complex viscosity increases with external magnetic field.

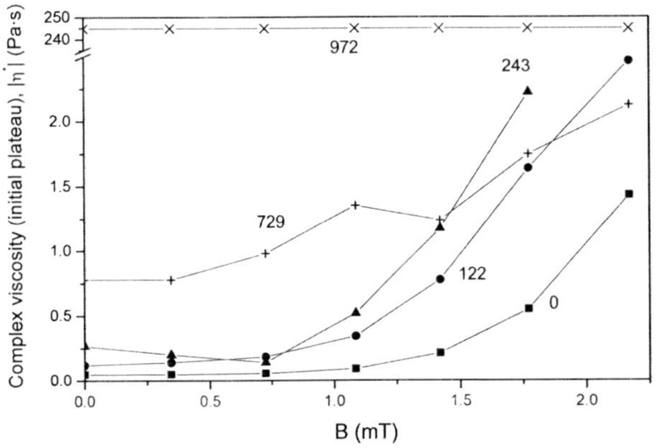

Figure 18. Absolute complex viscosity (corresponding to the initial plateau in $|\eta^*|$ vs $\sigma 0$ curves) as a function of the magnetic flux density for different silica contents (labeled in the Figure, in mmol/L). The iron volume fraction is 10 %. (Reprinted with permission from [de Vicente et al (2003)]).

However, for high silica concentration the complex viscosity practically does not change with the field. This confirms the results obtained in the viscometry experiments: for high enough silica concentration the MR effect is considerably hindered because of the silica gel network that prevents iron chaining, and the silica cover around the iron particles that limits field-induced particles approximation.

5.4. Analysis in the Time Domain. Creep-Recovery

These experiments were carried out as follows: firstly, samples were presheared for 30 s at a shear rate in the post-yield regime and zero magnetic flux density, to ensure a uniform distribution of particles and reproducible initial conditions. Then, a waiting time was allowed with no shear applied and at zero magnetic field. Finally, the rheological measurements were started: a constant stress σ (for $\Phi = 10$ % Fe: $\sigma = 0.06$ Pa for suspensions without silica, and $\sigma = 0.1$ Pa for 486 mmol/L silica concentration; for $\Phi = 30$ % Fe: $\sigma = 0.2$ Pa for suspensions without silica, and $\sigma = 3$ Pa for 486 mmol/L silica) was applied, and the compliance, or strain per unit stress, $J(t) = \gamma(t)/\sigma$, was measured during 10 s. Immediately afterwards, the stress was removed and the recovery curve was determined during another 10 s. Because the thixotropic behavior of the suspensions containing silica particles implies a time dependence of their rheological properties, the creep-recovery behavior strongly depends on the waiting time. Thus, the effect of different waiting times (WT = 2, 20 and 40 s) was studied.

Figure 19 shows the results of the creep-recovery tests for suspensions containing 10 % (Figure 19a) and 30 % (Figure 19b) iron volume fraction in the absence of silica. For the lowest iron content (Figure 19a), firstly, we observe that as the waiting time is increased the suspension experiences a more liquid-like behavior (the compliance increases), both in the presence and in the absence of magnetic field. This is a consequence of the gravitational particle settling. Secondly, for equal waiting time the application of a field implies a more solid-like behavior (the compliance decreases), which is due to the field-induced particles structuration. The behavior of the suspension with the highest concentration of iron (30 vol%, see Figure 19b) is completely different. As can be observed, the curves for different waiting times in the absence of field collapse. This is due to the fact that sedimentation rate is low in very loaded suspensions and, therefore, a similar sedimentation happens for the three waiting times studied. On the other hand, the application of an external magnetic field implies a sharp increment of the compliance (more liquid-like behavior). This contradictory behavior (the application of a field is expected to "harden" the magnetorheological suspension) is due to the field-induced formation of large flocculi that sediment faster under the effect of the gravitational field. A similar flocculation effect was previously reported on the stability of dilute aqueous suspensions of cobalt ferrite [de Vicente *et al* (2000)].

The results for suspensions containing a high concentration of silica (486 mmol/L) are plotted in Figure 20. As can be seen, for low iron content (10 vol%, see Figure 20a) the curves for the different waiting times collapse. This suggests that the silica network is completely built after only 2 s. It is also observed that the application of an external magnetic field provokes a much more solid-like behavior (much lower compliance), as a consequence of the field-induced structuration. However, for the highest concentration of iron (30 vol%; see Figure 20b) there is no effect of the application of the magnetic field. It seems that in this loaded suspension the field is not able to align the iron particles, which are entrapped by the silica network. In relation to the effect of waiting time on this suspension, it is observed that the compliance decreases (more solid-like behavior) (Figure 20b) as this variable increases.

This is a consequence of the very slow building process of the silica gel towards its final configuration, which is caused by the very high iron content.

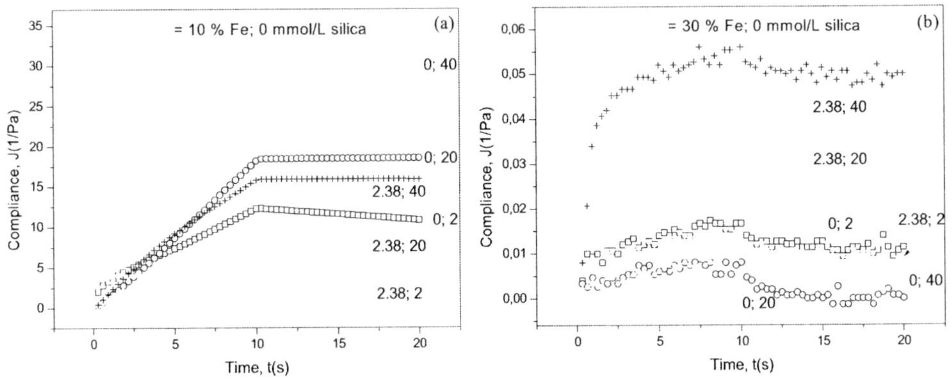

Figure 19. Creep-recovery results in suspensions of 10 % (a) and 30 % (b) iron volume fraction for different magnetic fields (*B*, mT) and waiting times (WT, s) in the absence of silica particles. Data labelled as: *B*; WT. (Reprinted with permission from [de Vicente *et al* (2003)]).

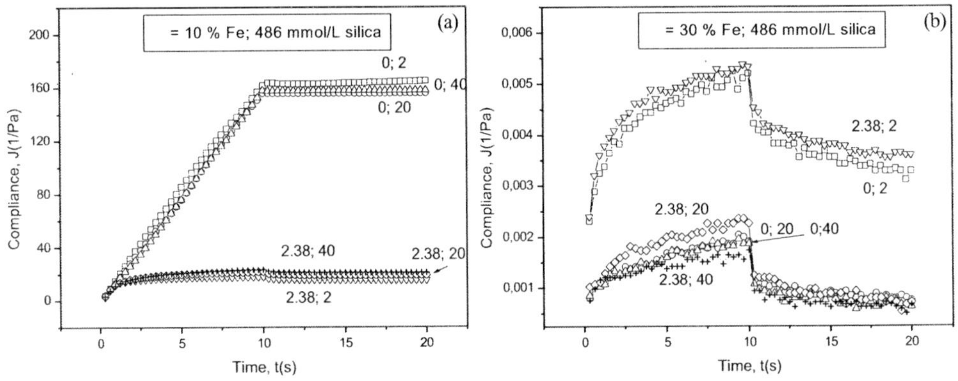

Figure 20. Same as Figure 19, but in the presence of 486 mmol/L silica particles. (Reprinted with permission from [de Vicente *et al* (2003)]).

These results show, again, that for high enough silica content the magnetorheological effect is strongly masked, as a consequence of the silica gel network, which entraps the iron particles and, therefore, hinders field-induced iron aggregation.

6. Rheology of Magnetic Fluid Composites

6.1. Experimental

Oleate-covered magnetite ferrofluids and pure kerosene (viscosity 16.6 ± 1.2 mPa·s) were used as dispersing media. Micron-sized iron particles were dispersed in both magnetite/kerosene ferrofluids and kerosene. The suspensions were prepared as described in paragraph 4. The iron volume fraction, Φ, ranged from 10 to 31.6 % while magnetite volume fraction, ϕ, was from 0 to 21.6 %. After preparing the suspensions, approximately 110 μL of them were placed between the plates of the magnetorheometer.

Magnetorheological properties of the suspensions were measured at 25.0 ± 0.1 °C in a controlled rate magnetorheometer (MCR 300, Physica, Germany). The measuring system geometry was a 20 mm diameter parallel-plate set for gap width of 0.35 mm. The steady-state regime was investigated as follows: samples were subjected to a shear rate ramp and the corresponding shear stress, and dynamic viscosity, were measured. In order to ensure a uniform distribution of the particles and reproducible initial conditions, samples were initially presheared for 30 s at a shear rate in the post-yield regime. Afterwards, an external magnetic field (up to H = 343 kA/m) was applied during a 30 s waiting time without shearing. Finally, rheological measurements were started in the presence of the same magnetic field strength applied during the waiting time.

6.2. Magnetoviscous Effect: Chain Model

As mentioned above, the free chain model of magnetorheology [Bossis *et al* (2002); Martin and Anderson (1996)] is based on the formation of a cubic network of infinite chains of particles aligned in the direction of the field. These chains are supposed to deform affinely with the strain. The model predicts the following simple relation between the dimensionless field-specific viscosity, η_F, and the Mason number, Mn [Martin and Anderson (1996)]:

$$\eta_F \equiv \frac{\eta - \eta_\phi}{\Phi \eta_c} = \frac{1}{5}\left(\frac{3}{2}\right)^{5/2} Mn^{-1} \tag{7}$$

where η is the apparent viscosity, η_c is the carrier viscosity, η_ϕ is the suspension viscosity at infinite Mn and Φ is the volume fraction of multidomain particles. The Mason number is defined as the ratio between hydrodynamic and magnetostatic forces:

$$Mn = \frac{\eta_c \dot{\gamma}}{2\mu_0 \mu_c \beta^2 H^2} \tag{8}$$

where $\dot{\gamma}$ is the shear rate and β the magnetic contract factor. μ_0 and μ_c are the relative permeability of vacuum and the carrier, respectively.

As an example, Figure 21 shows the dependence of η/Φ on the shear rate $\dot{\gamma}$ scaled by H^2 for an iron suspension (Figure 21a) and for and iron-magnetite suspension (Figure 21b). As can be seen, the curves corresponding to different magnetic field strengths collapse both for iron and for iron-magnetite suspensions, although the coincidence is slightly better in the iron-magnetite suspensions (Figure 21b). In addition, it can be observed that the values of the reduced viscosity, η/Φ, are slightly higher in the iron-magnetite suspension than in the iron suspension. Similar results were found for the other suspensions, not shown here for simplicity. The better collapse in the presence of magnetite, together with the higher values of the reduced viscosity, indicate the presence of stronger chain-like structures in iron-magnetite suspensions than in iron suspensions.

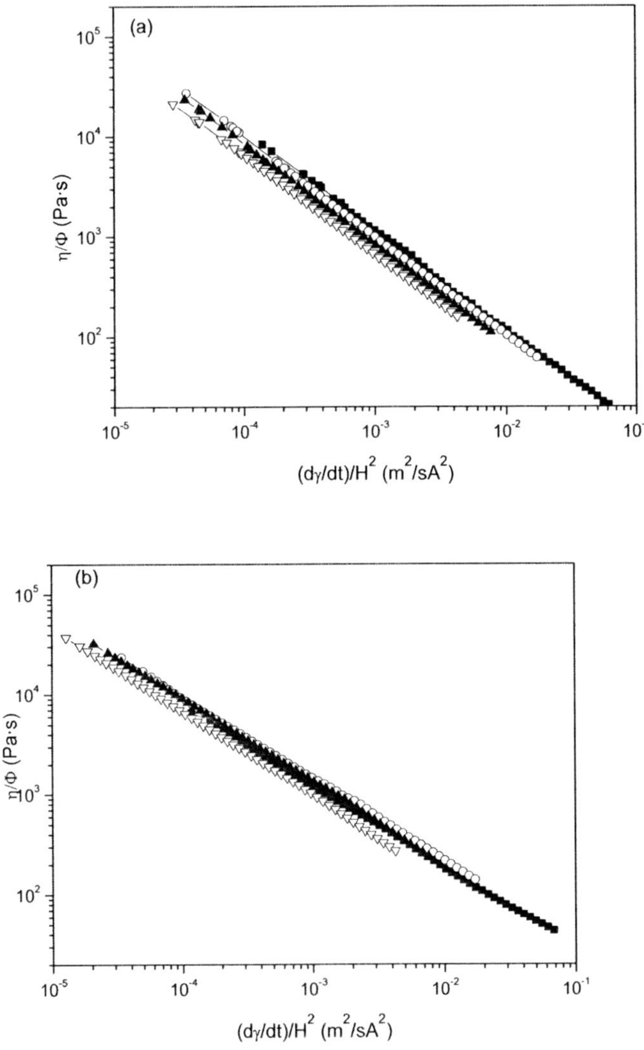

Figure 21. Viscosity scaled by iron volume fraction, η/Φ, as a function of shear rate scaled by the square of the external magnetic field strength, $(d\gamma/dt)/H^2$. (a) Suspension containing 31.6 % of iron; (b) suspension containing 10 % of iron and 21.6 % of magnetite. Magnetic field strength: ■: 85.7 kA/m; O: 171 kA/m; : 122; ▲: 257 kA/m; ∇: 343 kA/m.

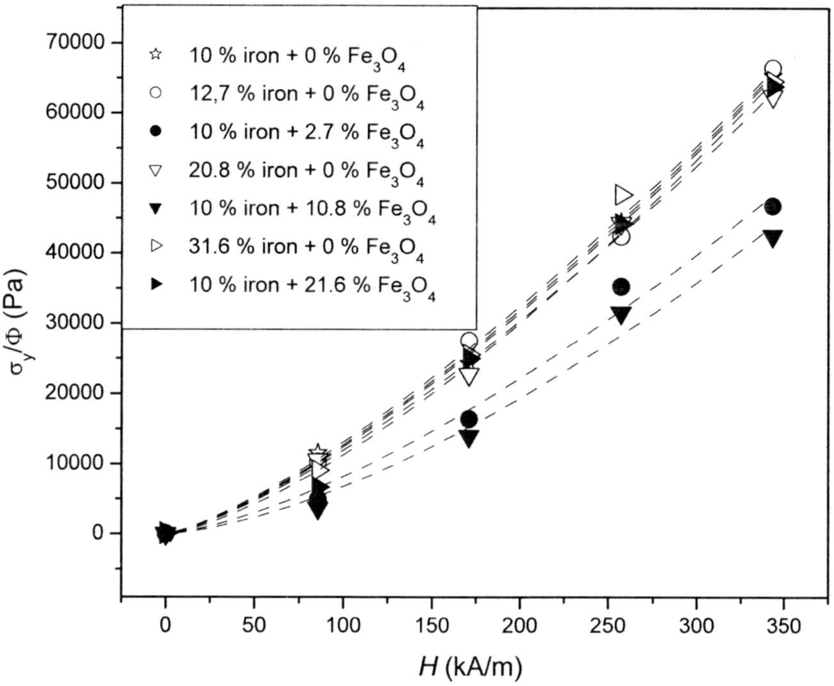

Figure 22. Scaled static yield stress, σ_y/Φ, as a function of the external magnetic field strength for different iron-magnetite suspensions. Dash lines correspond to the best fit to $\sigma_y/\Phi = bH^p$.

6.3. Magnetorheological Effect: Static Yield Stress

The chain model predicts a proportionality between the static yield stress, σ_y, and $H^{3/2}$ (see equation 6). In Figure 22 the static yield stress scaled by iron volume fraction (σ_y/Φ) is plotted as a function of the external magnetic field strength. First of all, note that there exists a linear dependence between the static yield stress and the iron volume fraction in iron suspensions, since the curves corresponding to different iron suspensions collapse. Secondly, Figure 22 shows that the addition of low to moderate magnetite content (2.7 % or 10.8 %) implies a decrease in the scaled yield stress (σ_y/Φ) when it is plotted as a function of the external magnetic field. Finally, the addition of a high magnetite amount (21.6 %) does not produce any significant change in the magnitude of the scaled yield stress (see Figure 22), as compared with the suspensions that only contain iron. At this point, it is important to mention that different trends are obtained when this plot is performed as a function of the internal magnetic field (see [López-López *et al* (2006b)]).

The experimental points in Figure 22 were fitted to $\sigma_y/\Phi = bH^p$ and p values corresponding to the best fits are shown in Table 4. From these values two conclusions can be reached. Firstly, all p values are close to that predicted by the model (1.5; see equation 6); secondly, the observed differences between p values in iron-magnetite suspensions and iron suspensions are not significant: a Student t test, at 95 % of confidence, demonstrated that no significant differences exist. Nevertheless, in the absence of magnetite all p values are significantly smaller than 1.5, whereas in the presence of magnetite they are very close or even higher than 1.5.

Table 4. p values obtained by fitting the equation $\sigma_y / \Phi = bH^p$ to the experimental data in Figure 22

Suspension	10% Fe + 0% Fe_3O_4	12.7% Fe + 0% Fe_3O_4	10% Fe + 2.7% Fe_3O_4	20.8% Fe + 0% Fe_3O_4	10% Fe + 10.8% Fe_3O_4	31.6% Fe + 0% Fe_3O_4	10% Fe + 21.6% Fe_3O_4
p	1.34±0.06	1.33±0.09	1.44±0.20	1.35±0.09	1.52±0.22	1.31±0.12	1.42±0.10

These results suggest that in the presence of magnetite "regular" chains are built whereas they are not in the absence of these ferromagnetic nanoparticles.

6.4. Magnetorheological Effect: Dynamic Yield Stress

Figure 23 shows the dynamic yield stress scaled by iron volume fraction as a function of the external magnetic field strength. If this graph is compared with that corresponding to the static yield stress (Figure 22), two important differences can be found. Firstly, there is no linear dependence between the dynamic yield stress and the iron volume fraction in iron suspensions, since the curves corresponding to different iron suspensions do not collapse in this case. This is probably due to the irreversible aggregation between iron particles in the absence of magnetite, which hinders normal iron chaining upon application of an external magnetic field. This irreversible aggregation is more significant the higher the volume fraction of solids, and so, the smallest values of the scaled dynamic yield stress are obtained for the most loaded suspension (31.6 % of iron).

The second difference is related to the magnitude of the yield stress for iron-magnetite suspensions in respect to iron suspensions. Whereas in Figure 22, the values of the scaled static yield stress for iron-magnetite suspensions are smaller than or equal to those corresponding to iron suspensions, in Figure 23, the scaled dynamic yield stresses for suspensions in the presence of moderate and high magnetite content are higher than those corresponding to iron suspensions. This is a consequence of the fact that dynamic yield stress gives a quantitative measurement of the shear stress required to break all internal structures in suspension (suspensions flow like a Newtonian liquid), while static yield stress corresponds to the point where suspensions start to flow. Therefore, this last difference is due to the fact that in iron-magnetite suspensions the internal structure is more complex than in iron suspensions (see Figure 13). Finally, as can be seen in Figure 23, the addition of a small amount of magnetite implies a decrease in dynamic yield stress and then, as magnetite concentration increases, dynamic yield stress increases.

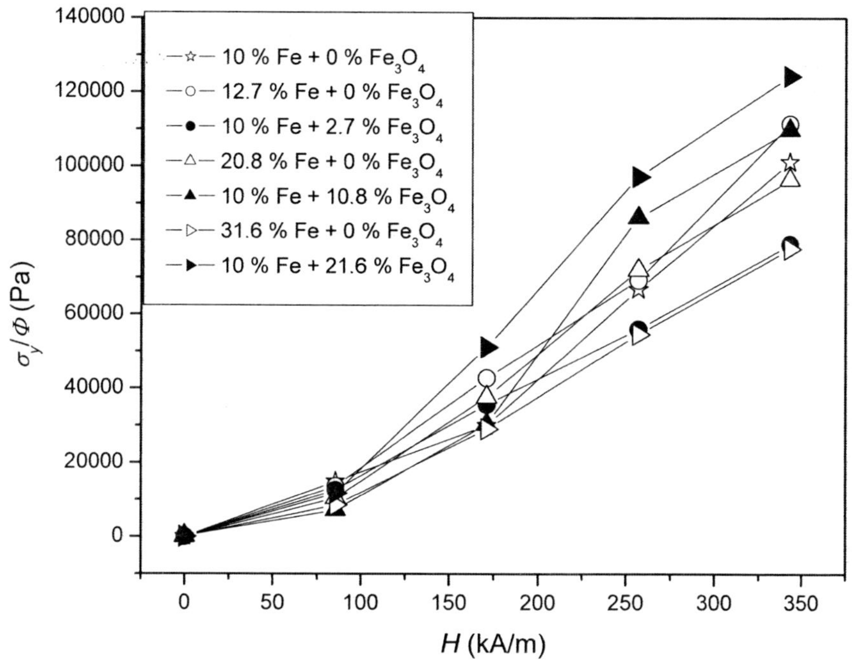

Figure 23. Scaled dynamic yield stress, σ_y/Φ, as a function of the external magnetic field strength for different iron-magnetite suspensions.

This behavior is probably due to the formation of heterogeneous iron-magnetite aggregates that avoid irreversible aggregation between particles and facilitate iron chaining. Finally, it is important to remark that in a previous work [López-López *et al* (2006b)] finite element calculations were used for calculating theoretically the values of the yield stress, predicted by the chain model, for MFC similar to these studied in this chapter. In this article, a comparison of theoretical values with experimental ones was performed, and it was found that this theory did not predict the experimental trends.

7. Conclusion

In this work, we firstly described the preparation of stable ferrofluids in non-polar carriers and the evaluation of their surface thermodynamic and magnetic properties. Secondly, magnetorheological fluids were prepared, using surfactants and gel-forming agents as stabilizing additives, and their stability properties were investigated. Afterwards, magnetic fluid composites constituted by micron-sized iron particles dispersed in the previously prepared ferrofluids were prepared and their stability properties investigated. Finally, the magnetorheological properties of both MR fluids stabilized by the addition of gel-forming agents (silica nanoparticles) and magnetic fluid composites were studied.

The most relevant conclusions of this work are summarized as follows:

a) It is possible to prepare stable ferrofluids in highly non-polar liquid carriers by chemical co-precipitation of ferric and ferrous ions in the presence of oleic acid. The thermodynamic analysis demonstrated that the loss of the stability when the dielectric constant of the carrier liquid is increased, is a consequence of the combined action of van der Waals and solvation (lyophobic) forces. The magnetization analysis of the ferrofluids indicates that there exists some degree of particle structuration, induced by both interfacial and magnetic attractions between the particles dispersed in the liquid phase, even in carriers with a very low dielectric constant.
b) A method for studying the sedimentation of opaque magnetic suspensions has been described. This method is based on the evolution of the electromotive force induced in a sensing coil, placed around the settling suspension, as the magnetic particles sediment and their concentration decreases in the region surrounded by the sensing coil.
c) It is feasible to stabilize concentrated iron suspensions against irreversible aggregation processes in oil media by means of the adsorption of fatty acids (oleic acid) or salts (aluminum stearate) on the magnetic particles. The adsorbed molecules impart the needed steric barrier to hinder agglomeration among iron particles. Although this steric repulsion does not avoid particle settling in concentrated magnetorheological fluids ($\Phi \sim 10$ %), the rheological tests showed that the redispersibility of these suspensions is considerably enhanced.
d) The addition of silica nanoparticles to stabilize concentrated iron suspensions in oil carriers against aggregation and sedimentation processes is a very efficient mechanism under rest conditions. Unfortunately, under shearing, particle settling is facilitated by the breakage of the silica network and compact sediments, which make difficult the redispersion, are created.
e) It is possible to stabilize magnetorheological fluids against aggregation and sedimentation processes by using a ferrofluid as carrier fluid. The magnetic attraction between the nanoparticles (magnetically single-domain) and the micron-sized particles (magnetically multi-domain) favors the formation of clouds of magnetite nanoparticles around each iron one.
f) When silica nanoparticles are used as stabilizing additives in magnetorheological fluids, the response of the suspensions to external magnetic fields can be considerably hindered. Indeed, for high silica content the gel network formed by the SiO_2 particles could be stiff enough to completely prevent field-induced aggregation. Additionally, the existence of heterogeneous iron-silica adhesion contributes to diminish the MR response of these fluids, since the physical approximation of iron particles is limited.
g) The use of appropriate ferrofluids as carriers for the preparation of iron-based magnetorheological fluids improves the MR response of these fluids. This is probably due to the formation of heterogeneous iron-magnetite structures that avoid irreversible aggregation between particles and facilitate iron chaining by decreasing the number of defects in the chain.

Therefore, it can be concluded that the use of magnetorheological fluids constituted by micron-sized iron particles dispersed in stable oleate-covered magnetite ferrofluids is the best alternative to obtain stable magnetorheological fluids that experience a high magnetorheological effect.

ACKNOWLEDGMENTS

Financial support by Ministerio de Educación y Ciencia (Spain) and FEDER funds (EU) under Project No. MAT2005-07746-C02-01, and Junta de Andalucía (Spain) under Project FQM410 are gratefully acknowledged. One of the authors (M.T. López-López) also acknowledges financial support by Secretaría de Estado de Universidades e Investigación (Ministerio de Educación y Ciencia, Spain) through its Postdoctoral Fellowship Program (No. EX2006-0467).

REFERENCES

Bossis, G.; Volkova, O.; Lacis, S.; Meunier, A. In *Ferrofluids*; Odenbach, S.; Ed.; Springer: Berlin, Germany, 2002; Vol. 1, pp 202-230.

Charles, S. W. In *Ferrofluids*; Odenbach, S.; Ed.; Springer: Berlin, Germany, 2002; Vol. 1, pp 3-18.

Chen, Z. Y.; Tang, X.; Zhang, G. C.; Jin, Y.; Ni, W.; Zhu, Y. R. In *Proc. 6th Int. Conference on Electro-rheological Fluids, Magneto-rheological Suspensions and their Applications*; Nakano, M.; Koyama, K.; Ed.; World Scientific: Singapore, 1998; Vol. 1, pp 486-493.

Chin, B. D.; Park, J. H.; Kwon, M. H.; Park, O. O. *Rheol. Acta* 2001, 40, 211-219.

Cullity, B. D. *Introduction to Magnetic Materials*; Addison-Wesley: Reading, Massachusetts, 1974.

Dang, A.; Ooi, L.; Fales, J.; Stroeve, P. *Ind. Eng. Chem. Res.* 2000, 39, 2269-2274.

De Gans, B. J.; Hoekstra, H.; Mellena, J. *Faraday Disc.* 1999, 112, 209-224.

De Vicente, J.; Delgado, A. V.; Durán, J. D. G.; González-Caballero, F. *Langmuir* 2000, 16, 7954-7961.

De Vicente, J.; López-López, M. T.; González-Caballero, F.; Durán, J. D. G. *J. Rheol.* 2003, 47, 1093-1109.

Garnett J. C. M. *Philos. Trans. R. Soc. London,* 1904, 203, 385-420.

Ginder, J. M. In *Encyclopedia of Applied Physics*; Trigg, G. L.; Ed.; VCH: Weinheim, Germany, 1996; Vol. 16, pp 487-503.

Ginder, J. M. *MRS Bulletin* 1998, 23, 26-29.

Jolly, M. R.; Carlson, J. D.; Muñoz, B. C. *Smart Mater. Struct.* 1996, 5, 607-614.

Kaiser, R.; Miskolczy, G. *J. Appl. Phys.* 1970, 41, 1064-1072.

López-López, M. T.; Durán, J. D. G.; Delgado, A. V.; González-Caballero, F. *J. Colloid Interf. Sci.* 2005a, 291, 144-151.

López-López, M. T.; de Vicente, J.; González-Caballero, F.; Durán, J. D. G. *Colloid Surface A* 2005b, 264, 75-81.

López-López, M. T.; de Vicente, J.; Bossis, G.; González-Caballero, F.; Durán, J. D. G. *J. Mater. Res.* 2005c, 20, 874-881.

López-López, M. T.; Zugaldía, A.; González-Caballero, F.; Durán, J. D. G. *J. Rheol.* 2006a, 50, 543-560.

López-López, M. T.; Kuzhir, P.; Lacis, S.; Bossis, G.; González-Caballero, F.; Durán, J. D. G. *J. Phys.: Condens. Matter* 2006b, 18, S2803-S2813.

Martin, J. E.; Anderson, R. A. *J. Chem. Phys.* 1996, 104, 4814-4827.

Massart, R. *IEEE T. Magn.* 1981, MAG-17, 1247-1248.

Odenbach, S. *Colloid Surface A* 2003, 217, 171-178.

Park, J. H.; Chin, B. D.; Park, O. O. *J. Colloid Interf. Sci.* 2001, 240, 349-354.

Phulé, P. P.; Ginder, J. M. In *Proc. 6th Int. Conference on Electro-rheological Fluids, Magneto-rheological Suspensions and their Applications*; Nakano, M.; Koyama, K.; Ed.; World Scientific: Singapore, 1998a; Vol. 1, pp 445-453.

Phulé, P. P.; Ginder, J. M. *MRS Bulletin* 1998b, 23, 19-21.

Rankin, P. J.; Ginder, J. M.; Klingenberg, D. J. *Curr. Opin. Colloid In.* 1998, 3, 373-381.

Rankin, P. J.; Horvath, A. T.; Klingenberg, D. J. *Rheol. Acta* 1999, 38, 471-477.

Rosensweig, R. E. *Sci. Am.* 1982, 247, 136-and.

Van Ewijk, G. A.; Philipse, A. P. *Langmuir* 2001, 17, 7204-7209.

Van Ewijk, G. A.; Vroege, G. J.; Philipse, A. P. *J. Magn. Magn. Mater.* 1999, 201, 31-33.

Van Oss, C. J. *Interfacial Forces in Aqueous Media*; Marcel Dekker: New York, New York, 1994.

Volkova, O.; Bossis, G.; Guyot, M.; Bashtovoi, V.; Reks, A. *J. Rheol.* 2000, 44, 91-104.

In: New Colloid and Surface Science Research
Editor: Lawrence V. Schwartz, pp. 197-219
ISBN: 978-1-60021-594-0

Chapter 5

EFFECT OF SURFACE CHARGE MIGRATION ON THE STABILITY OF EMULSION AND COLLOIDAL PARTICLES IN WATER

Tsuneki Ichikawa
Division of Materials Chemistry, Graduate school of Engineering,
Hokkaido University, Sapporo, 060-8628 Japan

ABSTRACT

Although the DLVO theory is a fundamental theory for explaining the stability of charged emulsion and colloidal particles in water, the treatments of the theory developed so far have a fundamental imperfection. The theory shows that the force preventing the coagulation and coalescence of charged particles is the double layer force that is generated by the overlap of diffuse electric double layers around the particles. The double layer force depends on the surface charge density and the separation distance between the surfaces of adjacent particles, whereas the surface charge density depends on the separation distance. The self consistency should therefore be attained between the surface charge density and the separation distance. However, except for our recent study, no attempt has been made for attaining the self consistency.

Here we have obtained the double layer potential energy for the front surfaces of approaching two charged particles by taking the migration of surface charge into account. The potential energy including the van der Waals energy is given by

$$U = \frac{\varepsilon_l \kappa \varsigma^2 \mathrm{e}^{-\kappa w}}{8(h+1)} \left[\frac{[4(h+1)+3h(a_1-a_2)E_0\cos\eta/\varsigma]^2}{h+1+(h-1)\mathrm{e}^{-\kappa w}} - \frac{[3h(a_1+a_2)E_0\cos\eta/\varsigma]^2}{h+1-(h-1)\mathrm{e}^{-\kappa w}} \right] - \frac{A_H}{12\pi w^2}$$

where $h = z_s e\zeta\,(s_{+,0} + s_{-,0})/\,[kT\,(s_{+,0} - s_{-,0})]$; z_s, the valence of surface ions; e, the unit charge; $s_{+,0}$ and $s_{-,0}$, number densities of positive and negative surface ions; kT, the thermal energy at temperature T; ε_l, the dielectric constant of water; κ, the Debye reciprocal length; ζ, the zeta potential; w, the separation distance between the front surfaces of approaching two spheres with radii a_1 and a_2; E_0, the intensity of an external electric field; η, the angle between the axis of approach and the external field; and A_H, the Hamaker constant of the particles, respectively. The above equation indicates that emulsion and colloidal systems can be destabilized by applying external electric fields. Experimental evidences of the efficient destruction of an oil-in-water emulsion by low external electric fields are also shown.

1. INTRODUCTION

Emulsion and colloidal systems have been widely used in our daily life as foods, cosmetics, medicines and so on. They are composed of tiny liquid and solid particles suspended in liquids. Three types of forces are known to prevent the coagulation of the particles in water. They are the double layer force acting between charged particles in ionic solutions [1, 2], the short-range hydration force acting between hydrated surfaces in water [3, 4], and the steric force acting between surfaces adsorbing macromolecules [5, 6]. The double layer force is generated by the overlap of diffuse electric double layers of adjacent particles and is the most fundamental force for stabilizing charged particles in water. The energy barrier preventing the coagulation of two charged particles is estimated by the DLVO theory [1, 2]. According to the theory, the barrier height is determined by the attractive van der Waals force and the repulsive double layer force that is composed of attractive Maxwell's electric field stress and repulsive osmotic pressure. The strength of the double layer force is determined by the concentration of ions in water and the surface charge densities of adjacent particles.

Although the DLVO theory is a fundamental theory for explaining the stability of charged emulsion and colloidal particles in water, as far as we know, the theory has not been strictly applied to particles with mobile surface charges. The difficulty in the application arises from the correct estimation of the surface charges of approaching particles. The surface charge densities are uniform and the same for all the particles, as long as separation distances between the particles are much longer than the thickness of diffuse electric double layers generated around the particles. Approach of two particles closer than the thickness of the double layers causes electrostatic interactions between the two surfaces, which results in the migration of surface charges. The surface charge densities should therefore be determined as a function of the separation distance. However, no such an attempt has been made except for our recent work [7].

Three methods have been proposed so far for obtaining the double layer force without calculating the surface charge densities as a function of separation distance. They are the fixed potential approximation, the fixed surface charge approximation, and the linear superposition approximation. The fixed potential approximation assumes that the surfaces of approaching two particles are completely conducting like metal-coated charged particles [8]. The surface electrostatic potentials of conducting particles are homogeneous, so that the surface potentials are kept constant during the approach. The fixed potential approximation necessitates very high density of charge carriers on the surfaces, which may not be the case for regular emulsion and colloidal systems. The fixed charge approximation assumes that the charges are immobilized on the surfaces. The problem of the fixed charge approximation is that the two surfaces never coagulate, since the surface electrostatic potential and therefore the double layer force increase infinitely with decreasing separation distance [9]. The linear superposition approximation assumes that the electrostatic potential. and the distribution of ions between two charged surfaces are given by the superposition of those for the isolated particles. Although this approximation gives more reasonable results, since the approximation lies between the former two extreme ones, there is no physical base for justifying the approximation. To determine the surface charge densities as a function of separation distance has therefore been expected.

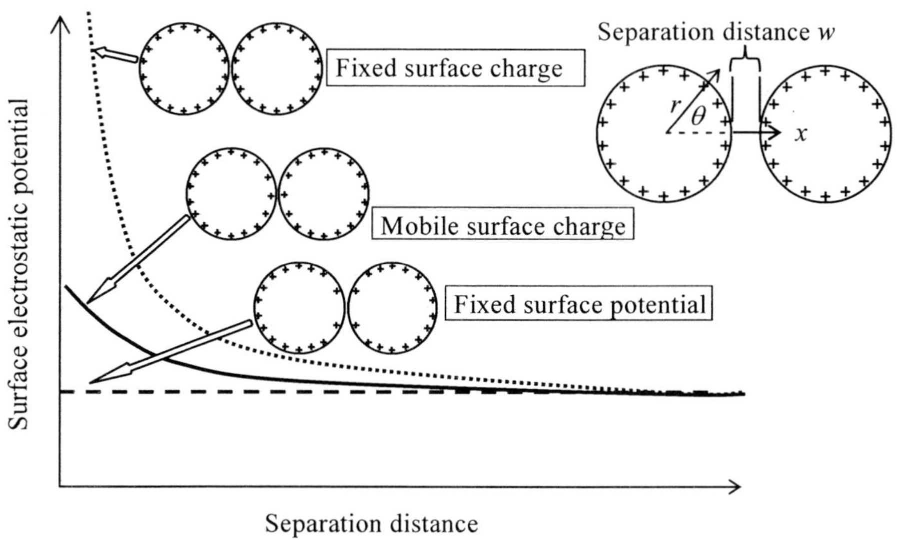

Figure 1. Schematic representation of the change of surface charge densities and potentials during the approach of two charged particles. The coordinate used for the calculation of surface electrostatic potential is also shown.

To know the effect of surface charge migration on the stability of emulsion and colloidal systems is also important from the practical point of views. Formation of emulsions in solvent extraction processes is troublesome since the processes are stopped until the emulsions are separated into oil and water. To find an efficient method of demulsification without contaminating the systems has therefore been expected. We have recently found that dense oil-in-water emulsions are demulsified very quickly by applying a low electric field of less than 10 V/cm [11, 12]. Based on the experimental and theoretical analyses of this new phenomenon, we have concluded that the applied field induces the rearrangement of surface charge on each oil droplet to compensate the external field-induced potential gradient, which causes the polarization of surface electrostatic potentials and therefore results in the hetero-coagulation and coalescence of oil droplets in homogeneous emulsion systems.

The rigorous estimation of the stability of approaching charged particles necessitates the determination of electrostatic potential in water by taking the mutual interaction of surface charges into consideration.

In the present chapter, the electrostatic potential and the distribution of surface charges are determined as a function of the separation distance by applying the Poisson-Boltzmann equation to ions both in water and on the surfaces of charged particles. The potential energy barrier preventing the coagulation and coalescence of the particles is then estimated by applying the DLVO theory. The results show how the migration of surface charge drastically changes the stability of emulsion and colloidal systems.

2. Fundamental Equation

The free energies of positive and negative ions, F_+ and F_-, respectively, in an ionic solution under no flow of ions are given by

$$F_+ = (U_+ + ez_+\phi) + kT\log c_+, \quad F_- = (U_- - ez_-\phi) + kT\log c_- \tag{1}$$

where $U_\pm$, $\pm z_\pm$, and $c_\pm$ are solvation energy, valence and number density of positive or negative ions, respectively; e, the unit charge; ϕ, the electrostatic potential; k, the Boltzmann constant; and T, the absolute temperature. The free energies under the equilibrium do not depend on the location of ions. Assuming that the solvation energies are constant, Eq. (1) gives

$$\nabla F_+ = ez_+\nabla\phi + \frac{kT}{c_+}\nabla c_+ = 0, \quad \nabla F_- = -ez_-\nabla\phi + \frac{kT}{c_-}\nabla c_- = 0 \tag{2}$$

The general solution of Eq. (2) satisfying the neutral condition of $z_+c_+ - z_-c_- = 0$ at $\phi = 0$ is given by the Boltzmann distribution

$$c_+ = c_{+,0}\exp\left(-\frac{z_+e\phi}{kT}\right), \quad c_- = c_{-,0}\exp\left(\frac{z_-e\phi}{kT}\right) \tag{3}$$

where $c_{+,0}$ and $c_{-,0}$ are average number densities of positive and negative ions.

The electrostatic potential is related to the charge density ρ of the solution through the Poisson's equation

$$\nabla^2\phi = -\frac{\rho}{\varepsilon_l} = -\frac{e(z_+c_+ - z_-c_-)}{\varepsilon_l} \tag{4}$$

where ε_l is the dielectric constant of the solution. Substitution of Eq. (3) into Eq. (4) gives the following Poisson-Boltzmann equation

$$\nabla^2\phi - \frac{1}{\varepsilon_l}\left[z_-ec_{0,-}\exp\left(\frac{z_-e\phi}{kT}\right) - z_+ec_{0,+}\exp\left(-\frac{z_+e\phi}{kT}\right)\right] = 0 \tag{5}$$

The second term in the left side of Eq. (5) attenuates the electrostatic potential by shielding it with ions in water.

The electrostatic potential on the surface of charged particles is determined by the surface charge density σ through the boundary condition of

$$\sigma \equiv e(z_{s,+}s_+ - z_{s,-}s_-) = \left[\varepsilon_i \frac{\partial \phi_i}{\partial n} - \varepsilon_l \frac{\partial \phi}{\partial n}\right]_{surface} \approx -\left[\varepsilon_l \frac{\partial \phi}{\partial n}\right]_{surface} \tag{6}$$

where ϕ_i and ε_i are the electrostatic potential and the dielectric constant of the particle; n, normal to the surface; and $\pm z_{s,\pm}$ and $s_\pm$, the valence and the number density of positive or negative ions adsorbed on the surface of the particle. The $\varepsilon_i \partial \phi_i / \partial n$ term in Eq. (6) is negligible, since ε_l is usually much higher than ε_i and, because of the shielding of electric field by ions in water, $\partial \phi / \partial n$ is much larger than $\partial \phi_i / \partial n$.

Assuming that the density of surface ions is low enough to neglect interactions among the ions other than electrostatic ones, the charge distribution is given by the Boltzmann distribution

$$\sigma = \left[z_{s,+}eS_+ \exp\left(-\frac{z_{s,+}e\phi}{kT}\right) - z_{s,-}eS_- \exp\left(\frac{z_{s,-}e\phi}{kT}\right)\right]_{surface} \tag{7}$$

Since the total numbers of positive and negative surface ions are conserved, the constants S_+ and S_- are defined by

$$s_{+,0}S = S_+ \int \exp\left(-\frac{z_{s,+}e\phi}{kT}\right) ds, \quad s_{-,0}S = S_- \int \exp\left(\frac{z_{s,-}e\phi}{kT}\right) ds \tag{8}$$

where S is the surface area of the particle; ds, the surface element; and $s_{\pm,0}$, the number density of positive or negative ions on the surface of the isolated particle.

Assuming that the size of the particles is much larger than the thickness of their double layers, the number densities are related to the surface electrostatic potential ς as [6]

$$\varsigma \approx \frac{e(z_{s,+}s_{+,0} - z_{s,-}s_{-,0})}{\varepsilon_l \kappa} \tag{9}$$

where κ is the Debye reciprocal length defined by

$$\kappa = \sqrt{c_{0,+}z_+(z_+ + z_-)e^2 / (\varepsilon_l kT)} = \sqrt{c_{0,-}z_-(z_+ + z_-)e^2 / (\varepsilon_l kT)} \tag{10}$$

The Debye length $1/\kappa$ is the same as the double layer thickness. The surface electrostatic potential ζ is approximately the same as the zeta potential that can be determined by measuring the electrophoresis rate of the particles. Substitution of Eq. (7) into Eq. (6) gives

$$-\left[\varepsilon_l \frac{\partial \phi}{\partial n}\right]_{surface} = \left[z_s e S_+ \exp\left(-\frac{z_s e \phi}{kT}\right) - z_s e S_- \exp\left(\frac{z_s e \phi}{kT}\right)\right]_{surface} \quad (11)$$

The problem of calculating the double layer force acting between particles with mobile surface charges has been reduced to solve Eq. (5) under the boundary condition of Eq. (11).

3. Surface Charge Migration under No External Electric Field

3.1. Electrostatic Potential

For making the problem simpler, we treat two charged particles in liquid that dissolves positive and negative ions of number density c_0 and valences $\pm z$, respectively. Eq. (5) is then reduced to

$$\nabla^2 \phi - \kappa^2 \frac{kT}{ze} \sinh\left(\frac{ze\phi}{kT}\right) = 0 \quad (12)$$

The valences of mobile positive and/or negative ions adsorbed on the surface of the particles are assumed to be $\pm z_s$, respectively.

Although the approach of two particles perturbs the electrostatic potential, the influence is limited within a very small region where the electric double layers overlap with each other. It is therefore plausible to assume that the surface electrostatic potentials are the same as that of the isolated particles except for a small area where the overlapping of the double layers is significant. ϕ in Eq. (8) is then replaceable with ζ, and Eq. (7) is simplified as

$$\sigma \approx \left[\varepsilon_l \kappa \varsigma \cosh\left(\frac{z_s e \varsigma}{kT}\right) + z_s e s_0 \sinh\left(\frac{z_s e \varsigma}{kT}\right)\right] \cosh\left(\frac{z_s e \phi_s}{kT}\right)$$
$$-\left[\varepsilon_l \kappa \varsigma \sinh\left(\frac{z_s e \varsigma}{kT}\right) + z_s e s_0 \cosh\left(\frac{z_s e \varsigma}{kT}\right)\right] \sinh\left(\frac{z_s e \phi_s}{kT}\right) \quad (13)$$

where $s_0 = s_{+,0} + s_{-,0}$ is the sum of the number densities of positive and negative surface ions, and ϕ_s is the surface electrostatic potential. Eq. (11) is then expressed as

$$\left[\frac{\partial \phi}{\partial n}\right]_{surface} = -\left[\kappa\varsigma\cosh\left(\frac{z_s e\varsigma}{kT}\right) + \frac{z_s e s_0}{\varepsilon_l}\sinh\left(\frac{z_s e\varsigma}{kT}\right)\right]\cosh\left(\frac{z_s e\phi_s}{kT}\right) + \left[\kappa\varsigma\sinh\left(\frac{z_s e\varsigma}{kT}\right) + \frac{z_s e s_0}{\varepsilon_l}\cosh\left(\frac{z_s e\varsigma}{kT}\right)\right]\sinh\left(\frac{z_s e\phi_s}{kT}\right) \quad (14)$$

Shown in Figure 2 are the surface potential and the surface charge density of one of two spherical particles of radius a, respectively, that are obtained by solving Eq. (12) numerically under the boundary condition of Eq.(14). It is evident that the approach of two particles causes both the changes of surface electrostatic potential and surface charge distribution.

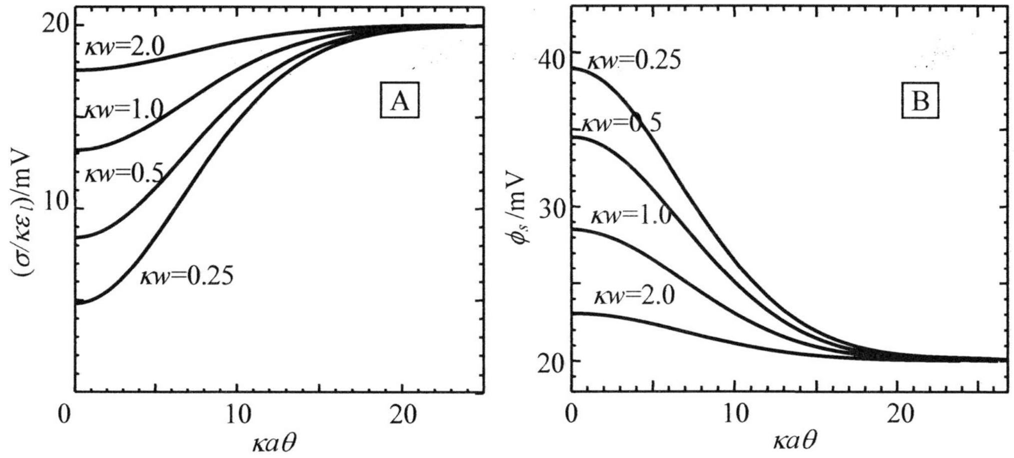

Figure 2. Electric charge (A) and electrostatic potential (B) on the surface of one of two identical spheres of normalized radius $\kappa a = 10^4$ and surface electrostatic potential $\zeta = 20$ mV. κ is the Debye reciprocal length; w, the separation distance between the front surfaces of approaching spheres of radius a; and $\kappa a\theta$, the normalized distance on the surface measured from ($r = a$, $\theta = 0$).

3.2. Double Layer Force Acting between Front Surfaces of Approaching Particles

Because the sizes of charged particles are generally much larger than the thickness of the double layers, for calculating the double layer potential, the front surfaces of the particles to be attached at first by the approach can be regarded as two parallel planar plates separated by distance w. For understanding the effect of surface charge reorientation on the stability of charged particles, we compare the solutions of Eq. (12) for two parallel plates with fixed surface charges, mobile surface charges, and fixed surface electrostatic potentials, respectively.

Taking x as the axis of approach, Eq. (12) is rewritten as

$$\frac{d^2\phi}{dx^2} - \kappa^2 \frac{kT}{ze}\sinh\left(\frac{ze\phi}{kT}\right) = 0 \tag{15}$$

Since ϕ is symmetrical with respect to the middle point of the two plates, a boundary condition for Eq. (15) is given by

$$\left[\frac{d\phi}{dx}\right]_{x=w/2} = 0 \tag{16}$$

Integration of Eq. (15) from $x = x$ to $x = w/2$ gives

$$\frac{d\phi}{dx} = -\frac{\phi}{|\phi|}\frac{\sqrt{2}\kappa kT}{ze}\left[\cosh\left(\frac{ze\phi}{kT}\right) - \cosh\left(\frac{ze\phi_{w/2}}{kT}\right)\right]^{1/2} \tag{17}$$

where $\phi_{w/2}$ is the electrostatic potential at $x = w/2$. Further integration of Eq. (17) from $x = 0$ to $x = w/2$ gives

$$\kappa w = 2\sqrt{2}\left[\frac{(F_s - F_{w/2})^{1/2}}{F_s^{1/2}(F_s+2)^{1/2}} + \int_{F_{w/2}}^{F_s}\frac{(F+1)(F-F_{w/2})^{1/2}\,dF}{F^{3/2}(F+2)^{3/2}}\right] \tag{18}$$

where $F = \cosh(ze\phi/kT) - 1$, $F_{w/2} = \cosh(ze\phi_{w/2}/kT) - 1$ and $F_s = \cosh(ze\phi_s/kT) - 1$.

The value of F_s is determined from Eq. (17) and the boundary condition of Eq. (6), that is

$$F_s = \frac{1}{2}\left(\frac{ze\varsigma}{kT}\right)^2 + F_{w/2} \tag{19}$$

for fixed surface charge,

$$\sqrt{2}\left(F_s - F_{w/2}\right)^{1/2} = \left[\frac{ze\varsigma}{kT}\cosh\left(\frac{z_s e\varsigma}{kT}\right) + \frac{z_s ze^2 s_0}{\varepsilon_l \kappa kT}\sinh\left(\frac{z_s e\varsigma}{kT}\right)\right](F_s + 1)$$
$$- \left[\frac{ze\varsigma}{kT}\sinh\left(\frac{z_s e\varsigma}{kT}\right) + \frac{z_s ze^2 s_0}{\varepsilon_l \kappa kT}\cosh\left(\frac{z_s e\varsigma}{kT}\right)\right]F_s^{1/2}(F_s + 2)^{1/2} \tag{19'}$$

for mobile surface charge, or

$$F_s = \cosh\left(\frac{ze\varsigma}{kT}\right) - 1 \tag{19''}$$

for fixed surface potential.

According to the DLVO theory, the pressure P acting in an ionic solution is given by a sum of the Maxwell's electric field stress and the osmotic pressure as

$$P = -\frac{\varepsilon_l}{2}\left(\frac{d\phi}{dx}\right)^2 + kTc_0\left[\exp\left(-\frac{ze\phi}{kT}\right) + \exp\left(-\frac{ze\phi}{kT}\right) - 2\right] \tag{20}$$

The pressure is the same everywhere between the plates [7], even at $w/2$ where the Maxwell's electric field stress is zero, so that Eq. (20) is simplified to

$$P = kTc_0\left[\exp\left(-\frac{ze\phi_{w/2}}{kT}\right) + \exp\left(-\frac{ze\phi_{w/2}}{kT}\right) - 2\right] = 2kTc_0 F_{w/2} \tag{21}$$

Integration of Eq. (21) from $w = w$ to $w = \infty$ gives the double layer potential U_D as

$$U_D = -2kTc_0 wF_{w/2} - 2kTc_0 \int_{F_{w/2}}^{0}\left(w\frac{dF_{w/2}}{dw}\right)\frac{dw}{dF_{w/2}}dF_{w/2}$$
$$= -\frac{4\sqrt{2}kTc_0}{\kappa}F_{w/2}\left[\frac{(F_s - F_{w/2})^{1/2}}{F_s^{1/2}(F_s + 2)^{1/2}} + \int_{F_{w/2}}^{F_s}\frac{(F+1)(F - F_{w/2})^{1/2}dF}{F^{3/2}(F+2)^{3/2}}\right]$$
$$+ \frac{4\sqrt{2}kTc_0}{\kappa}\int_0^{F_{w/2}}\left[\frac{(F_s - F_{w/2})^{1/2}}{F_s^{1/2}(F_s + 2)^{1/2}} + \int_{F_{w/2}}^{F_s}\frac{(F+1)(F - F_{w/2})^{1/2}dF}{F^{3/2}(F+2)^{3/2}}\right]dF_{w/2} \tag{22}$$

The value of double layer potential is given by numerical integration of Eq. (22). Figure 3 shows the double layer potentials for two plates with fixed surface charge, mobile surface charge and fixed surface electrostatic potential, respectively, as a function of normalized

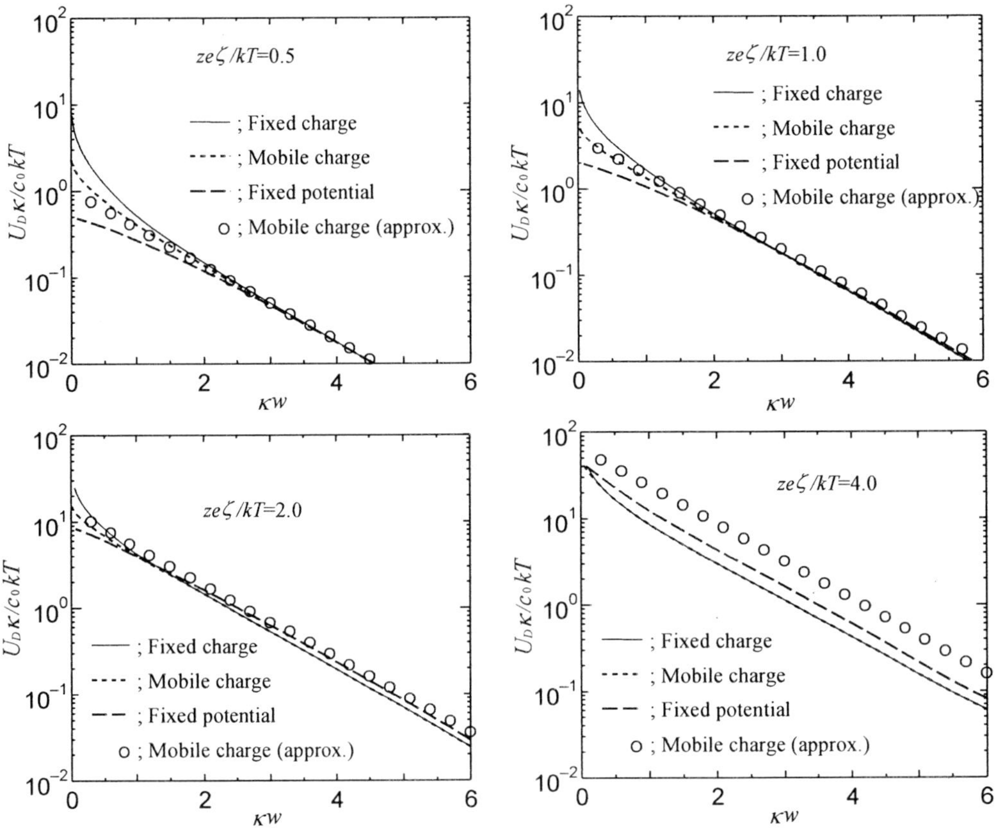

Figure 3. Double layer potential energies U_D for front surfaces of two approaching particles with fixed surface charge, mobile surface charge, and fixed surface potential. The surface charge is assumed to be composed of positive ions with the valence z same as ions in the aqueous phase; c_0, the concentration of ions in the aqueous phase; κ, the Debye reciprocal length; kT, the thermal energy at temperature T; w, the separation distance of the surfaces; and ζ, the zeta potential. The energy obtained under the Debye-Hückel approximation is also shown with open circles.

separation distance κw. It is evident that the difference between the three potentials increases significantly with decreasing the separation distance. The correct estimation of surface electrostatic potential and surface charge is therefore very important for discussing the stability of emulsion and colloidal systems.

The total potential energy U for the approach is the sum of the double layer potential and the van der Waals energy that is given by

$$U = U_D - \frac{A_H}{12\pi w^2} \tag{23}$$

where A_H is the Hamaker constant [13]. The total energy for two charged particles can be obtained by dividing the surfaces of the particles with tiny parallel plates and applying Eq. (23) to each segment [14].

The analytical solution of the Poisson-Boltzmann equation is obtained by applying the Debye-Hückel approximation of $\sinh(ze\phi/kT) = ze\phi/kT$ to Eq. (12) as

$$\frac{d^2\phi}{dx^2} - \kappa^2\phi = 0 \tag{24}$$

The solution of Eq. (24) satisfying the boundary condition of Eq. (14) is

$$\phi = B\left[e^{-\kappa x} + e^{\kappa(x-w)}\right] \tag{25}$$

where B is a constant defined by

$$B(1 - e^{-\kappa w}) = \varsigma \cosh\left(\frac{z_s e[B(1 + e^{-\kappa w}) - \varsigma]}{kT}\right) - \frac{z_s e s_0}{\varepsilon_l \kappa}\sinh\left(\frac{z_s e[B(1 + e^{-\kappa w}) - \varsigma]}{kT}\right) \tag{26}$$

Using the approximation of $\exp[z_s e(\phi-\zeta)/kT] \approx 1 + z_s e(\phi-\zeta)/kT$, Eq. (26) is reduced to

$$B = \frac{\varsigma(h+1)}{h + 1 + (h-1)e^{-\kappa w}} \tag{27}$$

where

$$h = \frac{z_s^2 e^2 s_0}{\varepsilon_l \kappa kT} \equiv \frac{z_s e \varsigma}{kT}\frac{s_{+,0} + s_{-,0}}{s_{+,0} - s_{-,0}} \tag{28}$$

Substitution of Eq. (27) and Eq. (25) into Eq. (22) gives the analytical expression of the double layer force

$$P = \frac{2\varepsilon_l \kappa^2 \varsigma^2 (h+1)^2 \mathrm{e}^{-\kappa w}}{[h+1+(h-1)\mathrm{e}^{-\kappa w}]^2} \tag{29}$$

The double layer potential is given by integrating Eq. (29) from w to infinite as

$$U_D = \frac{2\varepsilon_l \kappa (h+1)\varsigma^2 \mathrm{e}^{-kw}}{h+1+(h-1)\mathrm{e}^{-kw}} \tag{30}$$

The double layer potential obtained from Eq. (30) is also shown in Figure 3 with open circles. Comparison of the approximate and the rigorous potentials indicates that the Debye-Hückel approximation has enough accuracy as long as $|ze\zeta/kT| \le 2$. Eq. (30) is therefore applicable to charged particles with $\zeta \le 50$ mV in a monovalent ionic solution at ambient temperature.

The total energy for the approach of two charged surfaces is the sum of the double layer potential and the van der Waals energy, and is given by

$$U = \frac{2\varepsilon_l \kappa (h+1)\varsigma^2 \mathrm{e}^{-\kappa w}}{h+1+(h-1)\mathrm{e}^{-\kappa w}} - \frac{A_H}{12\pi w^2} \tag{31}$$

Eq. (31) is reduced to

$$U = \frac{2\varepsilon_l \kappa \varsigma^2 (z_s e|\varsigma| + kT)\mathrm{e}^{-\kappa w}}{(z_s e|\varsigma| + kT) + (z_s e|\varsigma| - kT)\mathrm{e}^{-\kappa w}} - \frac{A_H}{12\pi w^2} \tag{32}$$

for systems composed of only positive or negative surface ions.

Figure 4 shows the shape of potential energies that prevent the contact of two charged particles. It is noted that the potential energy barrier depends not only on the surface charge density but also on the number density of surface ions. Increase of the number density causes increase of the surface electric conductivity, which suppresses the increase of the surface electrostatic potentials during the approach and therefore decreases the double layer potential.

It is interesting to compare the present results with the potential energy barrier obtained from the conventional linear superposition approximation. The linear superposition approximation assumes that the electrostatic potential between two charged surfaces is expressed by the superposition of the potentials for isolated surfaces as

$$\phi = \frac{2kT}{ze}\log\left[\frac{1+\gamma\exp(-\kappa x)}{1-\gamma\exp(-\kappa x)}\right] + \frac{2kT}{ze}\log\left[\frac{1+\gamma\exp[-\kappa(w-x)]}{1-\gamma\exp[-\kappa(w-x)]}\right] \tag{33}$$

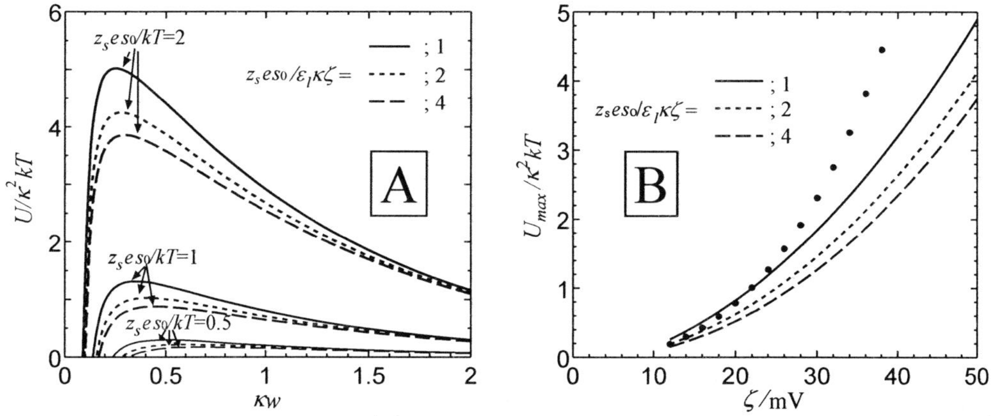

Figure 4. Potential energies (A) and the height of potential energy barriers (B) preventing the contact of two charged particles. w, κ, ζ and s_0 are the separation distance of two approaching spheres, the Debye reciprocal length, the zeta potential and the total density of surface ions, respectively. Parameters used for the calculations are $kT = 4\times10^{-21}$ J, $\kappa = 10^8$/m, $z = z_s = 1$, $\varepsilon_l = 80\varepsilon_0$ and $A_H = 5\times10^{-21}$ J. The solid circles in the left denote the barrier heights estimated by the linear superposition approximation.

where

$$\gamma = \frac{\exp(ze\varsigma/2kT)-1}{\exp(ze\varsigma/2kT)+1} \tag{34}$$

The double layer potential is then expressed as

$$U_D = \frac{64c_0kT\gamma^2}{3\kappa[1-\gamma^2\exp(-\kappa w)]^3}\left[3+\gamma^4\exp(-2\kappa w)\right]\exp(-\kappa w) \tag{35}$$

The total potential energy is a sum of the double layer potential and the van der Waals energy, so that

$$U = \frac{64c_0kT\gamma^2}{3\kappa[1-\gamma^2\exp(-\kappa w)]^3}\left[3+\gamma^4\exp(-2\kappa w)\right]\exp(-\kappa w) - \frac{A_H}{12\pi w^2} \tag{36}$$

Figure 4-B compares the height of the potential energy barriers for the contact of two charged particles that are obtained from Eq. (31) and Eq. (36), respectively, as a function of ζ. It is evident from the figure that the linear superposition approximation is valid only in the limiting case of $|\zeta| \approx 20$ mV and $s_0=\varepsilon_l\kappa|\zeta|/z_se$.

4. Surface Charge Migration under External Electric Field

4.1. Demulsification of Oil-in-Water Emulsion

We have shown in the previous section that the migration of surface ions during the approach of two charged particles changes the stability of emulsion and colloidal systems in ionic solutions. The migration of surface ions is induced not only by the approach of adjacent charged particle but also by an external electric field. Application of an external electric field to emulsion and colloidal systems is therefore possible to change the stability of the systems.

Because of its simplicity, electrical demulsification is one of the best methods for eliminating undesired emulsions [16]. However, the electrical demulsification of oil-in-water emulsions has scarcely been studied. Although several researchers reported the acceleration of demulsification by low external electric fields, they considered the acceleration as due to the increase of ionic concentration in water by the electrolysis of electrodes, the condensation of oil droplets near an electrode by electrophoresis or elimination of ions from the surface of oil droplets by electrolysis.

We have found that dense oil-in-water emulsions under external electric fields are destroyed not by electrolysis or electrophoresis but by the electric field itself. This is a general phenomenon as long as the emulsions are stabilized by the double layer force. Figure 5 shows the effect of external electric fields on the stability of oil-in-water emulsion that is prepared by shaking the 3 : 2 mixture of oil (1 : 1 mixture of benzyl alcohol and tetrahydropyran) and water (containing 2×10^{-4} mol/dm^3 of sodium 2-naphthyl sulfonate as a surfactant) in a separatory funnel more than 100 times. The radii and the zeta potential of the oil droplets are 25~30 μm and 60 mV, respectively. The emulsion layer thus prepared has a lifetime of more than 1 h.

As shown in Figure 5-A, application of a DC electric field of higher than 4 V/cm shorten the life time to less than 20 s. The acceleration of demulsification is never observed on water-in-oil emulsions and on oil-in-water emulsions stabilized with nonionic surfactants. The acceleration of demulsification is induced neither by electrophoresis nor electrolysis, since the fusion of the droplets takes place not only near the electrodes but everywhere in the emulsion layer [11].

As shown in Figure 5-B, the effect of external electric field on the demulsification decreases with increasing frequency of the applied square-wave AC electric field, since the coalescence of oil droplets is not completed within a half cycle and the deviation of the distribution of surface ions is compensated during the next half cycle with reverse polarity.

Application of a 6 V/cm pulsed square wave field of 50 % duty cycle gives a result quite different from the AC field. The time of demulsification increases with increasing frequency of the field, though it is always much shorter than that without applied electric field. The time of demulsification reaches a plateau value of 1 minute at a frequency of more than 1 kHz. Acceleration of the demulsification by the pulsed field is induced by the deviation of the distribution of surface charges that is induced during a 6 V/cm half cycle and is partially remained during the next 0 V half cycle. The difference between the AC and the pulsed fields can not be explained by mechanisms other than the migration of surface ions, such as electrolysis and electrophoresis.

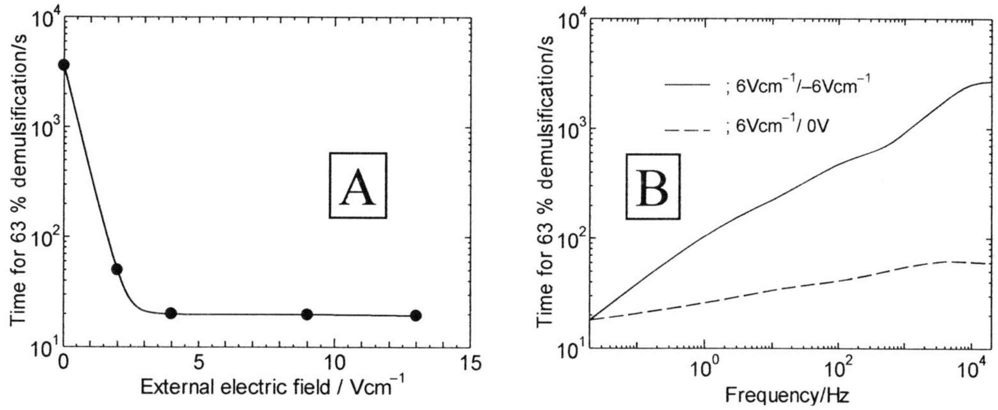

Figure 5. Effect of DC (A), pulsed (B) and AC (B) external electric fields on the life time of oil-in-water emulsion.

Figure 6 depicts the mechanism of external field-induced destruction of oil-in-water emulsion. The distribution of surface charge is homogeneous when the separation distances between the surfaces of adjacent droplets are much longer than the thickness of the double layers. Increase of the density of oil droplets causes the overlapping of the double layers and therefore modifies the distribution of the surface charges. However the modification is not enough for inducing the demulsification. Application of an external electric field causes the inhomogeneity of electrostatic potential on the surface of each oil droplet. The surface ions then migrate to compensate the inhomogeneity, which changes the surface charge densities. The electrostatic potentials of adjacent two surfaces arising from the surface charges are no more equal. It is well known that the mixing of two kinds of colloidal particles with different surface electrostatic potentials causes the hetero coagulation of the particles due to the decrease of the repulsive double layer force between the particles [15]. The same phenomenon happens between the oil droplets, so that the droplets coagulate and coalesce into large oil droplets. The external electric field thus breaks the emulsion. In the next section, we will translate this picture to a mathematical formula [17].

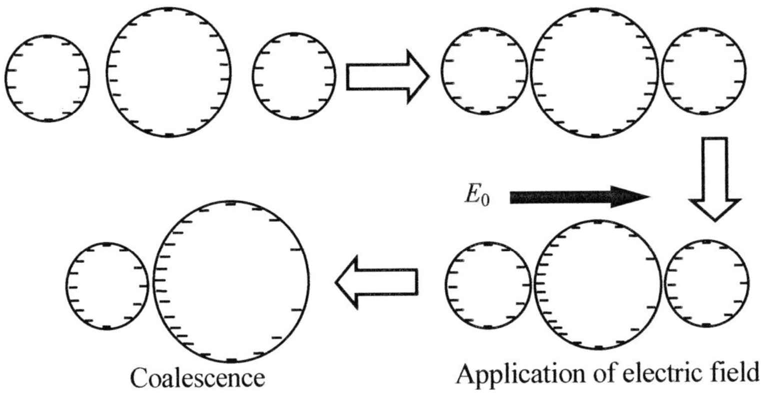

Figure 6. Schematic representation of the mechanism of electrical demulsification.

4.2. DLVO Theory under External Electric Field

The spatial distribution of ions in water under an external electric field can not be given by the Boltzmann distribution, since the external field induces the continuous flow of ions. Instead, the distribution is determined from the Smoluchowski diffusion equation

$$\frac{\partial c_{\pm}}{D_{\pm}\partial t} = \nabla\left(\nabla c_{\pm} \pm \frac{zec_{\pm}}{kT}\nabla\phi\right) \tag{37}$$

where $D_{\pm}$ is the diffusion constant of positive or negative ions. Under a steady-state condition, Eq.(37) is reduced to

$$\nabla\left(\nabla c_{\pm} \pm \frac{zec_{\pm}}{kT}\nabla\phi\right) = 0 \text{ or } \nabla c_{\pm} \pm \frac{zec_{\pm}}{kT}\nabla\phi = J_{\pm} \tag{38}$$

Here $J_{\pm}$ is the steady-state flow of positive or negative ions. Under no flow of ions, Eq. (38) is reduced to Eq. (4) and therefore gives the Boltzmann distribution. Although an external electric field induces the flow of ions and therefore changes the distribution of ions, the detailed analysis have revealed that the external field affects the distribution only through the change of the surface charge density, as long as the intensity of the field is less than 10^4 V/cm [12]. The Poisson-Boltzmann equation is therefore able to be used for determining the distribution instead of the Poisson-Smoluchowski equation.

Although an accurate numerical solution of the Poisson-Boltzmann equation can be obtained by solving Eq. (5) under the boundary condition of

$$\left[\frac{\partial \phi}{\partial n}\right]_{surface} = -\frac{z_s e}{\varepsilon_l}\left\{S_+ \exp\left[-\frac{z_s e}{kT}(\phi_s + \phi_{ext})\right] - S_- \exp\left[\frac{z_s e}{kT}(\phi_s + \phi_{ext})\right]\right\} \tag{39}$$

where ϕ_{ext} is the electrostatic potential generated by the external electric field.

For understanding the effect of the external field on the stability of charged particles in water, we will use several approximations in the following treatment.

We treat two charged spheres under a constant external electric field of E_0. Using the Debye-Hückel approximation of $\sinh(ze\phi/kT) = ze\phi/kT$, the Poisson-Boltzmann equation under a polar coordinate with the origin at the center of one of the spheres is expressed as

$$\frac{1}{r}\frac{\partial^2}{\partial r^2}(r\phi) + \frac{1}{r^2 \sin\theta}\frac{\partial}{\partial \theta}\left(\sin\theta \frac{\partial \rho}{\partial \theta}\right) = \kappa^2 \phi \tag{40}$$

Using the relation of $\partial\phi/\partial r >> \partial\phi/\partial\theta$, the solution of Eq. (40) for an isolated sphere of radius a is given by

$$\phi \approx \phi_s \frac{a e^{-\kappa(r-a)}}{r} \tag{41}$$

Substitution of Eq. (41) into Eq. (39) gives

$$\kappa\phi_s = \frac{z_s e G_+}{\varepsilon_l}\exp\left[-\frac{z_s e}{kT}(\phi_s - \varsigma - Ea\cos\theta)\right] - \frac{z_s e G_-}{\varepsilon_l}\exp\left[\frac{z_s e}{kT}(\phi_s - \varsigma - Ea\cos\theta)\right] \tag{42}$$

Here $Ea\cos\theta$ is the intensity of electrostatic potential generated on the surface by the external field, and is defined by

$$E = \frac{3\varepsilon_l E_0}{\varepsilon_i + 2\varepsilon_l} \approx \frac{3E_0}{2} \tag{43}$$

Since the total numbers of positive and negative ions are conserved, the constants G_+ and G_- must satisfy the conditions of

$$2s_{+,0} = G_+ \int_{-1}^{1} \exp\left[-\frac{z_s e}{kT}\left(\phi_s - \varsigma - Ea\cos\theta\right)\right] d\cos\theta$$

$$2s_{-,0} = G_- \int_{-1}^{1} \exp\left[\frac{z_s e}{kT}\left(\phi_s - \varsigma - Ea\cos\theta\right)\right] d\cos\theta \tag{44}$$

Figure 7 shows the surface electrostatic potentials of an isolated sphere at $\theta = 0$ and $\theta = \pi$ that are obtained as numerical solutions of Eqs. (42) and (44). The advantage of using ϕ_s-ζ-$Ea\cos\theta$ in Eq. (42) and Eq. (44) instead of ϕ_s-$Ea\cos\theta$ is that the exponents become close to zero, so that the accurate numerical calculation can be achieved without using small increment of $\Delta\cos\theta$.

Using the approximation of $\exp[\pm z_s e(\phi_s\text{-}\zeta\text{-}Ea\cos\theta)/kT] \approx 1\pm z_s e(\phi_s\text{-}\zeta\pm Ea\cos\theta)/kT$, Eq. (42) gives

$$\phi_s = \frac{(G_+ - G_-) + (G_+ + G_-)\dfrac{z_s e}{kT}\left(\varsigma + Ea\cos\theta\right)}{\dfrac{\varepsilon_l \kappa}{z_s e} + (G_+ + G_-)\dfrac{z_s e}{kT}} \tag{45}$$

Using the relation of $\zeta = z_s e(s_{+,0}\text{-}s_{-,0})/\varepsilon_l\kappa$ and the approximation of $\exp[\pm z_s e(\phi_s\text{-}\zeta\text{-}Ea\cos\theta)/kT] \approx 1\pm z_s e(\phi_s\text{-}\zeta \pm Ea\cos\theta)/kT$, substitution of Eq. (45) into Eq. (44) gives

$$G_+ = s_{+,0}, \quad G_- = s_{-,0} \tag{46}$$

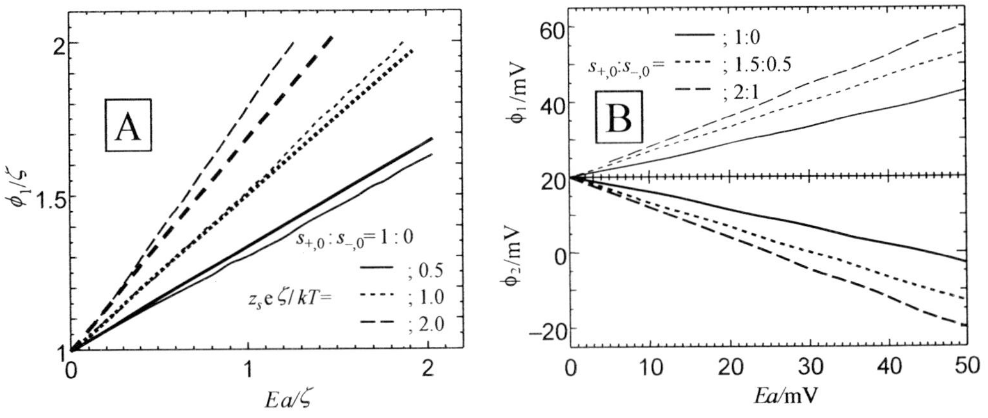

Figure 7. Effect of electric field E on the surface potentials at θ=0 ($\phi = \phi_1$) and θ=π ($\phi = \phi_2$) of charged sphere. ζ and a are the original surface electric potential and the radius of the sphere, respectively; $s_{+,0}$: $s_{-,0}$, the ratio of the total numbers of positive and negative surface ions with valence $\pm z_s$. The thick lines in Figure 7-A shows the approximate values obtained from Eq. (47).

which results in

$$\phi_s = \varsigma + \frac{h}{1+h} Ea\cos\theta \tag{47}$$

As shown in Figure 7-A, the values of electrostatic potentials obtained from Eq. (47) are in good agreement with exact ones.

The Double layer force acting between the interacting surfaces of two spheres with radii a_1 and a_2 is then obtained by using the DLVO theory. The solution of the Poisson-Boltzmann equation along the axis of approach ($\theta = 0$) under the Debye-Hückel approximation is given by

$$\phi = Ce^{-\kappa x} + De^{\kappa(x-w)} \tag{48}$$

Although the approach of the two spheres modifies the surface charge distribution, the effect is limited within a very small region where the double layers overlap with each other. The constants G_+ and G_- are then scarcely changed by the overlap of the double layers, and the boundary conditions at the surfaces are given by

$$\frac{\varepsilon_l \kappa}{z_s e}(C - De^{-\kappa w}) = s_{+,0}\exp\left[-\frac{z_s e}{kT}\left(C + De^{-\kappa w} - \varsigma - Ea_1\right)\right] - s_{-,0}\exp\left[\frac{z_s e}{kT}\left(C + De^{-\kappa w} - \varsigma - Ea_1\right)\right] \tag{49}$$

$$\frac{\varepsilon_l \kappa}{z_s e}(D - Ce^{-\kappa w}) = s_{+,0}\exp\left[-\frac{z_s e}{kT}\left(Ce^{-\kappa w} + D - \varsigma + Ea_2\right)\right] - s_{-,0}\exp\left[\frac{z_s e}{kT}\left(Ce^{-\kappa w} + D - \varsigma + Ea_2\right)\right] \tag{50}$$

Using the approximation of $\exp[\pm z_s e(\phi_s - \zeta \pm Ea)/kT] \approx 1 \pm z_s e(\phi_s - \zeta \pm Ea)/kT$, Eqs. (49) and (50) give the solutions of

$$C = \frac{(h+1)\varsigma}{(h+1)+(h-1)e^{-\kappa w}} + \frac{(h+1)a_1 + (h-1)a_2 e^{-\kappa w}}{(h+1)^2 - (h-1)^2 e^{-2\kappa w}} hE \tag{51}$$

$$D = \frac{(h+1)\varsigma}{(h+1)+(h-1)\mathrm{e}^{-\kappa w}} - \frac{(h-1)a_1\mathrm{e}^{-\kappa w}+(h+1)a_2}{(h+1)^2-(h-1)^2\mathrm{e}^{-2\kappa w}} hE \tag{52}$$

Since the double layer force is the same everywhere between $x = 0$ and $x = w$, by taking $x = 0$, the force given by Eq. (20) is expressed as

$$P = -\frac{\varepsilon_l \kappa^2}{2}\left[C - D\mathrm{e}^{-\kappa w}\right]^2 + 4kTc_0 \sinh^2\left(\frac{ze(C + D\mathrm{e}^{-\kappa w})}{2kT}\right) \tag{53}$$

Application of the Debye-Hückel approximation of $\sinh(ze\phi/kT) = ze\phi/kT$ to Eq. (53) gives

$$P = 2\varepsilon_l\kappa^2\mathrm{e}^{-\kappa w}\left\{\left[\frac{(h+1)\varsigma + h(a_1 - a_2)(E/2)}{(h+1)+(h-1)\mathrm{e}^{-\kappa w}}\right]^2 - \left[\frac{h(a_1 + a_2)(E/2)}{(h+1)-(h-1)\mathrm{e}^{-\kappa w}}\right]^2\right\} \tag{54}$$

which is rewritten as

$$P = \frac{[(h+1)\varsigma - ha_2E][(h+1)\varsigma + ha_1E][2\varepsilon_l\kappa^2\mathrm{e}^{-\kappa w}(h-1)^2]}{[(h+1)+(h-1)\mathrm{e}^{-\kappa w}]^2[(h+1)-(h-1)\mathrm{e}^{-\kappa w}]^2}$$
$$\times\left[\frac{(h+1)[(h+1)\varsigma + ha_1E]}{(h-1)[(h+1)\varsigma - ha_2E]} - \mathrm{e}^{-\kappa w}\right]\left[\frac{(h+1)[(h+1)\varsigma - ha_2E]}{(h-1)[(h+1)\varsigma + ha_1E]} - \mathrm{e}^{-\kappa w}\right] \tag{55}$$

Analysis of Eq. (55) under the condition of $h > 1$ gives that the double layer force is always attractive if $a_2E/\zeta. > (h+1)/h$. Since $h > 1$ for most of emulsion systems stabilized by the double layer force, the critical external electric field for inducing the coagulation and coalescence of two spheres is given by

$$E_0 \geq \frac{2[z_s e\varsigma(s_{+,0} + s_{-,0}) + kT(s_{+,0} - s_{-,0})]}{3z_s e(s_{+,0} + s_{-,0})a} \tag{56}$$

It should be noted that the critical electric field is determined by the radius of larger droplet. Addition of large oil droplets to an emulsion system composed of tiny oil droplets is therefore very effective for destroying the system by a low external electric field.

Integration of Eq. (54) gives the double layer potential of

$$U_D = \frac{\varepsilon_l \kappa [4(h+1)\varsigma + 3h(a_1 - a_2)E_0]^2 e^{-\kappa w}}{8(h+1)[h+1+(h-1)e^{-\kappa w}]} - \frac{9\varepsilon_l \kappa h^2 (a_1 + a_2)^2 E_0{}^2 e^{-\kappa w}}{8(h+1))[h+1-(h-1)e^{-\kappa w}]} \tag{57}$$

It is noted that substitution of $E_0 = 0$ into Eq. (57) gives Eq. (30) that has been obtained under no external electric field. Figure 8-A shows the double layer potential as a function of separation distance. The repulsive double layer potential is lowered by the external electric field, and changes to attractive force when E_0 exceeds $2(1+1/h)\zeta/3a$.

In more general case that the axis of approach is declined at angle η from the external electric field, the double layer potential is obtained simply by replacing E_0 in Eq. (57) with $E_0\cos\eta$. The total potential energy is finally given by

$$U = \frac{\varepsilon_l \kappa \varsigma^2 e^{-\kappa w}}{8(h+1)} \left[\frac{[4(h+1) + 3h(a_1 - a_2)E_0 \cos\eta / \varsigma]^2}{h+1+(h-1)e^{-\kappa w}} - \frac{[3h(a_1 + a_2)E_0 \cos\eta / \varsigma]^2}{h+1-(h-1)e^{-\kappa w}} \right] - \frac{A_H}{12\pi w^2} \tag{58}$$

Figure 8-B shows the shape of the potential energy barrier between the front surfaces of two spheres. The external electric field significantly reduces the height of the barrier.

The oil-in-water emulsion used for the electrical demulsification is characterized by ζ = 60 mV, T =298 K, $h = e\zeta/kT$, $\varepsilon_l = 80\varepsilon_0$, $c_0 = 2\times10^{-4}$ mol/dm^3, $A_H = 5\times10^{-21}$ J [18] and the critical electric field of $E_0 > 4$ V/cm. Substitution of these values into Eq. (56) gives the radius of a = 140 μm. Although the microscope image of the diluted emulsion shows that the radius of most of the droplets lies between 25 μm and 30 μm, a small number of droplets with the radius larger than 140 μm may act as seed droplets for the demulsification. The electrical demulsification is accelerated once the coalescence starts, since the external field-induced coalescence is more effective for droplets with larger radius.

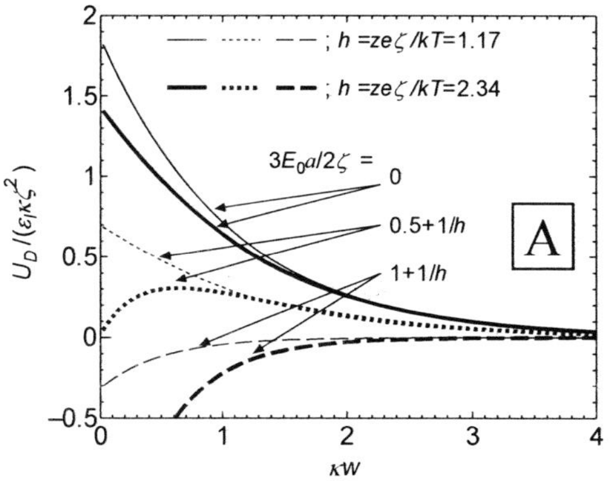

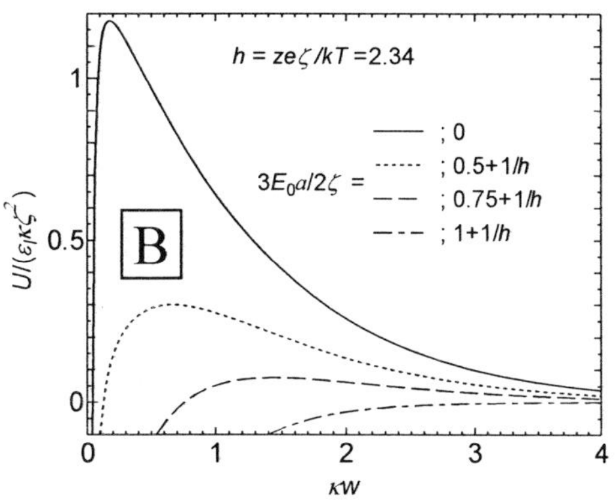

Figure 8. Double layer potential (A) and potential energy barrier (B) or the front surfaces of approaching two identical charged spheres. w, κ, ζ and a are the separation distance of the two surfaces, the Debye-Hückel reciprocal length, the electrostatic potential of the isolated spheres, and the radius of the spheres, respectively. Parameters used for the calculations are $kT = 4.1\times10^{-21}$ J, $\kappa = 4.6\times10^{7}$/m, $z = z_s = 1$, $\varepsilon_l = 80\varepsilon_0$, $\eta = 0$ and $A_H = 5\times10^{-21}$ J.

5. Summary and Conclusion

It has been shown experimentally and theoretically that the migration of ions adsorbed on the surface of colloidal and emulsion particles in water plays a significant role in the stability of the systems. The migration of surface ions opens a new way of destabilizing the systems by simply applying an external electric field.

The fundamental equation for controlling the migration of ions both on the surfaces and in water is the Poisson-Boltzmann equation. Application of the Poisson-Boltzmann equation to oil-in-water emulsions has revealed that a weak external electric field of the order of ζ/a destroys the emulsion.

The potential energy preventing the approach of two particles with mobile surface charges is given by

$$U = \frac{\varepsilon_l \kappa \varsigma^2 e^{-\kappa w}}{8(h+1)} \left[\frac{[4(h+1) + 3h(a_1 - a_2)E_0 \cos\eta / \varsigma]^2}{h+1+(h-1)e^{-\kappa w}} - \frac{[3h(a_1 + a_2)E_0 \cos\eta / \varsigma]^2}{h+1-(h-1)e^{-\kappa w}} \right] - \frac{A_H}{12\pi w^2}$$

where $h = z_s e\zeta (s_{+,0} + s_{-,0}) / [kT (s_{+,0} - s_{-,0})]$; z_s, the valence of surface ions; e, the unit charge; $s_{+,0}$ and $s_{-,0}$, number densities of positive and negative surface ions; kT, the thermal energy at temperature T; ε_l, the dielectric constant of water; κ, the Debye reciprocal length; ζ, the zeta potential; w, the separation distance between the front surfaces of approaching two spheres with radii a_1 and a_2; E_0, the intensity of an external electric field; η, the angle between the axis of approach and the external field; and A_H, the Hamaker constant of the particles, respectively.

References

[1] B.V. Derjaguin, L.D. Landau, *Acta Physicochim.* URSS. 14 (1941) 633.
[2] E.J.W. Verway, J.T.G. Overbeek, *Theory of the Stability of Lyophobic Colloids*; Elsevier, Amsterdam, 1948.
[3] B.V. Derjaguin, N.V. Churaev, M.V. Muller, *Surface Forces*, Plenum, New York, 1987.
[4] M. Manciu, E. Ruckenstein, *Langmuir,* 17, (2001) 7061.
[5] R. Evans, D.H. Napper, *J. Colloid Interface Sci.,* 52 (1975) 260.
[6] E.J. Clayfield, E.C. Lumb, *J. Colloid Interface Sci.*, 22 (1966) 269.
[7] T. Ichikawa, T. Dohda, Y. Nakajima, *Colloids and Surfaces* A, 279 (2006) 128.
[8] R. Hogg, T.W. Healy, D.W. Fuerstenau, *Trans Faraday* Soc.62 (1966) 1638.

[9] J. Gregory, *JCS Faraday* II, 69 (1973) 1723.
[10] J. Gregory, *J. Colloid Interface Sci.,* 51 (1974) 44.
[11] T. Ichikawa, K. Itoh, S.Yamamoto, M. Sumita, *Colloids and Surfaces* A 242 (2004) 21.
[12] T. Ichikawa, Y. Nakajima, *Colloids and Surfaces* A 242 (2004) 27.
[13] H.C. Hamaker, *Physica.* 4 (1937) 1058.
[14] B.V. Deryagin, *Kolloid Z.*, 69 (1934) 155.
[15] A. Bierman, *J. Colloid Sci.,* 10 (1955) 231.
[16] J.S. Eoe, M.Ghadiri, O.A. Sharif and T.J. Williams, *Chem. Eng. J.*, 84 (2001) 73, and references therein.
[17] T. Ichikawa, *Colloids and Surfaces* A, (2007) *in press*.
[18] J. Gregory, *Adv. Colloid Interface Sci.*, 2 (1970) 396.

In: New Colloid and Surface Science Research
Editor: Lawrence V. Schwartz, pp. 221-245

ISBN: 978-1-60021-594-0

Chapter 6

MODELING OF HEAVY METAL BIOSORPTION BY AEROBIC GRANULES

Zhi-Wu Wang and Yu Liu
Division of Environmental and Water Resources Engineering
School of Civil and Environmental Engineering, Nanyang Technological University,
50 Nanyang Avenue, Singapore 639798

ABSTRACT

A first-order reversible kinetic model and a general isotherm model describing Cd^{2+} and Cu^{2+} biosorption by aerobic granules were developed and verified by biosorption data obtained at different initial metal and aerobic granules concentrations. Compared to the second-order kinetic model, the proposed first-order kinetic model could provide a more satisfactory description of experimental data. The general isotherm model derived from a thermodynamic approach could be reduced to the well known Langmuir and Freundlich isotherm equations in some special cases.

The overall biosorption rate constants and biosorption capacities of Cd^{2+} and Cu^{2+} by aerobic granules were determined by the first-order kinetic model, and were found to be closely related to initial metal and granules concentrations, i.e., a high initial metal concentration would favor biosorption, while an increased initial granule (biosorbent) concentration would hinder bisorption. To interpret these experimental observations, two theoretical models were further developed from the proposed kinetic and isotherm models. The model simulation is in good agreement with experimental data obtained at different initial granule, Cd^{2+} and Cu^{2+} concentrations. It is expected that the modeling system developed can offer in-depth sights into biosorption of heavy metals by aerobic granules.

1. INTRODUCTION

Conventional methods for soluble heavy metal removal, such as chemical precipitation, solvent extraction and ion-exchange processes, have been used in industry for decades. However, those techniques are either expensive or not effective enough to recover heavy metals present in the effluent wastewater (Valdman and Leite 2000). In this case, research has turned to developing the cost-effective heavy metals removal techniques. It has been well known that bacterial surfaces contain polarizable groups (sites), such as phosphate, carboxyl, hydroxyl and amino-group, which are capable of interaction with cations and contribute to the reversible metal binding. Hence, extensive attention has been given to bioremediation of heavy metal pollution. For this purpose, many biomaterials including digested sludge, activated sludge and so on, have been tested as metal biosorbents (Lodi et al. 1998; Marques et al. 2000; Valdman and Leite 2000).

So far, most biosorbents used for soluble heavy metal removal are basically in the forms of suspended microorganisms or bioflocs, while the post-separation of bioflocs from the treated effluent is extremely difficult in practice. To overcome this drawback, cell immobilization technique has thus been developed. Recent research showed that aerobic granules have compact, porous structure and excellent settling ability (Liu and Tay 2004), e.g., the settling velocity of aerobic granules used in this study (Figure 1) is about 36 m h^{-1}, which is much higher than that of bioflocs with a typical value of less than 5 to 10 m h^{-1} (Campos et al. 1999).

Figure 1. Morphology of aerobic granules.

Such a unique feature of aerobic granules can ensure a fast and effective separation of used aerobic granules from the treated effluent. It seems that aerobic granules would be a good candidate of biosorbent for soluble heavy metals removal from wastewater (Liu et al. 2002; Xu et al. 2005). Therefore, this chapter attempted to systematically look into the modeling of Cd^{2+} and Cu^{2+} biosorption by aerobic granules.

2. BIOSORPTION KINETICS

2.1. The First-Order Biosorption Kinetics

2.1.1. The Model Description

Liu et al. (2003a) proposed that biosorption by aerobic granules would follow a first-order reversible kinetics:

$$S + M \underset{k_2}{\overset{k_1}{\rightleftarrows}} SM \tag{1}$$

where S is the available site for metal binding on aerobic granules, M is free metal ion, and SM represents metal ion-granule complex, while k_1 and k_2 are the rate constants for sorption and desorption processes, respectively. According to Eq. 1, the overall biosorption rate can be written as

$$-\frac{dC}{dt} = k_1 C - k_2 C_b \tag{2}$$

where C is concentration of soluble metal ion at time t, C_b is apparent concentration of adsorbed metal ions at time t, i.e., the metal ions adsorbed by aerobic granule per unit volume of the solution. A mass balance on metal ions gives

$$C_b = C_0 - C \tag{3}$$

where C_0 is initial concentration of metal ions. When biosorption reaches its equilibrium, Eq. 3 becomes

$$C_{be} = C_0 - C_e \tag{4}$$

At biosorption equilibrium, Eq. 2 reduces to

$$\frac{C_{be}}{C_e} = \frac{k_1}{k_2} \tag{5}$$

where C_{be} and C_e are apparent concentration of adsorbed metal ions and concentration of free metal ions at biosorption equilibrium, respectively. Combining Eqs. 2 to 5 yields

$$-\frac{dC}{dt} = (k_1 + k_2)(C - C_e) \tag{6}$$

Integration of Eq. 6 shows that

$$\frac{C_0 - C_e}{C - C_e} = e^{K_1 t} \tag{7}$$

where $K_1 = k_1 + k_2$ is termed the overall biosorption rate constant. Substitution of Eqs. 3 and 4 into Eq. 7 leads to

$$C_b = C_{be} \frac{e^{K_1 t} - 1}{e^{K_1 t}} \tag{8}$$

Dividing both sides of Eq. 8 by aerobic granule concentration yields

$$Q = Q_e (1 - e^{-K_1 t}) \tag{9}$$

It should be pointed out that Eq. 9 is similar to the first-order Lagergren model.

2.1.2. The Model Validation

Biosorption of Cd^{2+} and Cu^{2+} at different C_0

Figures 2 and 3 show Cd^{2+} and Cu^{2+} biosorptions by aerobic granules at different initial Cd^{2+} and Cu^{2+} concentrations (C_0), respectively. It is obvious that Eq. 9 can provide a satisfactory description for Cd^{2+} and Cu^{2+} biosorption by aerobic granules at various initial Cd^{2+} and Cu^{2+} concentrations. At a constant initial granules concentration, increased initial Cd^{2+} and Cu^{2+} concentrations would lead to an increase in the respective Cd^{2+} and Cu^{2+} uptake capacity.

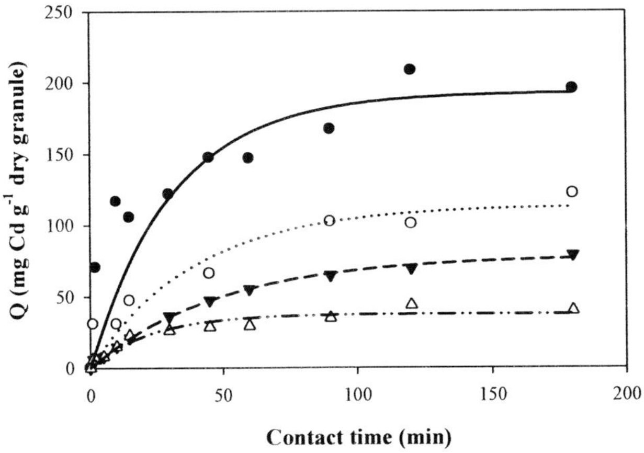

Figure 2. Cd^{2+} biosorption by aerobic granules at different initial Cd^{2+} concentrations. X_0 =150 mg l^{-1} and pH=7.0. △ 10 mg l^{-1}; ▼ 20 mg l^{-1}; ○ 50 mg l^{-1}; ● 100 mg l^{-1} (Liu et al. 2003a); and the

predictions by Eq. 9 are shown by various types of lines: ___ . . 10 mg l^{-1}; _ _ _ 20 mg l^{-1}; .... 50 mg l^{-1}; ______100 mg l^{-1}.

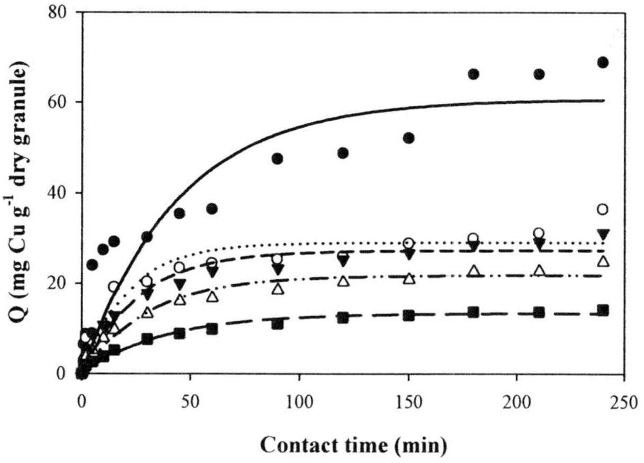

Figure 3. Cu^{2+} biosorption by aerobic granules at different initial Cu^{2+} concentrations. X_0 =150 mg l^{-1} and pH = 7.0, ■ 2 mg l^{-1}; △ 5 mg l^{-1}; ▼ 10 mg l^{-1}; ○ 25 mg l^{-1}; ● 100 mg l^{-1} (Liu et al. 2003a); and the predictions by Eq. 9 are shown by various types of lines: ___ ___ 2 mg l^{-1}; ___ . . 5 mg l^{-1}; _ _ _ 10 mg l^{-1}; .... 25 mg l^{-1}; ______100 mg l^{-1}.

Biosorption of Cd^{2+} and Cu^{2+} at Different X_0

Figures 4 and 5 show biosorption of Cd^{2+} and Cu^{2+} by aerobic granules at different initial concentrations of aerobic granules (X_0), while the initial Cd^{2+} and Cu^{2+} concentrations were fixed at 150 and 100 mg l^{-1}, respectively.

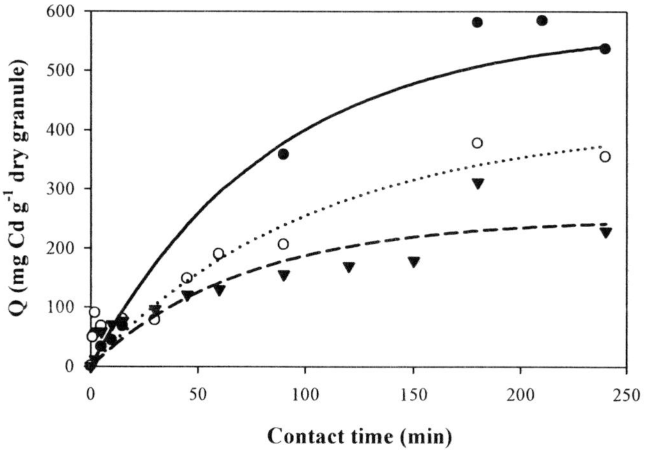

Figure 4. Cd^{2+} biosorption by aerobic granules at different initial aerobic granule concentrations. C_0 =150 mg l^{-1} and pH=7.0. ▼ 50 mg l^{-1}; ○ 100 mg l^{-1}; ● 200 mg l^{-1} (Liu et al. 2003a); and the predictions by Eq. 9 are shown by various types of lines: _ _ _ 50 mg l^{-1}; .... 100 mg l^{-1}; ______200 mg l^{-1}.

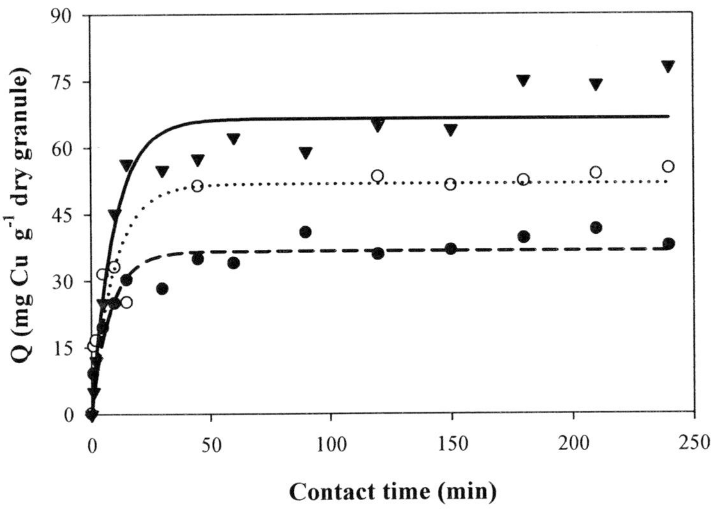

Figure 5. Cu^{2+} biosorption by aerobic granules at different initial aerobic granule concentrations. C_0 =100 mg l^{-1} and pH=7.0. ▼ 100 mg l^{-1}; ○ 200 mg l^{-1}; ● 300 mg l^{-1} (Liu et al. 2003a); and the predictions by Eq. 9 are shown by various types of lines: – – – 100 mg l^{-1}; .... 200 mg l^{-1}; ——— 300 mg l^{-1}.

It appears that the predictions by Eq. 9 are in good agreement with the experimental results, indicated by a coefficient of correlation greater than 0.93. Both the Cd^{2+} and Cu^{2+} biosorption capacities by aerobic granules are found to decrease with the increase of initial aerobic granules concentration.

2.2. The Second-Order Biosorption Kinetics

2.2.1. The Model Description

To describe adsorption kinetics, second-order kinetic model has been commonly used. According to Ho and McKay (1999), the second-order rate equation can be expressed as:

$$\frac{dQ}{dt} = K_2(Q_e - Q)^2 \tag{10}$$

where K_2 is the rate constant of second-order sorption. In fact, Eq. 10 is a kinetic analogue of the first-order Lagergren model. Integrated Eq. 10 shows that

$$Q = \frac{tQ_e^{\,2}K_2}{tK_2Q_e + 1} \tag{11}$$

Eq. 11 can be further rearranged to a linear form such that

$$\frac{1}{Q} = \frac{1}{Q_e^{\ 2}K_2} \cdot \frac{1}{t} + \frac{1}{Q_e} \tag{12}$$

Therefore, both K_2 and Q_e can be easily determined from Eq. 12.

2.2.2. The Model Validation

Biosorption of Cd^{2+} and Cu^{2+} at Different C_0

Figures 6 to 7 show comparison of the biosorption data obtained at different initial Cd^{2+} and Cu^{2+} concentrations over the Eq. 11 predictions. As compared to the first-order kinetic model (Figures 2 and 3), it appears that the second-order model can not offer a satisfactory description to Cd^{2+} and Cu^{2+} biosorption by aerobic granules, indicated by a correlation coefficient less than 0.88 for all data fittings.

Biosorption of Cd^{2+} and Cu^{2+} at Different X_0

Figures 8 and 9 show that for Cd^{2+} and Cu^{2+} biosorptions at different initial aerobic granules concentrations, the second-order model (Eq. 11) is not able to properly describe the obtained experimental data. Therefore, the first-order kinetic model will be used to study the effects of C_0 and X_0 on Cd^{2+} and Cu^{2+} biosorptions by aerobic granules in the subsequent sections.

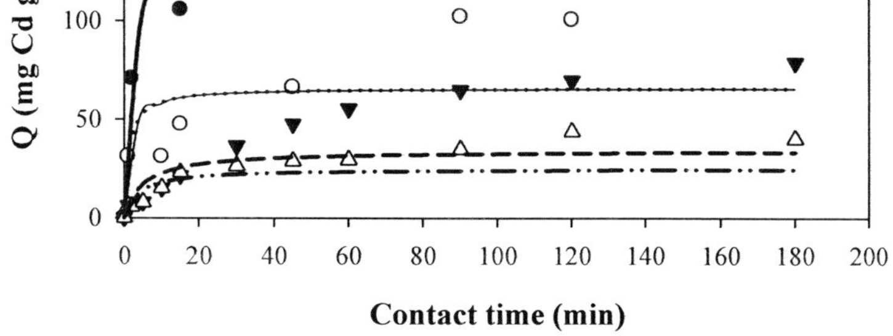

Figure 6. Cd^{2+} biosorption by aerobic granules at different initial Cd^{2+} concentrations. X_0 =150 mg l^{-1} and pH = 7.0. △ 10 mg l^{-1}; ▼ 20 mg l^{-1}; ○ 50 mg l^{-1}; ● 100 mg l^{-1} (Liu et al. 2003a); and the predictions by Eq. 11 are shown by various types of lines: —.. 10 mg l^{-1}; - - - 20 mg l^{-1}; .... 50 mg l^{-1}; ——100 mg l^{-1}.

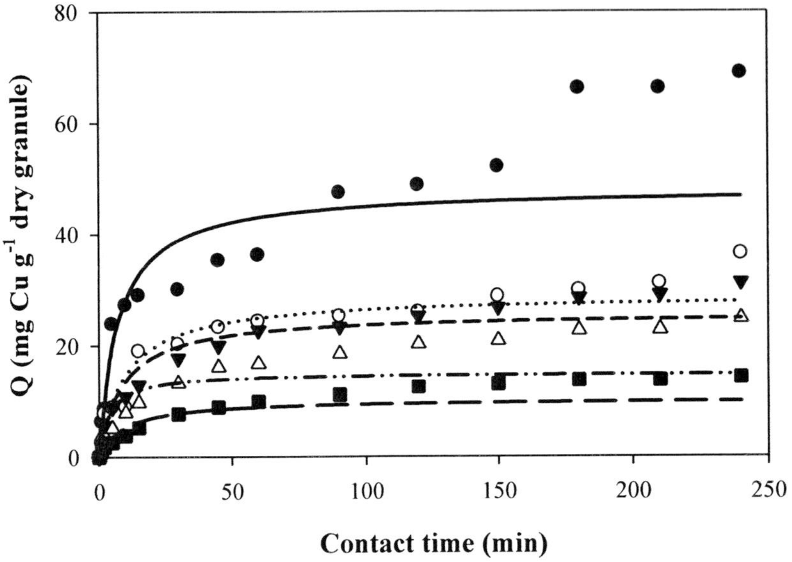

Figure 7. Cu^{2+} biosorption by aerobic granules at different initial Cu^{2+} concentrations. X_0 =150 mg l^{-1} and pH = 7.0. ■ 2 mg l^{-1}; △ 5 mg l^{-1}; ▼ 10 mg l^{-1}; ○ 25 mg l^{-1}; ● 100 mg l^{-1} (Liu et al. 2003a) and the predictions by Eq. 11 are shown by various types of lines: — — 2 mg l^{-1}; —.. 5 mg l^{-1}; - - - 10 mg l^{-1}; .... 25 mg l^{-1}; ——100 mg l^{-1}.

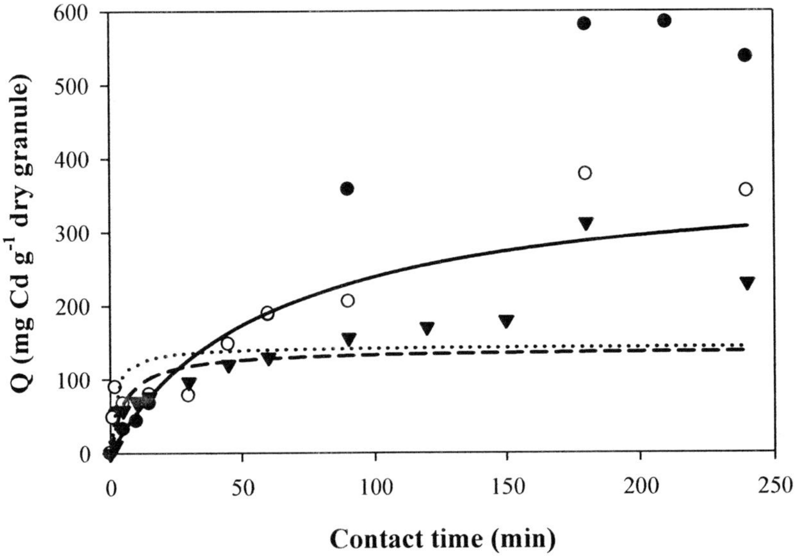

Figure 8. Cd^{2+} biosorption by aerobic granules at different initial aerobic granule concentrations. C_0 =150 mg l^{-1} and pH=7.0.▼ 50 mg l^{-1}; ○ 100 mg l^{-1}; ● 200 mg l^{-1} (Liu

et al. 2003a) and the predictions by Eq. 11 are shown by various types of lines: – – – 50 mg l^{-1}; .... 100 mg l^{-1}; —— 200 mg l^{-1}.

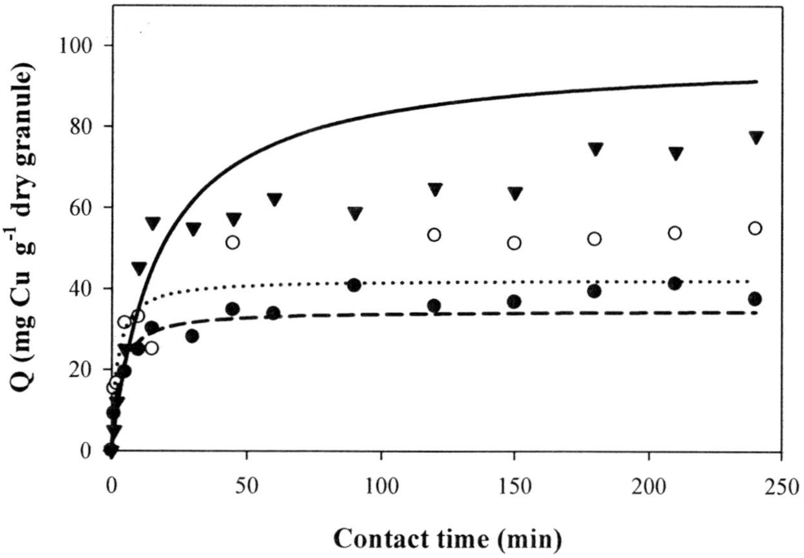

Figure 9. Cu^{2+} biosorption by aerobic granules at different initial aerobic concentrations. C_0 =100 mg l^{-1} and pH=7.0.▼ 100 mg l^{-1}; ○ 200 mg l^{-1}; ● 300 mg l^{-1} (Liu et al. 2003a) and the predictions by Eq. 11 are shown by various types of lines: – – – 100 mg l^{-1}; .... 200 mg l^{-1}; —— 300 mg l^{-1}.

3. Biosorption Isotherm

3.1. The Model Description

It has been proposed that the overall biosorption could be regarded as a simple change in the state of metal ion (Morel et al. 1993), i.e.,

$$C \rightarrow C_{ads} \qquad \Delta G^{o'} \tag{13}$$

where C and C_{ads} is respective metal concentration in bulk solution, and that adsorbed at time t, while $\Delta G^{o'}$ is change of the effective free energy of biosorption, which may vary with the biosorption reaction. Liu et al. (2003b) further proposed that $\Delta G^{o'}$ can be expressed as

$$\Delta G^{o'} = \Delta G^{o} - nRT\ln C \tag{14}$$

where n is positive coefficient. The real driving force of adsorption is the difference between the amount adsorbed by unit biosorbent (Q) at a given metal concentration and the theoretical amount that could be adsorbed by unit biosorbent at that concentration (Q_{th}). This driving force is disappearing when biosorption reaction gradually approaches its equilibrium state (Metcalf and Eddy 2003). As biosorption proceeds, the driving force decreases and adsorption resistance will increase. These seem to imply that the overall change of free energy of biosorption process (ΔG) would increase with the increase of the adsorption resistance, and decrease with the increase of the driving force of adsorption reaction. Liu et al. (2003b) suggested that the overall change of free energy of biosorption could be formulated as follows:

$$\Delta G = \Delta G^{o'} + RT\ln\frac{\text{Resistance}}{\text{Driving force}} \tag{15}$$

As pointed out earlier, adsorption becomes less favorable as adsorption proceeds, i.e., ΔG must increase accordingly. These seem to imply that Q would reflect the magnitude of adsorption resistance, while the difference between Q_{th} and Q may represent the actual driving force of biosorption. In this case, Liu et al. (2003b) thought that Eq. 15 can be translated to

$$\Delta G = \Delta G^{o'} + RT\ln\frac{Q}{Q_{th} - Q} \tag{16}$$

Eq. 16 shows that when $Q = 0.5Q_{th}$, $\Delta G^{o'}$ is equal to ΔG. This in turn implies that $\Delta G^{o'}$ can be defined as the overall free energy change at $Q = 0.5Q_{th}$, i.e., the driving force of biosorption is equal to the resistance. Substitution of Eq. 14 into Eq. 16 yields

$$\Delta G = \Delta G^{o} - nRT\ln C + RT\ln\frac{Q}{Q_{th} - Q} \tag{17}$$

When the adsorption reaches its equilibrium, ΔG becomes zero. Therefore,

$$0 = \Delta G^{o} - nRT\ln C_e + RT\ln\frac{Q_e}{Q_{th}^{e} - Q_e} \tag{18}$$

where C_e, Q_e and Q_{th}^e are respective values of C, Q and Q_{th} at equilibrium. Solving Eq. 18 for Q_e gives

$$Q_e = Q_{th}^e \frac{C_e^{\ n}}{\exp\left(\frac{\Delta G^o}{RT}\right) + C_e^{\ n}} \tag{19}$$

Eq. 19 can be rearranged to

$$Q_e = Q_{th}^e \frac{C_e^{\ n}}{K_{ads} + C_e^{\ n}} \tag{20}$$

where

$$K_{ads} = \exp\left(\frac{\Delta G^o}{RT}\right) \tag{21}$$

3.2. The Model Validation

The respective biosorption isotherms of Cd^{2+} and Cu^{2+} at a constant granule concentration of 150 mg l^{-1} are shown in Figs. 10 and 11. It can be seen that the proposed model (Eq. 20) can provide a satisfactory description for the experimental data, indicated by a respective correlation coefficient of 0.935 for Cd^{2+} and 0.989 for Cu^{2+}. The respective K_{ads} value obtained is 0.00347 for Cd^{2+} biosorption and 0.0221 for Cu^{2+} biosorption by aerobic granules. For Cd^{2+} biosorption by aerobic granules, the value of n estimated is 0.71, while 0.51 for Cu^{2+} biosorption.

The Langmuir and Freundlich adsorption isotherms have been commonly used to fit biosorption data. Thus, the experimental data were also fitted to these two isotherms (Figs. 10 and 11). The predictions by the Langmuir and Freundlich adsorption isotherms turn out to be not as good as that given by Eq. 20, which is evidenced by relatively small correlation coefficient, especially the prediction by the Langmuir isotherm seems poorer than those by Eq. 20 and the Freundlich model. In fact, Eq. 20 can also be applied to other biosorbents other than aerobic granules even though they are not shown here (Ruiz-Manriquez et al. 1998; Yang and Volesky 1999; Zhou 1999; Wang et al. 2001; Pagnartelli et al. 2003; Loukidou et al. 2004).

It should be realized than Eq. 20 can reduce to the following form when n equals 1, which is the same as the well known Langmuir adsorption isotherm:

$$Q_e = Q_{th}^e \frac{C_e}{K_{ads} + C_e} \tag{22}$$

Eq. 22 shows that the Langmuir adsorption isotherm would be a particular case of Eq. 20 when n has a value of 1. When the value of C_e^n is much less than the value of K_{ads}, Eq. 20 is simplified to the Freundlich adsorption isotherm:

$$Q_e = \frac{Q_{th}^e}{K_{ads}} C_e^{\ n} \tag{23}$$

Eq. 23 reveals that the Freundlich constant indeed equals Q_{th}^e / K_{ads}.

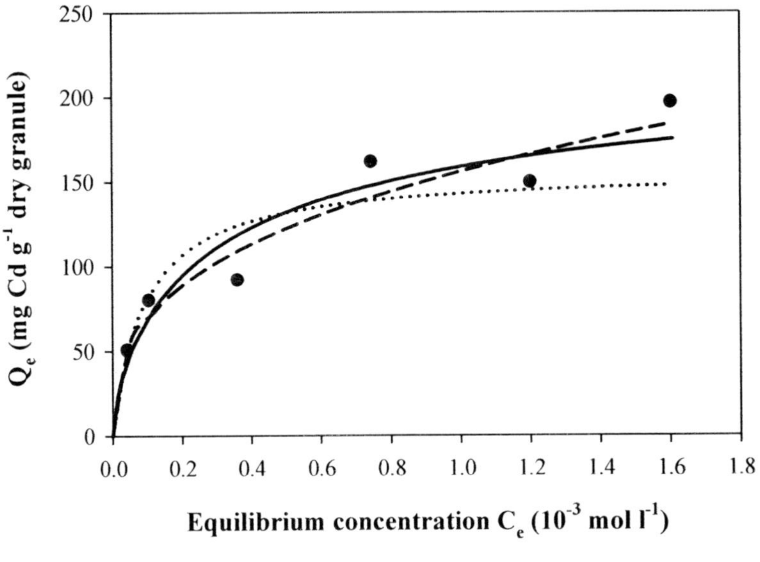

Figure 10. Biosorption isotherm of Cd^{2+} by aerobic granules. ——— prediction by Eq. 20 with R = 0.935 (Liu et al. 2003b); .... prediction by the Langmuir model with R= 0.912; – – – prediction by the Freundlich model with R= 0.929.

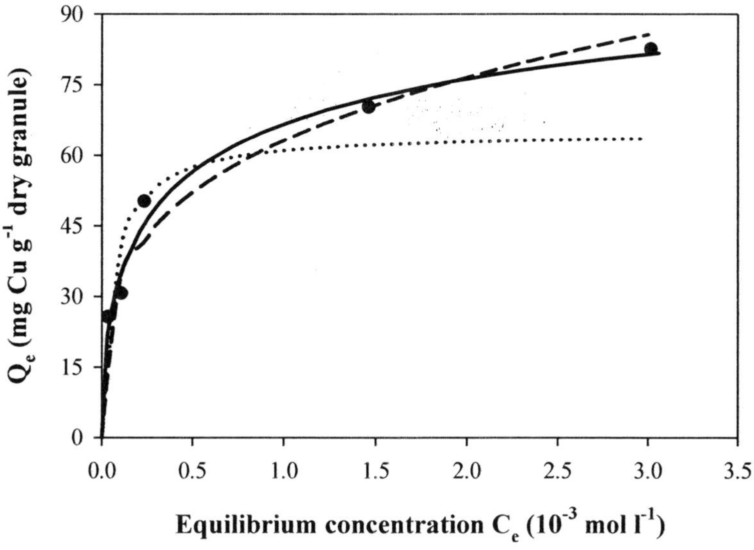

Figure 11. Biosorption isotherm of Cu^{2+} by aerobic granules. ——— prediction by Eq. 20 with R = 0.989 (Liu et al. 2003b). .... prediction by the Langmuir model with R= 0.842; – – – prediction by the Freundlich model with R= 0.944.

Eq. 21 shows that the value of $1/K_{ads}$ is related to the magnitude of the thermodynamic equilibrium constant, K_{eq}. When the value of K_{ads} is small, the thermodynamic equilibrium constant, K_{eq} is large. In this case, biosorption would proceed far towards completion, and the position of equilibrium lies far toward the biosorption of metal. On the other hand, when the value of K_{ads} is very large, i.e, K_{eq} is very small, the position of equilibrium lies far towards the soluble metal.

4. Effect of C_0 on Biosorption

4.1. Effect of C_0 on K_1

The relationships between the overall biosorption rate constant (K_1) and initial Cd^{2+} and Cu^{2+} concentrations (C_0) are shown in Figures 13 and 14, respectively. It is found that K_1 tends to increase with C_0 in both cases, e.g., in Cd^{2+} biosorption, K_1 increased from 0.021 min^{-1} at C_0 of 10 mg l^{-1} to 0.032 min^{-1} at C_0 of 100 mg l^{-1}; while in Cu^{2+} biosorption, K_1 increased from 0.032 min^{-1} at C_0 of 2 mg l^{-1} to 0.092 min^{-1} at C_0 of 100 mg l^{-1}. These in turn imply that a high initial metal concentration would favor Cd^{2+} and Cu^{2+} biosorption by aerobic granules, i.e., a fast biosorption would be expected at high C_0. Moreover, K_1 for Cu^{2+} biosorption is almost 3 times higher than that for Cd^{2+} biosorption at the same initial metal concentration, this means that Cu^{2+} biosorption should be faster than Cd^{2+} biosorption by aerobic granules.

The positive correlation of K_1 to the respective Cd^{2+} and Cu^{2+} concentrations seem to indicate that the concentration gradient-driven mass transfer would play an important role in Cd^{2+} and Cu^{2+} biosorption by aerobic granules. Microscopic observation clearly shows that aerobic granules have high porosity and channel structure (Figure 12). It is a reasonable consideration that aerobic granules may have a similar internal structure to granular activated carbon. Thus, Cd^{2+} and Cu^{2+} biosorptions by aerobic granules would be determined, at least, in part by external and internal mass transfer efficiencies.

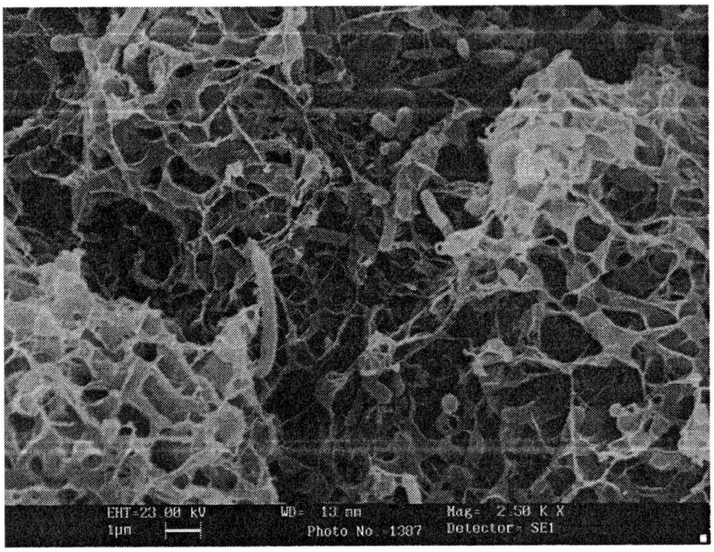

Figure 12. Porous structure of aerobic granule.

Obviously, the net rates of transport of Cd^{2+} and Cu^{2+} should be proportional to their concentrations in the bulk solutions. This may explain why K_1 is proportionally correlated to initial Cd^{2+} and Cu^{2+} concentrations as observed in Figures 13 and 14.

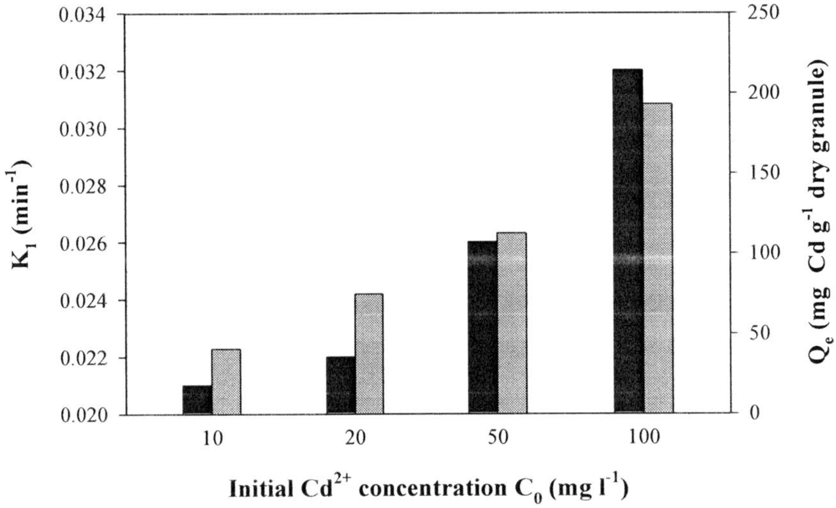

Figure 13. Effect of initial Cd^{2+} concentration on K_1 (*black*) and Q_e (*gray*).

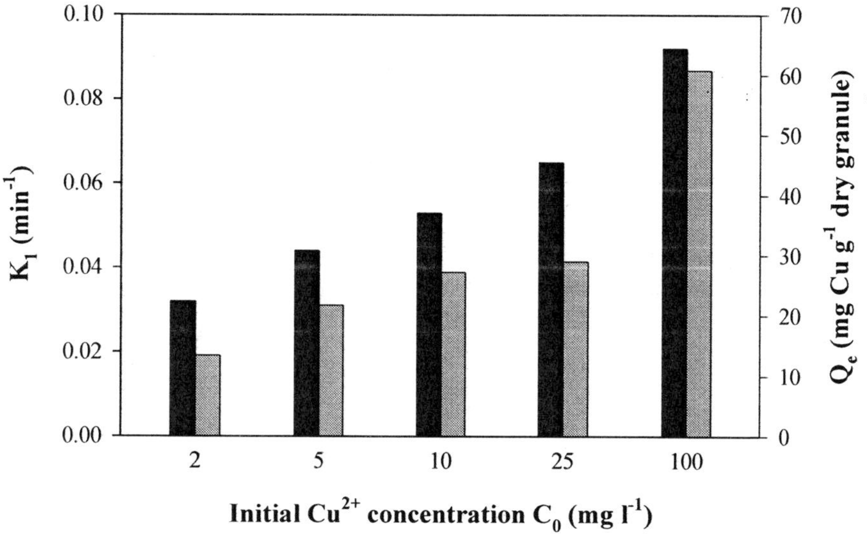

Figure 14. Effect of initial Cu^{2+} concentration on K_1 (*black*) and Q_e (*gray*).

4.2. Effect of C_0 On Q_e

Figures 13 and 14 exhibit the relationships between the initial Cd^{2+} and Cu^{2+} concentrations (C_0) and the equilibrium biosorption capacities (Q_e) of ions at the constant aerobic granules concentration. Q_e is found to be positively related to C_0, i.e., a relatively high C_0 would enhance the biosorption capacity. In Cd^{2+} biosorption, Q_e increased from 46 mg Cd^{2+} g^{-1} granules at C_0 of 10 mg l^{-1} to 193 mg Cd^{2+} g^{-1} granules at C_0 of 100 mg l^{-1}, while in Cu^{2+} biosorption, Q_e increased from 15 mg Cu^{2+} g^{-1} granules at C_0 of 2 mg l^{-1} to 60 mg Cu^{2+} g^{-1} granules at C_0 of 100 mg l^{-1}. Moreover, the biosorption capacity of Cd^{2+} by aerobic granules is much higher than that of Cu^{2+} at the same initial metal concentration. Puranik et al. (1999) also reported that the amounts of zinc and lead adsorbed onto *S. cinnamoneum, P. chrysogeum* and *Citrobater sp.* were increased with the increase of the initial metal concentration.

5. Effect of X_0 on Biosorption

5.1. Effect of X_0 on K_1

Relationships between the overall biosorption rate constants (K_1) and initial aerobic granules concentration (X_0) for Cd^{2+} and Cu^{2+} biosorptions are shown in Figures 15 and 16, respectively. K_1 tends to decrease with the increase of X_0 in both cases, e.g., K_1 decreased from 0.01 min^{-1} at X_0 of 50 mg l^{-1} to 0.0083 min^{-1} at X_0 of 300 mg l^{-1} in Cd^{2+} biosorption

(Figure 15); and likewise, K_1 decreased from 0.1 min^{-1} at X_0 of 100 mg l^{-1} to 0.051 min^{-1} at X_0 of 300 mg l^{-1} in Cu^{2+} biosorption (Figure 16).

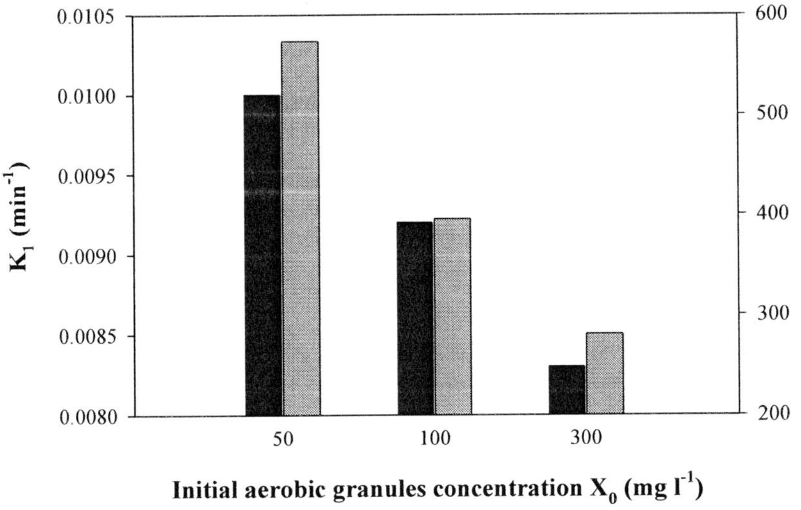

Figure 15. Effect of X_0 on K_1 (*black*) and Q_e (*gray*) for Cd^{2+} biosorption.

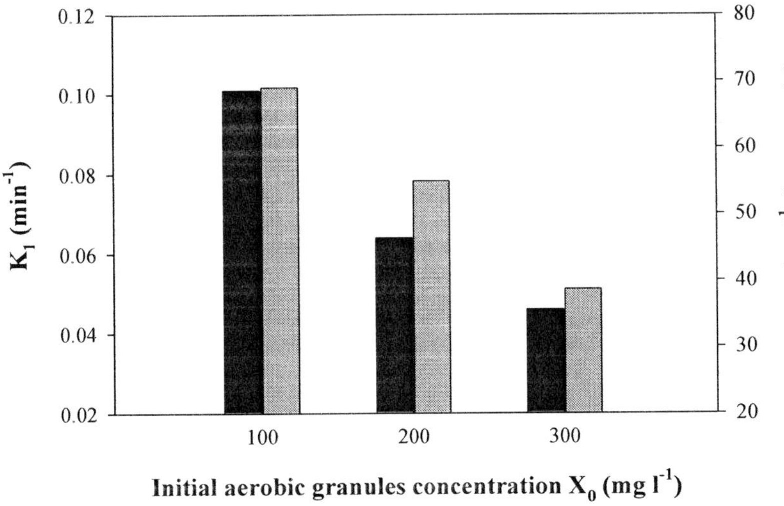

Figure 16. Effect of X_0 on K_1 (*black*) and Q_e (*gray*) for Cu^{2+} biosorption.

5.2. Effect of X_0 on Q_e

The effects of initial granule concentration (X_0) on the equilibrium biosorption capacities (Q_e) of Cd^{2+} and Cu^{2+} are shown in Figures 15 and 16, respectively. It appears that an increased X_0 would result in a decreased Q_e, .e.g., for Cd^{2+} biosorption, Q_e decreased from

573 mg Cd^{2+} g^{-1} granules at C_0 of 50 mg l^{-1} to 281 mg Cd^{2+} g^{-1} granules at C_0 of 300 mg l^{-1}; while for Cu^{2+} biosorption, Q_e decreased from 69 mg Cu^{2+} g^{-1} granules at C_0 of 100 mg l^{-1} to 39 mg Cu^{2+} g^{-1} granules at C_0 of 300 mg l^{-1}.

It is a reasonable consideration that the number of binding sites to Cd^{2+} and Cu^{2+} carried by aerobic granules is proportional to the amount of aerobic granules. This means that a higher granule concentration could result in a lower relative Cd^{2+} and Cu^{2+} concentration on the basis of unit mass of aerobic granules, which in turn is responsible for the observed decline of the biosorption rate constant and the equilibrium biosorption capacity at high granule concentrations (Figures 15 and 16). In fact, similar phenomenon has also been reported in metal biosorption study with marine algae as biosorbents (Aksu and Gulen 2002).

6. Theoretical Interpretation of Effects of C_0 and X_0 on K_1

Figures 13 to 16 reveal that K_1 increases with increasing C_0 and decrease with the increase of X_0. The likewise effects have also been observed by other researchers (Puranik et al. 1999; Aksu and Gulen 2002). However, there is still lack of a theoretical interpretation of such results. Just recently, Rudzinski and Plazinski (2006) developed a theoretical expression for the interfacial adsorption rate constant, based on the Statistical Rate Theory, in which the interfacial transfer rate and the number of active binding sites were both taken into consideration, i.e.,

$$K_1 = \frac{2K_{ls}C_e\left(1-\frac{Q_e}{Q^e_{th}}\right)(\varepsilon_m-\varepsilon_1)}{nRT} \tag{24}$$

where ε_1 to ε_m represent the dispersion of adsorption energies on interface. K_{ls} is a pure constant. For a defined biosorption reaction, Eq. 24 can be simplified to

$$K_1 = K'C_e\left(1-\frac{Q_e}{Q^e_{th}}\right) \tag{25}$$

where $K' = \frac{2K_{ls}(\varepsilon_m-\varepsilon_1)}{nRT}$, and can be determined by a simple least square regression.

Eq. 25 shows that the overall biosorption rate constant (K_1) is the function of both metal concentration (C_e) and biosorption capacity (Q_e) at equilibrium. According to Eq. 4, C_{be} can be expressed as

$$C_{be} = X_0 Q_e \tag{26}$$

Thus, C_e and Q_e can be solved by combining Eqs. 4, 20, 25 and 26. In this case, K_1 in Eq. 25 becomes an implicit function of the initial metal (C_0) and aerobic granules (X_0) concentrations, but without analytical expression, i.e.,

$$f(K_1, C_0, X_0) = 0 \tag{27}$$

Numerical solution of Eq. 27 can be obtained by using Matlab 7.1 and the predicted profiles of K_1 as the function of initial Cd^{2+} and Cu^{2+} and aerobic granules concentrations are plotted in Figures 17 and 18. It can be seen that the predicted trend of K_1 is reasonably matching with the experimental data, i.e., K_1 tends to increase with C_0 for both Cd^{2+} and Cu^{2+} biosorptions.

Figures 19 and 20 further show the effect of X_0 on K_1 obtained in Cd^{2+} and Cu^{2+} biosorptions by aerobic granules. Obviously, the predictions by Eq. 27 are in good agreement with the experimental data, i.e., an increased X_0 will result in a decreased K_1 at a constant initial metal concentration, and thus in turn slows down biosorption reaction.

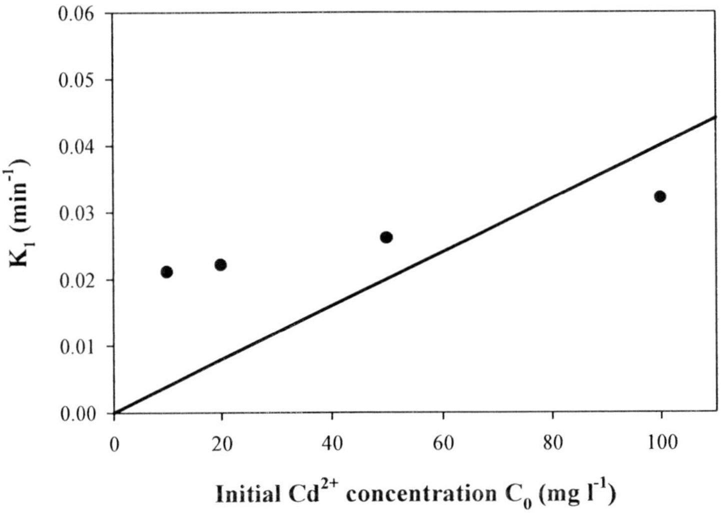

Figure 17. Dependence of K_1 on C_0 observed in biosorption of Cd^{2+}. The prediction by Eq. 27 is shown by solid line.

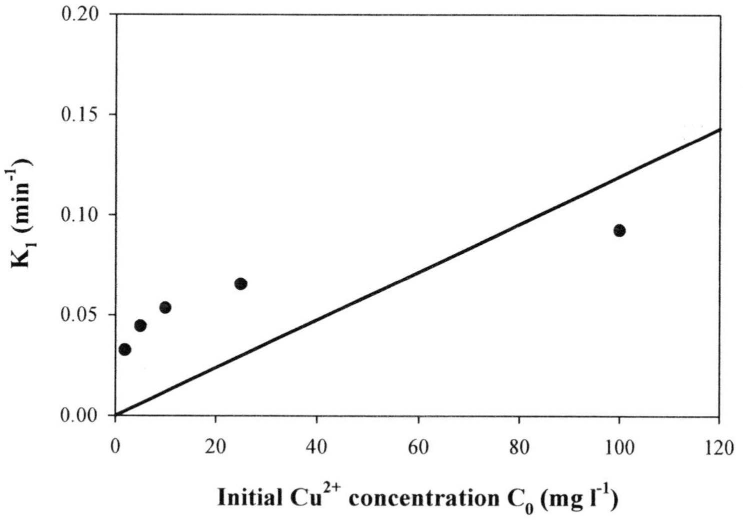

Figure 18. Dependence of K_1 on C_0 observed in biosorption of Cu^{2+}. The prediction by Eq. 27 is shown by solid line.

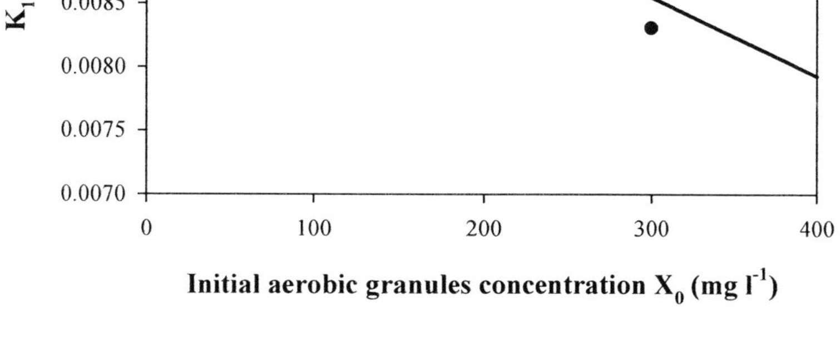

Figure 19. Dependence of K_1 on X_0 observed in biosorption of Cd^{2+}. The prediction by Eq. 27 is shown by solid line.

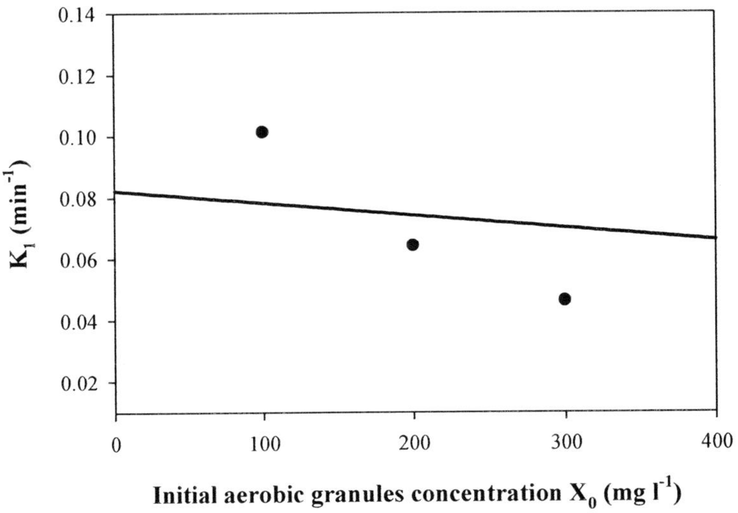

Figure 20. Dependence of K_1 on X_0 observed in biosorption of Cu^{2+}. The prediction by Eq. 27 is shown by solid line.

7. THEORETICAL INTERPRETATION OF EFFECTS OF C_0 AND X_0 ON Q_E

Figures 13 to 16 show that the initial Cd^{2+} and Cu^{2+} concentrations can significantly influence the equilibrium biosorption capacity of aerobic granules. Combination of Eqs. 4, 20 and 26 yields

$$\frac{Q_{th}^{e}(C_0 - Q_e X_0)^n}{K_{ads} + (C_0 - Q_e X_0)^n} - Q_e = 0 \qquad (28)$$

In this equation, Q_e is indeed an implicit function of the initial metal concentration (C_0) and initial aerobic granules concentration (X_0). The numerical solution of Eq. 28 for Q_e thus can be obtained by using Matlab 7.1. It appears from Figs. 21 and 22 that Eq. 28 can reasonably predict the experimental data obtained at different initial Cd^{2+} and Cu^{2+} concentrations. Eq. 28 shows that Q_e would be zero when initial metal concentration is nil; and when C_0 approaches infinite, the limit of Eq. 28 would become

$$Q_{th}^{e} - Q_e = 0 \qquad (29)$$

Therefore, Q_e in Figures 21 and 22 will gradually approach Q_{th}^{e} with the continuous increase of C_0. In addition, Figures 23 and 24 also exhibit the simulation by Eq. 28 on the effect of X_0 on Q_e. Good agreement between the model prediction and experimental data appears from these two figures. Consequently, Eq. 28 can offer a theoretical interpretation to the observed effects of C_0 and X_0 on Q_e.

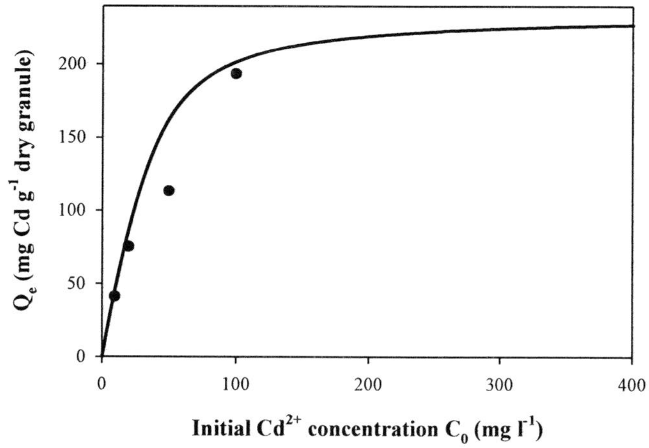

Figure 21. Relationship between Q_e and C_0 observed in Cd^{2+} biosorption. The prediction by Eq. 28 is shown by solid line.

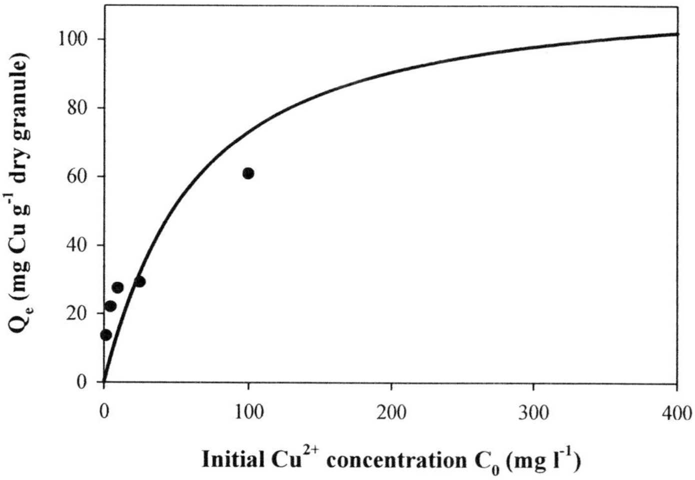

Figure 22. Relationship between Q_e and C_0 observed in Cu^{2+} biosorption. The prediction by Eq. 28 is shown by solid line.

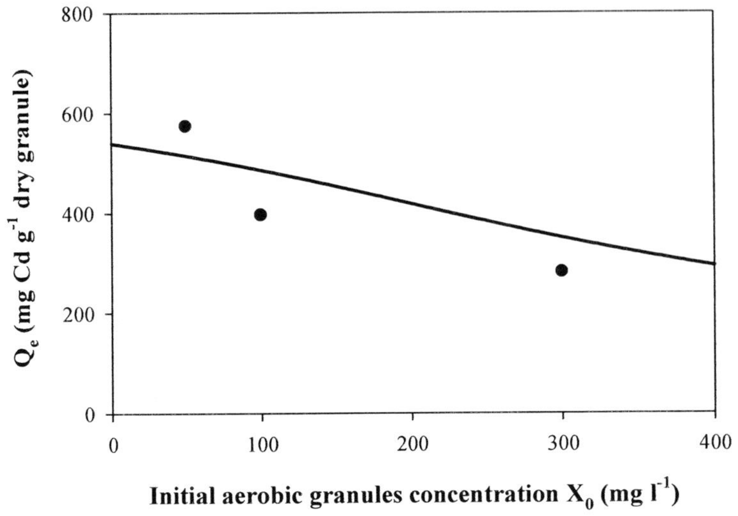

Figure 23. Relationship between Q_e and X_0 observed in biosorption of Cd^{2+}. The prediction by Eq. 28 is shown by solid line.

In fact, Eq. 28 can also be used to provide satisfactory prediction for heavy metal biosorption by other biosorbents. For example, data obtained from Cr^{6+} biosorption by *Aeromonas caviae* by Loukidou (2004) are used to fit Eq. 20 and following outcomes are generated from Eq. 20: Q_{th}^{e}=180.55 mg g^{-1}, K_{ads}=0.00184, and n=0.702. Using these values, Fig. 25 is thus plotted according to Eq. 28. It can be seen that Eq. 28 can fairly describe these experimental data. In addition, Eq. 28 can be rearranged to

$$C_0 = Q_e X_0 + \left(\frac{Q_e K_{ads}}{Q_{th}^{e} - Q_e} \right)^{\left(\frac{1}{n}\right)} \tag{30}$$

Eq. 30 shows that a wide spectrum of C_0 and X_0 satisfying Eq. 30 will generate the same Q_e. This implies that Eq. 30 can offer a useful numerical tool for predicting Q_e in biosorption experiments conducted at various initial metal and biosorbent concentrations.

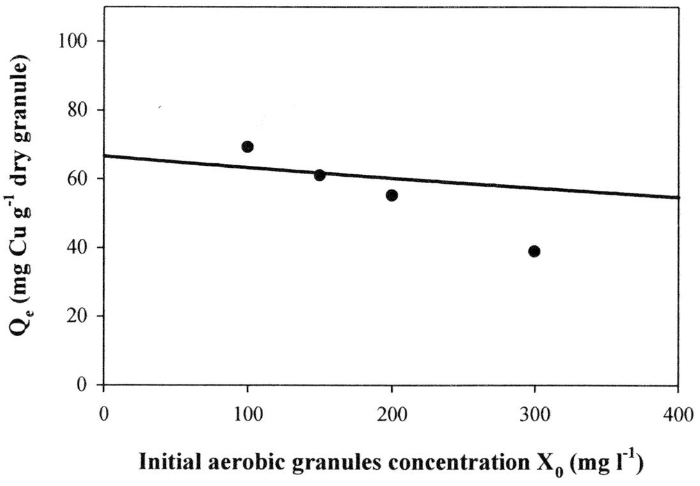

Figure 24. Relationship between Q_e and X_0 observed in biosorption of Cu^{2+}. The prediction by Eq. 28 is shown by solid line.

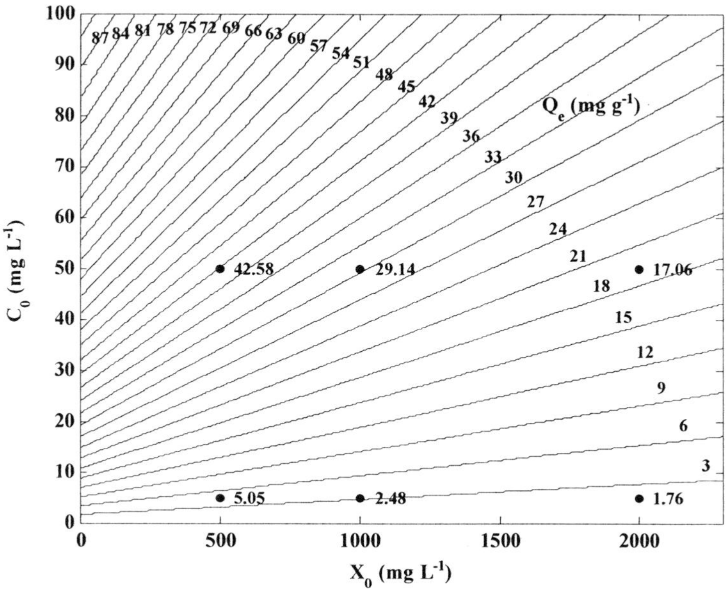

Figure 25. Simulation of Q_e as the functions of C_0 and X_0 in Cr^{6+} biosorption by *Aeromonas caviae*. •: experimental points taken from Loukidou et al. (2004) and the prediction by Eq. 28 is shown by lines.

8. CONCLUSION

A comprehensive modeling system including kinetics and isotherm is developed for description of Cd^{2+} and Cu^{2+} biosorption by aerobic granules. Compared to the second-order kinetic model, the first-order kinetic model can provide a more reasonable prediction of experimental data obtained at different initial Cd^{2+}, Cu^{2+} and aerobic granules concentrations. It is shown that the proposed isotherm equation derived from a simple thermodynamic approach is relatively better than the Langmuir and Freundlich isotherm equations in describing Cd^{2+} and Cu^{2+} biosorptions by aerobic granules. Moreover, it is demonstrated that the Langmuir and Freundlich isotherms are the special case of the proposed isotherm equation.

The overall biosorption rate constant and equilibrium biosorption capacity by aerobic granules are highly dependent on initial Cd^{2+}, Cu^{2+} and aerobic granules concentrations. In order to theoretically interpret the observed phenomena, two additional models were thus derived based on the proposed kinetic and isotherm models. These two models can be used to simulate biosorption under various initial conditions, and the outcomes of the models appear to be in good agreement with the experimental observations, i.e., the overall biosorption rate constant and biosorption capacity will increase with increasing initial metal concentration, but tend to decrease with the increase of initial aerobic granules concentration. It is expected that the modeling system presented here can offer a deeper understanding of heavy metal biosorption by aerobic granules.

REFERENCES

Aksu, Z. and Gulen, H. (2002). Binary biosorption of iron (III) and iron (III)-cyanide complex ions on *Rhizopus arrhizus*: modelling of synergistic interaction. *Process Biochemistry*, *38*, 161-173.

Campos, J. L., Garrido-Fernandez, J., Mendez, R. and Lema, J. M. (1999). Nitrification at high ammonia loading rates in an activated sludge unit. *Bioresource Technology*, *68*, 141-148.

Ho, Y. S. and McKay, G. (1999). Pseudo-second order model for sorption processes. *Process Biochemistry*, *34*, 451-465.

Liu, Y., Yang, S.F., Tan, S.F., Lin, Y.M. and Tay, J.H. (2002). Aerobic granules: a novel zinc biosorbent. *Letters in Applied Microbiology 35,* 548-551.

Liu, Y., Yang, S. F., Xu, H., Woon, K. H., Lin, Y. M. and Tay, J. H. (2003a). Biosorption kinetics of cadmium (II) on aerobic granular sludge. *Process Biochemistry*, *38*, 997-1001.

Liu, Y., Xu, H., Yang, S. F. and Tay, J. H. (2003b). A general model for biosorption of Cd^{2+}, Cu^{2+} and Zn^{2+} by aerobic granules. *Journal of Biotechnology*, *102*, 233-239.

Liu, Y. and Tay, J. H. (2004). State of the art of biogranulation technology for wastewater treatment. *Biotechnology Advances*, *22*, 533-563.

Lodi, A., Solisio, C., Converti, A. and Del Borghi, M. (1998). Cadmium, zinc, copper, silver and chromium(III) removal from wastewaters by *Sphaerotilus natans*. *Bioprocess and Biosystems Engineering*, *19*, 197-203.

Loukidou, M. X., Zouboulis, A. I., Karapantsios, T. D. and Matis, K. A. (2004). Equilibrium and kinetic modeling of chromium(VI) biosorption by *Aeromonas caviae*. *Colloids and Surfaces a-Physicochemical and Engineering Aspects*, *242*, 93-104.

Marques, P., Rosa, M. F. and Pinheiro, H. M. (2000). pH effects on the removal of Cu^{2+}, Cd^{2+} and Pb^{2+} from aqueous solution by waste brewery biomass. *Bioprocess Engineering*, *23*, 135-141.

Metcalf and Eddy (2003). *Wastewater engineering: treatment and reuse* (4th edn). Boston, McGraw-Hill.

Morel, F., Hering, J. G. and Morel, F. (1993). *Principles and applications of aquatic chemistry*. New York, Wiley.

Pagnartelli, F., Esposito, A., Toro, L. and Veglio, F. (2003). Metal speciation and pH effect on Pb, Cu, Zn and Cd biosorption onto *Sphaerotilus natans*: Langmuir-type empirical model. *Water Research*, *37*, 627-633.

Puranik, P. R., Modak, J. M. and Paknikar, K. M. (1999). A comparative study of the mass transfer kinetics of metal biosorption by microbial biomass. *Hydrometallurgy*, *52*, 189-197.

Ruiz-Manriquez, A., Magana, P. I., Lopez, V. and Guzman, R. (1998). Biosorption of Cu by *Thiobacillus ferrooxidans*. *Bioprocess Engineering*, *18*, 113-118.

Rudzinski, W. and Plazinski, W. (2006). Kinetics of solute adsorption at solid/solution interfaces: A theoretical development of the empirical pseudo-first and pseudo-second order kinetic rate equations, based on applying the statistical rate theory of interfacial transport. *Journal of Physical Chemistry B*, *110*, 16514-16525.

Valdman, E. and Leite, S. G. F. (2000). Biosorption of Cd, Zn and Cu by *Sargassum sp* waste biomass. *Bioprocess Engineering*, *22*, 171-173.

Wang, J. L., Zhan, X. M., Ding, D. C. and Zhou, D. (2001). Bioadsorption of lead (II) from aqueous solution by fungal biomass of *Aspergillus niger*. *Journal of Biotechnology*, *87*, 273-277.

Xu, H., Liu, Y. and Tay, J.H. (2005). Effect of pH on nickel biosorption by aerobic granular sludge. *Bioresource Technology 97,* 359-363.

Yang, J. B. and Volesky, B. (1999). Biosorption of uranium on *Sargassum* biomass. *Water Research*, *33*, 3357-3363.

Zhou, J. L. (1999). Zn biosorption by *Rhizopus arrhizus* and other fungi. *Applied Microbiology and Biotechnology*, *51*, 686-693.

INDEX

A

B

C

D

E

F

K

L

M

N

O

P

Q

R

U

V

W

X

Y

Z